P9-EKS-673

Library of Congress Cataloging-in-Publication Data

Hundal, Mahendra S., 1924—
Systematic mechanical designing: a cost and management perspective/ Mahendra S. Hundal
p. cm.
Includes bibliographical references and index.
ISBN 0-7918-0042-3
1. Design, Industrial—Management.
2. Design, Industrial—Costs.
I. Title
TS171.H83 1997 97-42670
658.5'75—dc21—dc21 CIP

Preface

The first part of the book looks at the overall product realization process, and then (in Chapter III) describes a method of systematic design. This method of design is based upon concepts and principles rather than on preexisting solutions and physical objects. The chapter describes different stages of the design method: (1) clarification of the task; (2) conceptual design; (3) preliminary, or embodiment, design; and (4) detail design. Methods of generating solutions are discussed, including the intuitive and the systematic methods. Two examples of the application are given. The first shows the use of different physical effects in the conceptual design of a temperature measurement device. A second example shows the use of embodiment and detail design principles in the design of a variable-speed power transmission.

The economic future of a technologically advanced country such as the United States depends on its manufacturing base. This base can be rapidly eroded by foreign competition in today's global economy. The economic health of the country can only be maintained by developing and manufacturing high-quality products at low cost and in a short time. Chapter IV, on the role of management, describes techniques and procedures companies may apply in order to attain this goal. This chapter also discusses the role of management in lowering product costs. Topics include accelerated product development, concurrent engineering, and approaches for providing cost information and using teamwork to lower costs. This is followed by Chapter V, on Quality Function Deployment. QFD is a tool, developed in Japan but now extensively used by major U.S.

companies, that consists of listening to the needs of the customer and translating these into dependable products. A brief section on Taguchi methods provides an appreciation of the importance of robust design.

Chapter VI discusses cost structures, with a focus on absolute and relative costs. A cost structure shows the breakdown of the product cost according to one of several criteria: parts, type of cost, functions, production process, etc. Such a structure brings into focus the proportion of cost contributed by various factors. Other topics in this chapter include how costs are influenced at the concept and embodiment stages, the effect of the number of parts, and life cycle costs.

Next in Chapter VI, cost models of various types are presented. Function and parametric costing and costs based on units, operations, and activities are discussed. Nearly all cost models make use of regression analysis. It can be used when costs can be expressed as a function of the various statistical parameters. Similarity laws, on the other hand, are based upon physical relationships and expressed in the form of power laws. These are discussed later in a brief chapter (Chapter X). Designing to cost is demonstrated through an example in Chapter VI. A cost goal for the product must be set by the company and provided to the design group. The designer then sets up cost structures according to parts or other criteria, based upon similar products. Only if cost calculation is carried out in parallel with design steps is there a reasonable chance to hold to the goal.

Chapter VII presents the principles of embodiment design, including methods of obtaining variants at the embodiment stage, part design principles, and shape design. Designing for easy assembly is discussed with the help of several examples.

Although materials and manufacturing engineering are separate disciplines in which there are several excellent textbooks, two chapters are included in this book on these topics (Chapters VIII and IX). The designer needs a knowledge of materials and manufacturing in order to design cost-effective products. The economic aspects of these fields are particularly emphasized.

Chapter X, on standardization, shows the advantages of standardization for reducing costs. Lower costs result from the use of standard and repeated parts, from manufacturing products in a series of size ranges, and from using modular design.

Chapter XI, on design and manufacturing under environmental constraints, provides some background on this subject, which is becoming increasingly important. It is at the design stage that a company has the greatest control over all aspects of the product including recycling, remanufacturing, and toxicity.

Nomenclature used in this book is listed in Appendix A. The next five appendices elaborate further on various aspects of systematic design. The last three appendices present computer aids in design, numerical details of regression analysis, and mathematical methods for optimization

Throughout the book, cost figures have been cited and used. Needless to say, they should not be used by the reader without verification. While relative costs remain relatively constant, absolute costs vary with place, time, quantity, and quality.

A book on design cannot be a comprehensive work without reaching an encyclopedic length. Such was not the intent in preparing this work. It does not include specialty topics such as mechanism design. None of the engineering science fields is discussed—stress analysis, heat transfer, etc. The reader is referred to the appropriate sources on many of the topics presented here.

AUDIENCE

The book is appropriate for all individuals engaged in product design, manufacturing, and management—designers, product planners, product designers, design engineers, project engineers, design managers, process planners, and manufacturing engineers. Among the practicing professionals this book should be particularly useful for

- Engineers in industry who have not been exposed to the systematic design method or to the importance of cost awareness during designing, and who need to realize the importance of communication with other departments involved in the product development process.
- Engineering managers, to whom it may be informative for many of the same reasons. For them, an even greater emphasis is placed on the importance of communication with other departments.

- Other product development professionals, so they may obtain an appreciation of the design process.

This book will also be useful for engineering design students in mechanical, manufacturing, and related engineering disciplines. These are the senior-level students who have completed the course in design of machine elements and are looking at mechanical design from a broader perspective. It should also be well suited for courses taught as "integrated product development" or under similar titles. These are the interdisciplinary courses in product design, which emphasize the integration of marketing, costing, design, and manufacturing into the development process.

ACKNOWLEDGMENTS

I am indebted to many whose influence has culminated in this work. My erstwhile colleagues in the Design Office at the Tata Iron & Steel Company in Jamshedpur, India. Much of my thinking is influenced by my early experiences of design of heavy equipment for the steel industry. Professor Klaus Ehrlenspiel and colleagues in the Department of Mechanical Engineering Design at the Munich University of Technology, where I spent a year on a Fulbright scholarship. My colleagues at the University of Vermont, with whom I have developed and team-taught a course in Integrated Product Development: Delcie Durham and Branimir von Turkovich in the Department of Mechanical Engineering; Larry Haugh of the Department of Statistics; Larry Shirland, Jacque Grinnell, Richard Jesse, and Lauck Parke of the School of Business Administration; and last but most important, Jerry Manock, an engineer/designer/entrepreneur who kept us in touch with the real world.

Mahendra S. Hundal
Burlington, Vermont
Summer 1996

Table of Contents

List of Tables

Table
Number

List of Figures

Figure
Number

Introduction

- Design Literature
- Management Issues
- Bibliography

The United States is a country with high labor costs, as are most Western European countries and Japan. Competitive market conditions require that product costs be kept low. But the designer often does not have a good understanding of customer requirements, and sufficient knowledge of materials, manufacturing processes, and the costs incurred therein. Boothroyd et al. (1994) quote from a survey of designers regarding this. Of those surveyed regarding manufacturing processes:

49% had little or no knowledge of injection molding.
73% had little or no knowledge of sintering.
94% had little or no knowledge of hot isostatic pressing.

Of those surveyed regarding polymer materials:

59% had little or no knowledge regarding nylon 6.
74% had little or no knowledge regarding polycarbonates.
84% had little or no knowledge regarding fluoropolymers.

A company is organized according to departments—sales, planning, design, purchasing, manufacturing, and so forth. Each of these may be staffed by specialists who have little knowledge of other departments' operations. This is especially true of designers, who generally do not have an understanding of manufacturing, purchasing, and costing. The cost of a product grows from conception, through design, development, and use, all the way to disposal or recycling. Product costs can be influenced the most in the early stages and to decreasing degrees as the process proceeds. The later in the product realization process any changes are made, the more expensive they get. For a *new* product, sufficient cost information is not available when the costs can be influenced the most—during designing. In order to rectify this situation, designers should be able to take advice from other departments and use it to do their own estimates of costs. To design for cost, designers must be able not only to work with physical properties, such as strength or wear, but also at the same time to calculate costs.

Costs are too important to be the domain of only cost estimating and industrial engineering personnel. The inputs and expertise of these specialists are of great importance in the design process. However, the technical knowledge of designers is the most important factor in determining product costs. Figure 1 shows that a company has the greatest control over a product at the early stages, at which time the product's major properties, such as cost and life expectancy, are being set. The designer determines both the factory cost and the total life cycle cost. Therefore the design department ought to have the most information about costs. Unfortunately, this is not the case in most companies. Product design and the designer have long been undervalued and misunderstood—the designer is regarded as a person who draws lines and circles on paper (now on a computer screen). Greater importance has been given to the analyst. But the analyst must have something to analyze—the design, i.e., shapes, materials, and motion. If the design is poor to begin with, no amount of analysis can fix it—it can only be fine-tuned in the final stages. Design forms the core of development and manufacture of competitive products. A competitive product must not only provide the desired functionality, but also be affordable to the customer, be of high quality to satisfy the customer, and be delivered on time.

There is potential for cost reduction in every industry; Munro (1995a) provided the figures shown in Table 1. In order to put the subjects of

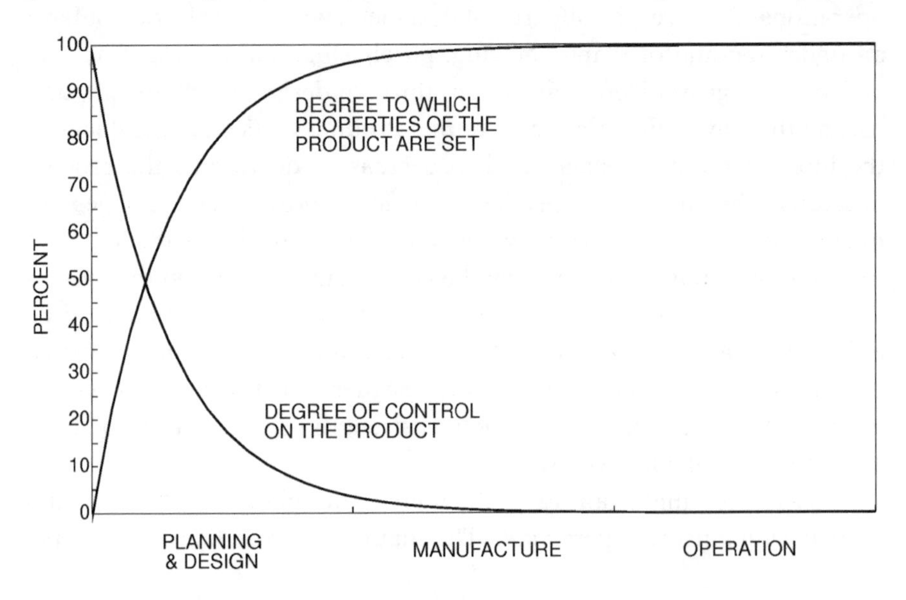

Figure 1. Control on a product and setting of its properties.

TABLE 1 Potential for Cost Reduction in Various Industries

Industry	Potential
Defense	80%
Aircraft	50–60%
Medical	40–50%
Machine tool	35–40%
Automobile	30%
Large appliance	15–20%
Small appliance	8%
Commodities	3–5%

(Munro, 1995a)

systematic design and design to cost in perspective, we will review the major works in these areas.

DESIGN LITERATURE

Over the years many good books on design have appeared, and more will continue to appear. Every design engineer should broaden his or her knowledge and perspective by reading as much as time permits on this and related subjects. In the following we give a brief review of design literature and by doing so also provide a glimpse of how the subject has evolved.

In the United States, the majority of books on mechanical engineering design deal with machine element design—e.g., those by Juvinall and Marshek (*Fundamentals of Machine Component Design*, 1991) and Shigley and Mischke (*Mechanical Engineering Design*, 1989). In the design of machine elements, shapes of objects are specified and the primary problem is that of determining the sizes, and secondarily choosing the material; manufacturing processes are seldom referred to. Engineering and particularly product design, on the other hand, are much more comprehensive topics. During the 1960s a number of books appeared in the United States that dealt with design from a wider viewpoint, such as

Dixon's *Design Engineering* (1966). For various reasons, however, engineering sciences have been emphasized to a much greater degree, at the expense of design, in both education and research. Since 1986, when the American Society of Mechanical Engineers (ASME) published a position paper on the importance of engineering design and how it must relate to manufacturing (Rabins, 1986), the picture has changed. The results, however, if at all observable, are variable. This contrasts sharply with activities in Europe. In Germany during the 1960s, for example, a concerted effort was launched for the improvement of design education, research, and practice. It should also be mentioned that in design education the influence of European educators as found in, for example, Hubka, Andreasen, and Eder's *Practical Studies in Engineering Design* (1988), Hubka and Eder's *Engineering Design* (1992), and Pahl and Beitz's, *Engineering Design—A Systematic Approach* (1988) has made some inroads, largely because these works are available in the English language.

The first book on the method of systematic design to appear in English was by Matousek, also entitled *Engineering Design—A Systematic Approach* (1963), a translation of his original work in German. Matousek brings together all the ideas of problem definition, the requirements list, and preparing the basic design before going on to the choice of materials and manufacturing methods. He provides an extensive discussion of shape (form) design for the different materials and manufacturing methods. This book appeared in 1963, and perhaps in light of what has been said above it was hardly taken note of in the United States. The book on systematic design by Pahl and Beitz (1988) provides a very thorough treatment of the subject. The authors emphasize that a designer must clarify the problem, work with an open mind, come up with a number of solutions at each stage, and evaluate and choose before proceeding to the next stage. The book covers the topics of design fundamentals, product planning, clarification of the task, conceptual design, embodiment design, and development of products in size ranges and as modular products. The chapter on embodiment design (which implies giving concrete shape to an abstract concept) includes ideas on designing for production and ease of assembly. The method is also discussed by Hundal (1990) from the point of view of computer application. Cross, in his book *Engineering Design Methods* (1989), provides several examples and applications of systematic designing. Roth, in *Designing with Design Catalogs* (1982), has developed a methodology characterized by systematically assembled design catalogs, i.e., collections of solutions. The second edition of his book appeared in 1994; it is a totally revised and greatly enhanced version. It is a pity this book is not available in an English translation. However, it is worth acquiring it for the figures alone, particularly the design catalogs. The English language version of VDI Guideline 2221, "Systematic Approach to the Design of Technical Systems and Products" (1987), sums up the German design philosophy.

Other English language works on engineering design deserve some description. Love's book *Planning and Creating Successful Engineered Designs* (1986) is written for practicing engineers by a practicing engineer. He follows the tried and tested procedure: talk to the customer, pin down the requirements, check feasibility, choose between alternatives, and optimize the design. He also puts weight on design management. Pugh, in *Total Design* (1990), describes the "design core" topics of market needs, product specifications, conceptual design, detail design, manufacture, and marketing. The more recently emerged topics of quality function deployment (QFD), failure mode and effect analysis (FMEA), and Taguchi methods are also presented. Ullman, in the first part of his 1992 text *The Mechanical Design Process*, discusses types of design and the human element in design—how humans design and interact with products. The main part of the book deals with the design process, including specifications, planning, concept generation and evaluation, detailed design, evaluation, design for assembly, and finalizing the design. A book by Ulrich and Eppinger, *Product Design and Development* (1995), addresses the complete product development cycle. The book includes chapters on organizations, customer needs, product specifications, concept generation and selection, product architecture, industrial design, design for manufacturing, prototyping, and the economics and management of development projects. Boothroyd, Dewhurst, and Knight, in their book *Product Design for Manufacture and Assembly* (1994), have provided a thorough coverage of design and manufacturing considerations in design, including a significant amount of information on costs of materials and manufacturing processes. The chapters in the book deal with design for manual, automatic, and robotic assembly; wire harness and printed circuit board design; and design for machining, injection molding, sheet metal, die casting, and powder metal processing. Finally, the most recent book on design to come to this author's attention is Dixon and Poli's *Engineering Design and Design for Manufacturing* (1995), a comprehensive, major work on design, with emphasis on manufacturability aspects.

There appears to be a lack of communication between manufacturing management professionals and design engineering professionals. It is hoped that the current thrust toward concurrent engineering will help to bring these groups closer. In design engineering and design for manufacturability literature, there has been considerable activity in the area of design for cost. Outstanding examples of this are the proceedings of the International Conference on Engineering Design (ICED), beginning in 1981, published by Heurista, in Zurich, Switzerland. Also, ASME has published the proceedings of a session on cost under *Design for Manufacturability — 1993*, DE-Vol. 52.

Although there is a considerable amount of literature available on product costing, the first comprehensive work relating to costs in mechanical design was published (in German) by Ehrlenspiel, *Cost-Effective Designing*, in 1985. The book provides guidelines and rules for lowering product costs, and methods for estimating these during the design process. Extensive reference to German language literature on the subject is provided in the book. A continual emphasis in the book is on the systematic design method. It points out that while product's costs are influenced at all stages of its life—from design order to sales—the most important factors are:

1. The concept, including physical effects, material type, and number and type of active surfaces
2. Size of the product, i.e., dimensions and amount of material
3. Number of parts, including standard, similar, and same parts

Figure 2, from Ehrlenspiel's book, shows what percentage of costs are already set at each stage of product development and what additional costs are incurred at each stage. The point made here is that at the design stage we have the most leverage on the product costs. It has been estimated from various sources that 70 to 80% of the costs are set at the design stage, whereas designing itself accounts for only 6% of the product cost. Trucks, in his book *Designing for Economical Production* (1987), gives helpful and practical hints for designers, with the aim of selecting the best production process, as well as for reducing manufacturing costs.

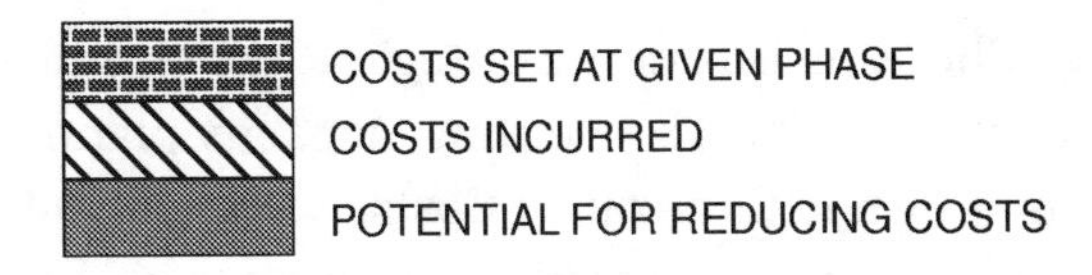

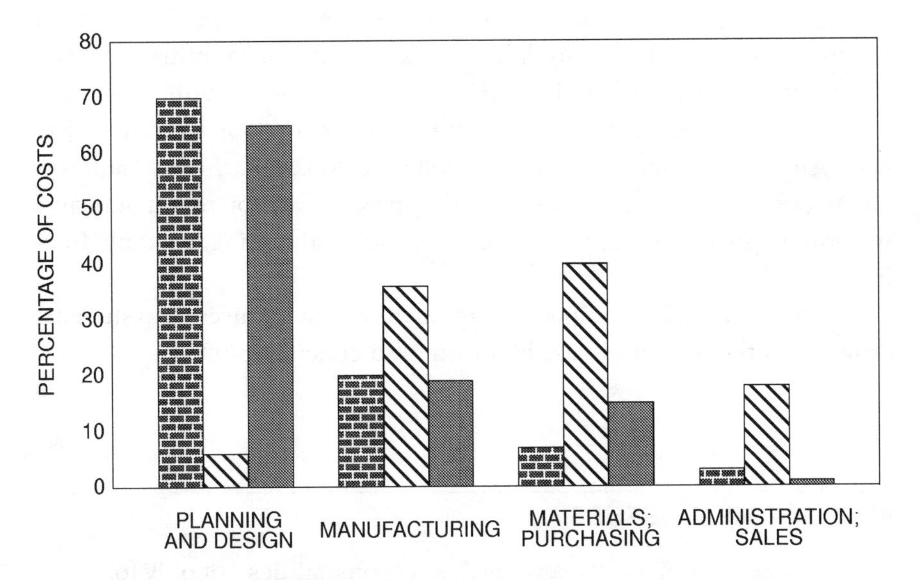

FIGURE 2. Product costs set and incurred in different activities. (Ehrlenspiel, 1985)

The first part of the book presents manufacturing methods for metallic materials. The second part of the book discusses the processing of polymers, and cost reduction methods, limits and fits, mechanical assemblies, and management tools.

A management perspective of designing to cost, oriented toward the defense industry, is given in a book by Michaels and Wood, *Design to Cost* (1989). Among the chapters in the book are "Getting Started," which addresses management decision profile, affordability analysis, cost-effectiveness analysis, value analysis, risk management, and design-to-cost goals; "Defining Activities and Responsibilities," which includes functional responsibilities, suppliers, specifications, and the review process; "Designing for Inherently Low Cost," which looks at product simplification, role of castings, simplification in production, testability, and uniformity; "Choosing Business Strategies," which addresses procurement, suppliers, make-or-buy, scheduling, capacity planning, and inventories; and "Estimating and Controlling Costs," which includes cost structures, learning curves, estimating confidence, time value of money, and elements of cost control. Each chapter includes case studies and exercises. Ostwald, in *Engineering Cost Estimating* (1992), provides a thorough treatment of cost estimating, including labor, material, and accounting analyses, forecasting, preliminary and detail methods of cost estimating, the pertinent topics of operation and product cost estimating, as well as project and system estimating. Winchell, in *Realistic Cost Estimating for Manufacturing* (1989), describes cost estimating procedures for manufacturing, including applications to machining, die and sand casting, welding, forging, etc.

The present book attempts to combine the best features of systematic design and the design-to-cost literature in a concise volume.

MANAGEMENT ISSUES

The management of a company carries responsibilities not only for product costs but for a host of other characteristics such as quality and development time, as well as for the activities which affect those characteristics. It is therefore worthwhile to review the history of U.S. manufacturing industry in order to understand the place of design and manufacturing today.

After World War II, U.S. manufacturing industry prospered up until the 1970s and then began a decline that is only now being reversed. American industry relied on innovation and short time to market for its success in producing goods of high quality and low cost. The decline has been attributed to management's attitudes toward manufacturing and technology, its capabilities, and the strategies it used in pursuing its aims, as pointed out by Hayes, Wheelwright, and Clark in *Dynamic Manufacturing* (1988). To be sure, there were other factors present, such as high interest rates, an overvalued dollar, and tax laws that encouraged personal debt. However, the success stories during this period (the 1980s) of companies such as Xerox (Bebb, 1989) and Chrysler show that actions of management were able to turn major companies' fortunes around in this same climate.

American industry entered the 1960s after the postwar boom flush with capital, with a large pool of skilled labor as well as management that had gained its experience during those boom years. American industry was in a position where it could make standard products at lower cost and of higher quality than foreign competition. The decline has been attributed to the fact that management began to neglect manufacturing technology; it became top-heavy and distanced itself from labor. However, by about this time Japanese manufacturers were moving in just the opposite direction. They had become adept in getting workers, managers, and personnel in engineering, manufacturing, and marketing to work together. They also made their suppliers part of their team and even went as far as to make their customers feel that they had a voice in what products were made and what features the products had.

The key to becoming competitive is not necessarily new plants and equipment but rather applying the resources of the company, both human and machines, to achieve a competitive edge in the marketplace by delivering high-quality products at low cost and providing dependable service. Management must rediscover the obvious fact that the most important function of a manufacturing company is manufacturing. The term *world-class manufacturing* has come about in the past few years. To be a world-class manufacturer, a company must be as good as the best of the competition in one or more measures: cost, quality, service, etc. These goals can be achieved by the company by matching the manufacturing process to the product, continual improvement, and learning from experience.

American industry grew in the nineteenth century by combining the old-world tradition of craftsmanship with the newly developed methods of scientific management. Hayes et al. (1988) cite the history of Mesta Machine Company, which produced industrial machinery for a worldwide market. Its success was based on superb engineering, innovative development of new products, and dependable service. Its management was in close touch with the factory floor. The head of Mesta, Lorenz Iversen, who guided the company through its growth for over 50 years, was a degreed engineer as well as a craftsman. He always looked for innovative solutions to problems, stayed at the cutting edge of technology, instilled a sense of pride and a sense of purpose in the workers, and demanded precision and excellence. After Iversen's retirement in the 1950s, the company's management changed and it lost its technological drive and the competitive advantage it had enjoyed based on excellence in engineering. The new management of this (for its time) high-tech company came from a financial background and had little knowledge of technology. It added more layers of management and became too distant from its staff and the people on the factory floor, losing touch with the engineers and workers. The company's sales declined, and it eventually declared bankruptcy. The story of Mesta was to repeat many times over in other companies and other industries such as steel, automobiles, and machine tools.

In the early days, U.S. industry was run in the "artisan" mode of management, which was characterized by individual responsibility, technical excellence, customer service, and continual improvement. On this scene came Frederick W. Taylor, who perfected and applied his ideas of scientific management. While his name is associated with time and motion studies, his vision encompassed the whole of the manufacturing enterprise. His ideas may be summarized by the following points, taken from his book *Scientific Management* (1947):

- Find, through scientific studies, the one best way of doing a certain task.
- Since different people have different talents and abilities, match people to the tasks.
- Put all of the planning and control in the hands of a trained staff.

Beginning with the early twentieth century, Taylor's methods produced a revolutionary change in U.S. manufacturing industry. Taylor introduced statistical process control and methods for planning and managing material flows, which helped mass production enterprises to form and flourish. The shortcomings of Taylor's methods became evident only decades later: the "best" way became the only way, fixed forever; there was no call for continual improvement; and managers became separated from the workers. Taylor's model for mass production systems is claimed by Waldner, in *CIM—Principles of Computer-Integrated Manufacturing* (1992), to resist change from a quantity-based to a quality- and variety-based culture.

To his credit, Taylor emphasized research and experimentation. However, he separated learning and improvement of the production processes by the staff from actual production by the workers. This broke the vital link necessary for two-way communication. Unless staff personnel are involved in actual production, they lose learning opportunities. Likewise, the workers are separated from the analysis and learning processes. Each loses sight of the real objective, to produce competitive products. The rigid, static system created by scientific management is just the opposite of the aim of a world-class manufacturer: to tap the reservoir of knowledge, skills, experience, technology—i.e., all the benefits of its human resources—to produce quality products.

A significant factor contributing to the prosperity of U.S. (or, for that matter, any) manufacturing industry was the fostering of organized industrial research. Science was applied by engineers at first to make better measurements, to design reliable and repeatable tests, and for identification of product properties. Only later was scientific research used for improvements in products and processes. The guiding factors in conducting industrial research that helped industry so much were:

- Realization that technological improvement is an evolutionary process, which proceeds in small steps
- Keeping laboratory research closely tied to industry's strategic goals
- Developing the product and the process to manufacture it concurrently
- A cross-functional, collaborative effort directed toward success

Part of the decline of U.S. manufacturing industry can be attributed to the distancing of industrial research from practice. More emphasis was put on "big science" at the expense of improving products and manufacturing methods in incremental steps. Instead of building quality into the products, rework stations and warranty contracts came into being.

As a side note, while industry was starting on its decline, U.S. engineering schools were becoming more theoretical. The format of today's engineering education in the United States derives in large part from the "Grinter Report" of 1955. The report made recommendations on changes in curricula in light of the conditions following the Second World War: (1) strengthen mathematics and basic sciences; (2) provide a common core of engineering sciences; (3) integrate study of engineering analysis, design, and systems; (4) include elective subjects to provide individuals flexibility; (5) integrate humanities and social sciences into the curriculum; (6) emphasize oral, written, and graphic communication; (7) encourage experiments; and (8) strengthen graduate education.

Over the years these goals have been met with varying degrees of success. After the Grinter Report, a further stimulus to the reshaping of the engineering curricula was the launching of *Sputnik* in 1957. The support of the National Science Foundation (NSF) and additional funding from NASA and the defense agencies led to outstanding research in the engineering sciences and the strengthening of the basic and engineering science content of the curricula. There was, however, no corresponding effort in engineering design, which in any case was only alluded to in the Grinter Report (item 3), nor in manufacturing. This contrasts sharply with Europe; in Germany during the 1960s, a concerted effort was launched for the improvement of design education, research, and practice. Education in manufacturing had a similar fate in the United States: since there was little science in manufacturing in the 1950s and early 1960s, it was taken out of many programs, when an effort should have been made to develop a science base in this critical activity.

Another concept that has come to the forefront of today's manufacturing is "just-in-time" delivery of materials. Actually it had been applied successfully by Henry Ford in the early years of the twentieth century. Ore arriving in boats on a given day was out after two days in the form of a Model T. By the time of the Second World War, mass production had been mastered by U.S. manufacturers. The combination of craftsmanship skills and scientific management had brought about a flowering of industrial enterprises of unprecedented productivity.

The onset of World War II required U.S. industry to:

- Switch from civilian to military production
- Increase its production capacity manyfold
- Manufacture products that were themselves rapidly developing
- Develop manufacturing processes that had never existed

Armaments of various types as well as aircraft were, of course, already being manufactured, but in small quantities. The need was that they be manufactured in tens or hundreds of thousands. Also, the new systems had to be developed quickly. Industry achieved these aims by:

- Standardizing parts
- Simplifying designs, and using innovative designs
- Managing the flow of materials
- Having the design and manufacturing teams work together
- Improving existing manufacturing processes and developing entirely new processes
- Developing manufacturing systems in which labor with little or no training could be introduced
- Galvanizing management and the workforce to work for a common goal

There are many examples of such advances from World War II, but perhaps none more dramatic than the development and manufacture of military aircraft. Here, in addition to all of the above needs, the sciences of materials, aerodynamics, control, and propulsion had to be concurrently advanced. There are several outstanding examples of rapid product development from this time. North American's Mustang fighter was in prototype stage 100 days after order placement, and the Grumman Hellcat was developed in seven months. The B-24 bomber is a classic example of continual product and process improvement. The number of labor hours to produce one of these was reduced from 40,000 in 1943 to 8000 in 1945. To make one of these craft, 500,000 parts were needed, whose flow had to be planned and controlled toward the final assembly (without the help of a computer). The company achieved its objectives, among other ways, by

- Reducing the number of parts through design improvements
- Developing better production methods and tools
- Involving factory floor personnel in improving the production processes

The outstanding achievements of Lockheed's famous "Skunk Works" are mentioned elsewhere in the book.

The Japanese industrial development following World War II owes much to the U.S. occupation authorities. American engineers and scientists were brought in to train the Japanese in manufacturing engineering methods, including statistical process control, scheduling, and planning. The key ideas from the American teachers were: use ingenuity, be consistent in quality, and strive for continual improvement. The slogan "Put quality ahead of everything else" came from an American shipbuilding company of that time.

In his 1946 book *Fundamentals of Successful Manufacturing* (note the date!), Hyde makes the following points for a manufacturing company to succeed:

- Product and process designers must work together.
- Be flexible. Nothing is set forever; everything can be improved.
- Tools and mechanization are meant not to replace labor, but rather to provide leverage to labor.
- Aim for continual improvement in products and manufacturing processes.
- Invest in research and laboratory facilities and in new product development.

What U.S. industry had in the years following World War II, it is beginning to regain by applying the same ideas that made it so prosperous then.

The ultimate test of good engineering and management is the customer's satisfaction with the product—be it a skateboard, a space shuttle, a computer, an artificial joint, or a building.

BIBLIOGRAPHY

Bebb, H. B. "Quality Design Engineering: The Missing Link to U.S. Competitiveness." Keynote address, NSF Design Engineering Conference, Amherst, Mass., June 1989.

Boothroyd, G., P. Dewhurst, and W. A. Knight. *Product Design for Manufacture and Assembly*. New York: Marcel Dekker, 1994.

Cross, N. *Engineering Design Methods*. New York: Wiley, 1989.

Dixon, J. R. *Design Engineering*. New York: McGraw-Hill, 1966.

Dixon, J. R., and C. Poli. *Engineering Design and Design for Manufacturing*. Conway, Mass.: Field Stone Publishers, 1995.

Ehrlenspiel, K. *Kostengünstig Konstruieren (Cost-Effective Designing)*. Berlin and New York: Springer Verlag, 1985.

Grinter, L. E., ed. "Report on Evaluation of Engineering Education." ASEE, 1955. (Summary appears in *J. Eng. Ed.* 83(1): 74–94, 1994.)

Hayes, R. H., S. C. Wheelwright, and K. B. Clark. *Dynamic Manufacturing*. New York: Free Press, 1988.

Hubka, V., M. M. Andreasen, and W. E. Eder. *Practical Studies in Engineering Design*. London: Butterworth, 1988.

Hubka, V., and W. E. Eder. *Engineering Design*. Zurich: Heurista, 1992.

Hundal, M. S. "A Systematic Method for Developing Function Structures, Solutions and Concept Variants." *Mechanisms and Machine Theory* 25(3): 243–256 (1990).

Hyde, G. G. *Fundamentals of Successful Manufacturing*. New York: McGraw-Hill, 1946.

Improving Engineering Design. Washington, D.C.: National Academy Press, 1991.

Juvinall, R. C., and K. M. Marshek. *Fundamentals of Machine Component Design*. New York: Wiley, 1991.

Love, S. F. *Planning and Creating Successful Engineered Designs*. North Hollywood, Calif.: Advanced Professional Development, 1986.

Matousek, R. *Engineering Design—A Systematic Approach*. London: Blackie, 1963.

Michaels, J. V., and W. P. Wood. *Design to Cost*. New York: Wiley, 1989.

Munro, A. S. (private communication). Norwalk, Conn.: Reed Exhibition Companies, 1995(a).

Munro, A. S. "Let's Roast Engineering's Sacred Cows." *Machine Design*, February 9, 1995(b), pp 41–46.

Nevins, J. L., and D. L. Whitney. *Concurrent Design of Products and Processes*. New York: McGraw-Hill, 1989.

Ostwald, P. F. *Engineering Cost Estimating*. 3rd ed. Englewood Cliffs, N.J.: Prentice Hall, 1992.

Pahl, G., and W. Beitz. *Engineering Design—A Systematic Approach*. Berlin and New York: Springer Verlag, 1988.

Pugh, S. *Total Design*. Reading, Mass.: Addison-Wesley, 1990.

Rabins, M. J. *Goals and Processes for Research on Theory and Methodology*. New York: ASME, 1986.

Roth, K. *Konstruieren mit Konstruktionskatalogen (Designing with design catalogs)*. Berlin and New York: Springer Verlag, 1982. Also 2nd ed. (2 vol.), 1994.

Shigley, J. E., and C. R. Mischke. *Mechanical Engineering Design*. 5th ed. New York: McGraw-Hill, 1989.

Taylor, F. W. *Scientific Management*. New York: Harper, 1947.

Trucks, H. *Designing for Economical Production*. 2nd ed. Dearborn, Mich.: Society of Manufacturing Engineers, 1987.

Ullman, D. G. *The Mechanical Design Process*. New York: McGraw-Hill, 1992.

Ulrich, K. T., and S. D. Eppinger. *Product Design and Development*. New York: McGraw-Hill, 1995.

"Systematic Approach to the Design of Technical Systems and Products." VDI Guideline 2221. Düsseldorf: VDI Verlag, 1987.

Waldner, J.-B. *CIM—Principles of Computer-Integrated Manufacturing*. Chichester and New York: Wiley, 1992.

Winchell, W. *Realistic Cost Estimating for Manufacturing*. 2nd ed. Dearborn, Mich.: Society of Manufacturing Engineers, 1989.

The Product Realization Process

The realization of a product consists essentially of two steps: design and manufacturing. However, in order to understand the process, we must look at the total life of a product. The life of a product begins with its planning and ends with its disposal, i.e., scrapping and/or recycling. Figure 3 shows the different life phases of a product. These will be described in brief.

STEPS IN THE PRODUCT REALIZATION PROCESS

A need for the product, whether real or imagined, must exist. This may come from external or internal sources. External forcing for a new product may be due to:

- A direct order from a customer
- Obsolescence of an existing product
- Availability of new technologies
- Change in market demands

Internal to the company, new product ideas may come from:

- New discoveries and developments within the company
- Need for a product identified by the marketing department

Once the need has been established, the product has to be designed and manufactured. We refer to these steps—from the go-ahead to when the product physically exists—as the *product realization process (PRP)*. This process includes the first four steps shown in Figure 3:

- Product planning
- Design
- Process planning
- Manufacturing

These steps will be briefly summarized here. Design and manufacturing are described in more detail in later chapters.

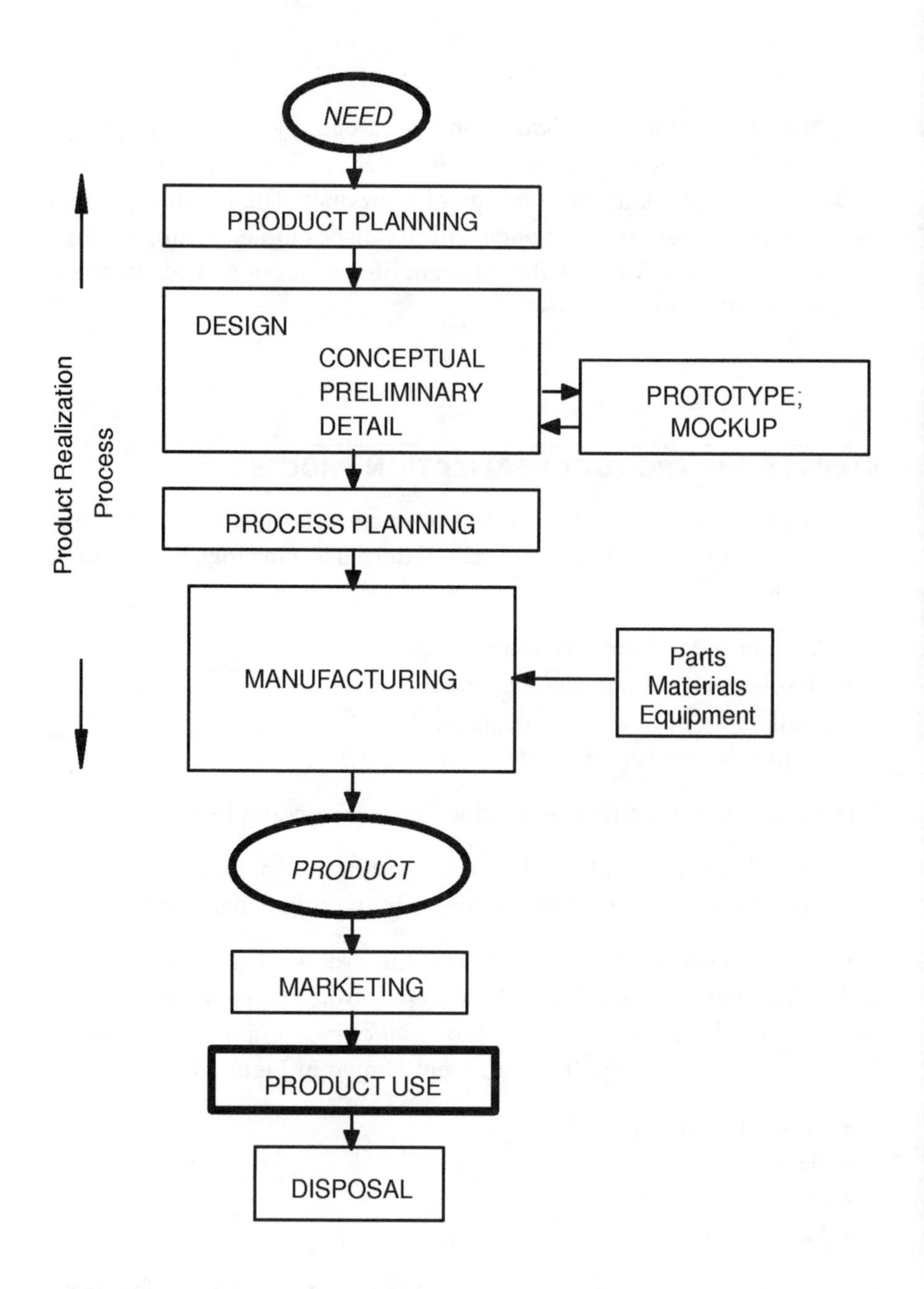

FIGURE 3. Life phases of a product.

It should be emphasized here that the above steps appear in Figure 3 as sequential. This is the so-called gateway model of project administration. In this process each step must be complete before the next step begins. As we shall see later, this procedure leads to delays, mistakes, poor quality, and high costs. A preferable product development procedure is one in which the activities in the successive stages are concurrent and partially overlap.

Product Planning

Product planning is the search for and selection and development of ideas for new products. A systematic approach to product planning will lead to a better meeting of the constraints of cost and time.

Product planning activities include:

- Establishing product goals
- Conducting market analyses
- Detailing the benefits the product will provide the customer
- Deciding on the features the product will have
- Establishing product performance
- Conducting an economic analysis and setting the cost target
- Establishing the expected sales volume
- Setting deadlines for completion of tasks, such as design, prototype building, and setting up the manufacturing line

The two most important entities involved in making the decision to develop a product are the company and the market. There are also secondary factors, such as government laws, economic policies, and state of the technology. This interaction is shown in Figure 4. Specifically,

- The company needs to define its objectives and examine its capabilities.
- The types of strengths that a company has are its personnel, its facilities, and its financial situation.

The personnel and facilities are distributed among various types of activities or departments, e.g., design, production, and marketing; and among different buildings, such as those for design, test, and production equipment and those for distribution systems. An evaluation of resources and

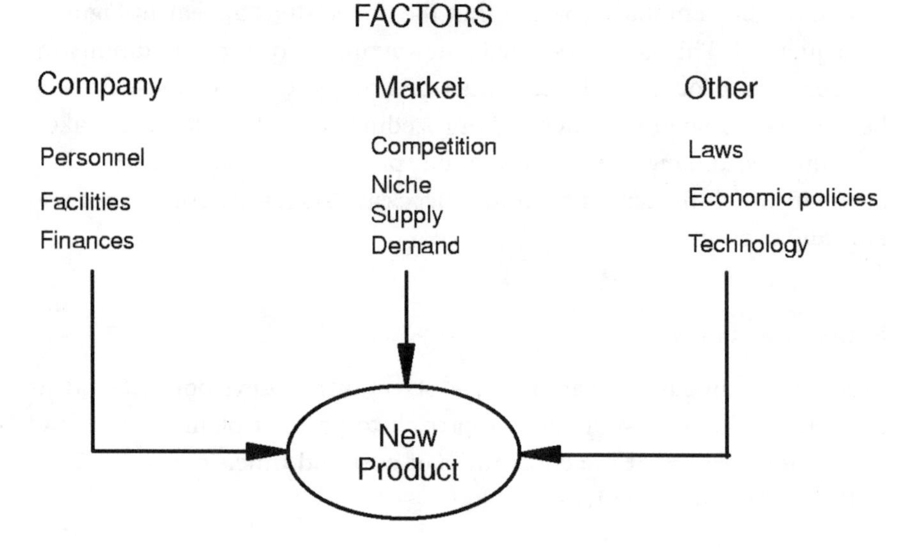

FIGURE 4. Factors influencing new product development.

objectives will help focus the company on the type of products it should develop.

The market is a moving target. The length of time it takes for product development is very critical. The longer the time to product introduction, the more uncertain will be the market forecast and therefore the greater the risk. Let us say it takes two years from the start to the time a product is introduced into the market. The market analysis will have to predict what the market will be like two years hence. Thus it pays to develop products in as short a time as possible. If we cut down the development time to one year, planning becomes simpler and there will be less risk involved. We shall elaborate further on this point in Chapter IV (Figure 36).

DESIGN

Product design includes activities from generating a requirements list for the product, to developing ideas on what it should look like and how it should operate, and on to producing complete drawings and documentation. The latter contain the complete information on the product, from which it may be manufactured. Steps in design are:

- Preparation of the requirements list
- Conducting a technology survey: determining what is feasible at the time using feasible technologies
- Conceptual design
- Preliminary or embodiment design
- Designing, building, and testing a prototype; building a mockup if appropriate
- Detail design
- Preparing documents such as a bill of materials and assembly, operating, and service instructions

Process Planning

Also called production or manufacturing planning, process planning involves decisions on how the product is to be manufactured—for example, what steps are required to manufacture the product; which manufacturing processes, machines, and tools are required; and how the parts are to be assembled. Steps in process planning are:

- Producibility analysis
- Initial process design
- Vendor and sourcing selection
- Tooling design
- Final process design

Manufacturing

Under manufacturing, we include materials handling, production of parts, assembly, quality control, and related activities. Many of the decisions regarding manufacturing have already been made during the embodiment design stage, knowingly or unknowingly.

Steps in manufacturing are:

- Tool and equipment procurement
- Production line setup
- Test runs
- Production runs

The steps in product realization are summarized in Figure 5.

Product planning

Establishing product goals
Conducting market analyses
Detailing the benefits this product will provide the customer
Deciding on the product features
Establishing product performance
Conducting economic analysis
Setting cost target
Establishing expected sales volume
Setting deadlines: design, prototype building, set–up of manufacturing line, etc.

Design

Preparing requirements list
Conducting technology survey: determine what is feasible with the available technologies
Concept devlopment
Preliminary, or embodiment design
Building and testing prototype
Detail design
Preparing documents: bill of materials, operating instructions, service instructions, etc.

Process planning

Producibility analysis
Initial process design
Vendor/source selection
Tooling design
Final process design

Manufacturing

Tool and equipment procurement
Production line set–up
Test runs
Production runs

FIGURE 5. Steps in product realization.

TYPES OF PRODUCTS

We can classify products in many different ways, for instance, consumer products versus commercial and industrial products. Consumer products are intended to be used by the general public, who have little or no training in their use. Commercial and industrial products are, as a rule, used and operated by trained personnel, in a given type of environment.

Much of what is said in this book is aimed at a "standard" or "generic" product. This is a type of product for which a market need has been recognized; the company then proceeds to fulfill this need. Products can be categorized into four other types, besides the standard or generic product (Ulrich and Eppinger, 1995):

- *Technology push product*. A company develops a new technology. It then puts this technology to work in a variety of different products.

An example of this type is Gore-Tex, an expanded fluorocarbon sheet produced by Gore Associates, which has been used for rain gear, artificial blood vessels, and insulation, among other applications.

- *Platform product*. A company develops a new technology that is tried, tested, and established. It is then used as a basis for other products. One example of this type of product is Polaroid instant film. Many electronic products also fall into this category.
- *Process-driven product*. This is a product whose very existence depends on a very specialized process developed for its manufacture. Examples are paper, semiconductors, electric wiring, and certain types of food products.
- *Customized product*. This is a variation on an existing, well-established product of the company. (See also Figure 14 and accompanying discussion on types of design in Chapter III.) Examples of this type of product are many "standard" items such as electric motors.

ORGANIZATIONAL STRUCTURES

Engineers, designers, manufacturing personnel, and others in a company are organized so that the company may carry out its functions in an optimal fashion. An organization establishes links between individuals in groups and between groups. The links, which enable and facilitate the flow of information in an organization, may be hierarchical, the result of physical proximity, or set up through electronic networks.

Organizational structures are of one of three types: functional, project-based, or matrix type. These will be discussed next.

FUNCTIONAL ORGANIZATION

In a functional organization, individuals are grouped together according to their training and expertise. This is the traditional type of organization of most manufacturing companies. The divisions or groups, as shown in Figure 6, are headed by managers or vice presidents. The actual division names vary with companies. For example, the group labeled "product engineering" in Figure 6 may be called "engineering" or "new product

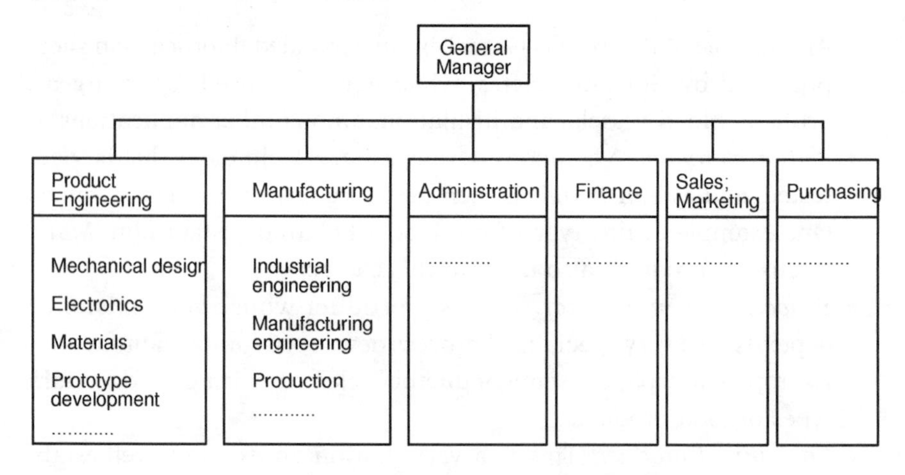

FIGURE 6. Functional organization.

development," and there might be a separate group called "research and development." Each division or group may have specialized subgroups, as shown under "product engineering" and "manufacturing." The people in each group perform the same or similar functions. Each individual may work on one or more projects at a time. The individuals develop a high degree of expertise in their fields. Their links with other groups are weak. It is difficult for such organizations to respond to rapidly changing market needs. Companies producing "customized products," such as power plants, tend to be functionally organized.

PROJECT ORGANIZATION

In a project organization, a number of individuals from different functions are brought together to form a group that will work on a project, led by a project manager. As shown in Figure 7, each project group has enough individuals from each specialty to carry through the given project. For the duration of the project these individuals from different fields have strong links to each other. When the project is completed, the individuals are reassigned to other projects. Construction companies are the best-known example of project-based organizations.

Project-based organizations are better suited to respond to rapidly changing market demands, because of their flexibility. There is better

PROJECT No. 1	PROJECT No. 2	PROJECT No. 3
Mechanical	Mechanical	Mechanical
Materials	Materials	Materials
Electrical	Electrical	Electrical
Manufacturing	Manufacturing	Manufacturing
....	....	....

FIGURE 7. Project-based organization.

cross-functional communication, and thus better coordination; and both technical and nontechnical aspects of a project can be optimized faster. A drawback in a project organization is the lack of opportunity to develop the same degree of expertise as in a functional organization.

MATRIX ORGANIZATION

A combination of functional and project organizations, known as a *matrix organization*, is shown in Figure 8. Individuals from the same function are physically located together and report to their group leader as in a

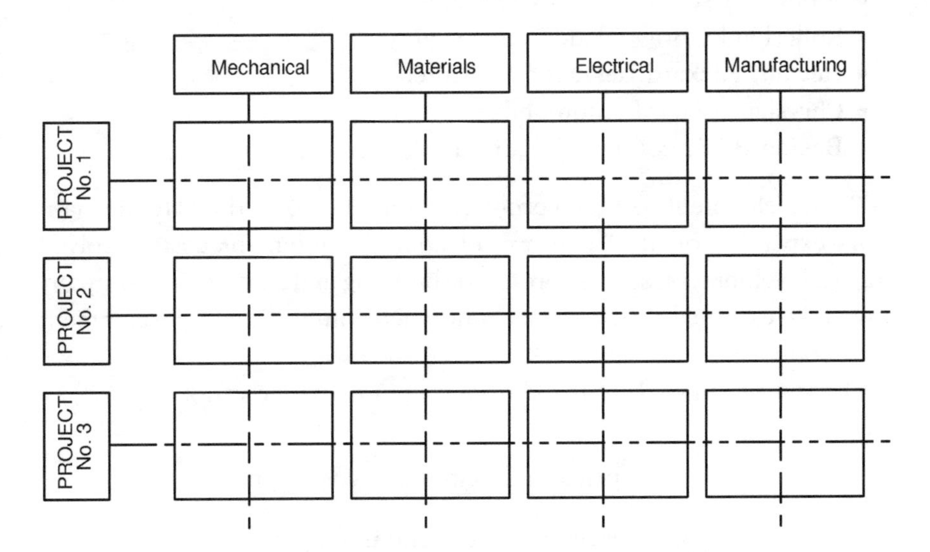

FIGURE 8. Matrix organization.

functional organization. However, they are assigned to a project, led by a project manager. While the latter is responsible for the completion of the project within time and money constraints, the functional manager is responsible for the resources needed for the various projects. A matrix organization appears to combine the desirable features of both the functional and the project organizations.

Theoretically, each individual in a matrix organization has two supervisors—the functional and the project supervisor. In reality, one or the other type of link tends to be stronger. Matrix organizations can accordingly also be divided into lightweight and heavyweight project organizations (Hayes, Wheelwright, and Clark, 1988), depending on whether the functional or the project manager plays the stronger role, i.e., has the budget authority.

PRODUCT DEVELOPMENT RESOURCE REQUIREMENTS

Figure 9 shows data from the development of five different products, which vary over wide ranges of cost, size, and complexity (Ulrich and Eppinger, 1995):

- Jobmaster screwdriver, from Stanley Tools
- Rollerblade, Bravoblade in-line skates
- Hewlett-Packard DeskJet 500 printer
- Chrysler Concord automobile
- Boeing B-777 commercial aircraft

The development cost for a product is roughly proportional to the man-years expended on it. There are, of course, equipment costs involved during development, such as prototype building and testing. The company must take expected sales into account when committing to development of a product.

The last bar in each of the charts in Figure 9 is the ratio:

$$\text{Sales price to development cost ratio}$$

$$= \frac{\text{(Sales price per unit)(Total units produced)}}{\text{Development cost}}$$

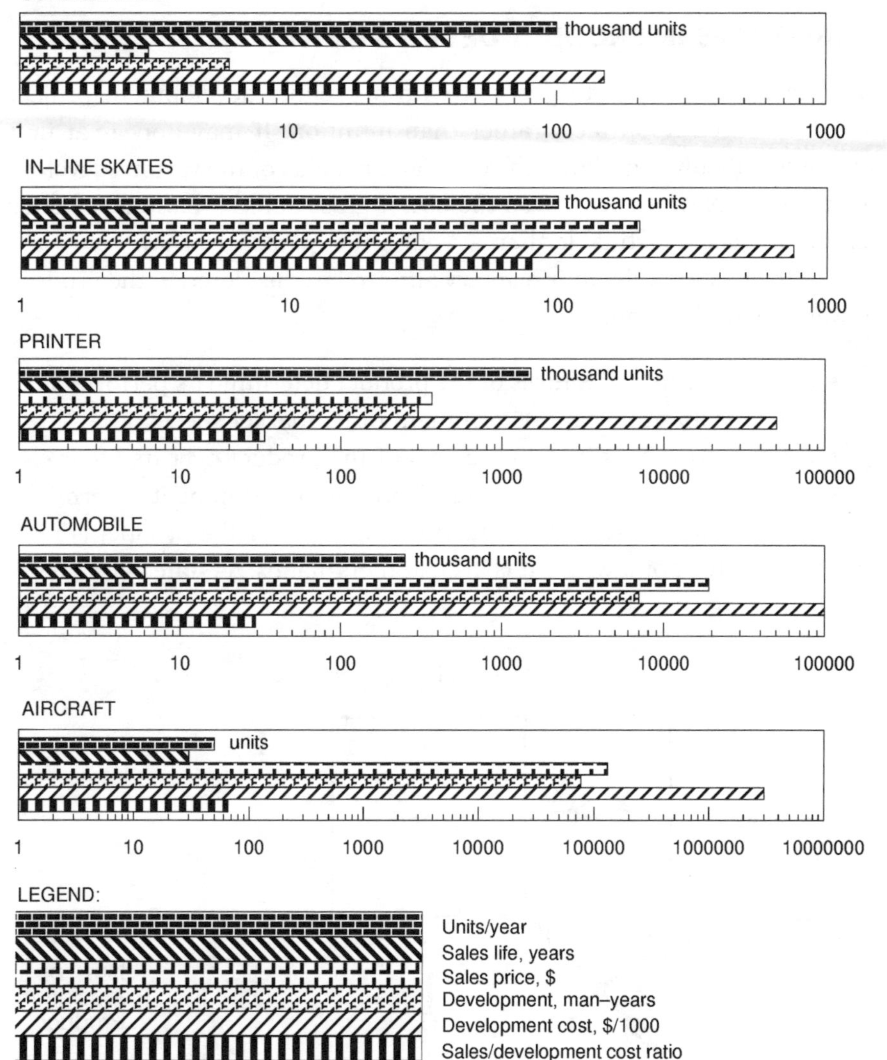

Figure 9. Development resources for different products. (Ulrich and Eppinger, 1995)

This ratio is of the same order of magnitude for all the products—in the range from 30 to 80. The investment required to set up the production facilities has been shown by Ulrich and Eppinger to be of the same order of magnitude as the development cost.

TRADE-OFFS IN PRODUCT DEVELOPMENT

The decision whether to proceed with a product development project must be based upon facts rather than intuition. It may appear at first that it is difficult to quantify the costs and benefits of the various development goals. Nevertheless, even the use of gross approximations is better and more easily justifiable than a "seat-of-the-pants" decision.

Figure 10 shows the four primary aims to be considered in the product development decision:

- *Product features.* Features of the product determine its performance, which is an important determinant of its market success.
- *Product cost.* This is the total cost of the product over its life cycle.
- *Development speed.* The speed of product development determines the time to market—from the time the need for the product is realized, to the point where it is in customers' hands. As stated elsewhere,

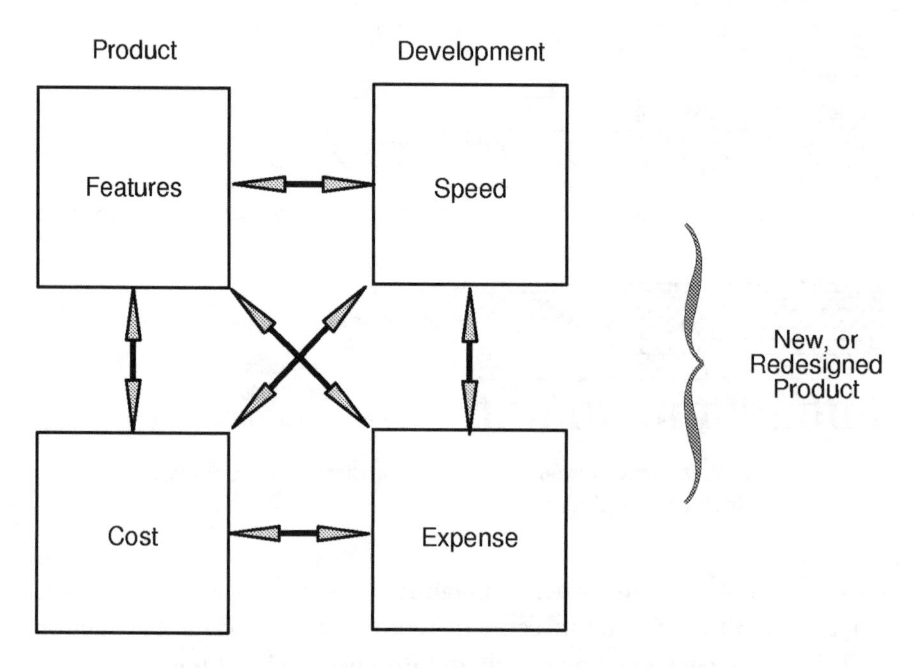

FIGURE 10. Product development aims and trade-offs. (Smith and Reinertsen, 1991)

it can be crucial to the success of the product. It involves all of the company's departments, not just design engineering.

- *Development expenses.* These are the one-time costs associated with development of the product. Although seldom an overriding concern, these expenses must be justified to upper management.

Each of these aims needs to be weighed against the other three on the basis of costs and benefits before the decision to proceed can be made.

• • • • •

EXAMPLE
LAWN TRIMMER DEVELOPMENT TRADE-OFFS

In order to illustrate these trade-offs, we consider the following example. A company makes lawn trimmers at the rate of 2500 units per year.

- Sale price of each unit = $1500.
- Over a period of 5 years, the gross sales are (2500) × ($1500) × (5) = $18,750,000.
- The gross profit margin is 32% (factory cost is 68% of the sales cost).
- Incremental profit (extra profit for each dollar of sales) is 17%.

The company has to decide whether to add a certain feature (say, a self-starter) to the lawn trimmer. In the following, we show the calculation of the costs and benefits as each of the development goals in Figure 10 is individually weighed against the others. After that we imagine the real-life situation in which all the goals are considered in combination.

I. Product Features versus Product Cost

Should we add the feature? The feature will:

- Increase sales by 15%.
- Increase cost by 20%.

A. Benefit

Increase sales by ($18,750,000) × (15%) = $2,812,500.

Increase profits @ 17% profit margin: ($2,812,500) × (17%) = $478,125.

B. Cost

Product cost: ($18,750,000) × (68%) = $12,750,000.

Increase in cost is ($12,750,000) × (20%) = $2,550,000.

Since the cost is higher than the benefit, the conclusion on the basis of this comparison is: Don't add the feature!

II. Product Features versus Development Cost

Is it worth spending $30,000 in development costs to add the feature?

A. Benefit

Increase sales by ($18,750,000) × (15%) = $2,812,500.

Increase profits @ 17% profit margin: ($2,812,500) × (17%) = $478,125.

B. Cost

Development cost = $30,000.

Since the benefits are higher than the cost, the conclusion on the basis of this comparison is: Add the feature!

III. Product Cost versus Development Cost

Should we spend $30,000 to reduce product cost by 2%?

A. Benefit

Product cost: ($18,750,000) × (68%) = $12,750,000.

Reduction in cost: ($12,750,000) × (2%) = $255,000.

B. Cost

Development cost = $30,000.

Since the benefits are higher than the cost, the conclusion on the basis of this comparison is: Add the feature!

IV. Product Cost versus Development Speed

Next, we look at the advantages of reducing the product cost by redesign. The redesign will reduce the factory cost by 20%. However, the redesign is expected to delay introduction of the product by 12 months. On the basis of the lost sales, it is estimated that each month of delay will cost $100,000.

A. Benefit

One year delay amounts to 20% of the 5-year period considered.

Thus the revenue affected = ($18,750,000) × (20%) = $3,750,000.

Product factory cost @ 68% = $2,550,000.

Cost decrease (20% of $2,550,000) = $510,000.

B. Cost

Each month of delay costs: $100,000.

Thus 12 months of delay costs: $1,200,000.

Since the cost is higher than the benefit, the conclusion on the basis of this comparison is not to add the feature. We should continue with the existing design for the next 12 months. Of course, the sooner the change is made, the better will be the benefit-to-cost relationship.

V. Product Features versus Development Speed

Should we add a feature that will add 1% to the sales but delay product introduction by 3 months?

A. Benefit

Feature adds 1% to sales: ($18,750,000) × (1%) = $187,500.

At 17% profit margin, profit = $31,875.

B. Cost

Each month of delay costs: $100,000.

Thus 3 months of delay costs: $300,000.

Since the cost is higher than the benefit, the conclusion on the basis of this comparison is not to add the feature under these conditions.

VI. Development Cost versus Development Speed

Next, we look at the advantages of spending extra money (overtime, or outside services) to speed up development. Use of outside services will save 1 month of product development time. Its cost is $5400.

A. Benefit

Savings @ $100,000 per month = $100,000.

B. Cost

Outside services cost: 180 hours @ $30 per hour = $5400.

Since the benefits are higher than the cost, the conclusion on the basis of this comparison is: Spend the extra money to speed up development.

Combination of Trade-Offs

Thus far we have looked at the trade-off parameters in pairs. Now we consider a more realistic situation where a combination of trade-offs must be weighed.

The company is considering adding a feature which will:

- Increase sales by 15%.
- Increase the product cost by 1%.
- Delay development by 1 month.
- Add $30,000 to the development cost.

A. Benefit

The increased sales result in increased profits:

Increase sales by ($18,750,000) × (15%) = $2,812,500.

Increase profits @ 17% profit margin: (\$2,812,500) × (17%)
= \$478,125.

B. Cost

The cost components are as follows:

Product cost = (\$18,750,000) × (68%) = \$12,750,000
Increase in cost = (\$12,750,000) × (1%) = \$127,500
One month of delay cost = \$100,000
Development cost = \$30,000
Total cost = \$257,500

Since the benefits are higher than the cost, the conclusion on the basis of this comparison is: Add the feature!

• • • • •

PREREQUISITES TO ENGINEERING DESIGN

The total engineering design effort requires the knowledge and appreciation of a number of fields, including the basic and engineering sciences, the humanities, the social sciences, and the fine arts. Engineering design is put in perspective in Figure 11, which is adapted from Dixon (1966). The different fields are arranged into four separate stems:

1. *The basic and engineering sciences stem.* The primary knowledge we need for engineering design derives from the basic or natural sciences: physics, chemistry, biology. The knowledge from the basic sciences has evolved further in the form of engineering or applied sciences, which are more directly useful to engineers. Engineering sciences provide an understanding of human-made devices and phenomena and help us optimize designs and operations. As an example, the engineering design of a water heater would require the application of thermodynamics, heat transfer, fluids, materials science, controls, and combustion or electrical science. The manufacture of all objects requires a knowledge of materials and manufacturing sciences. We should also mention here the medical sciences,

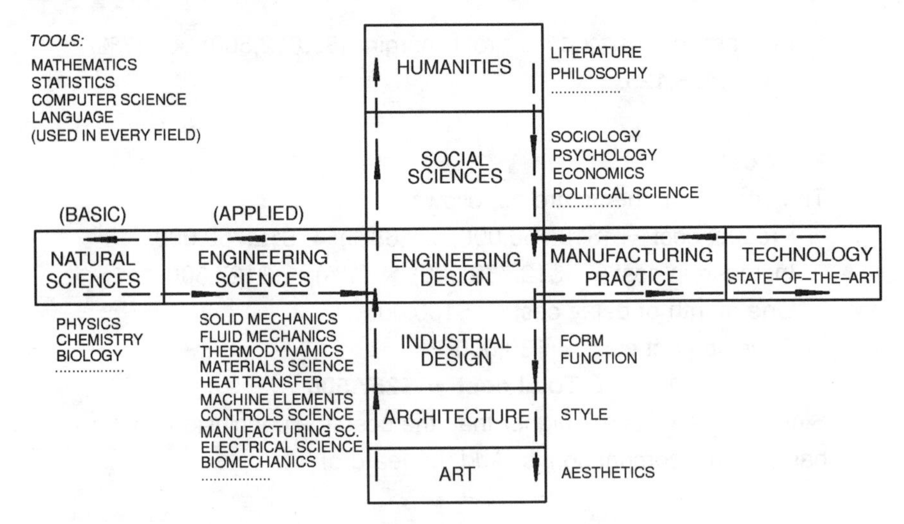

FIGURE 11. Place of engineering design in the technological and cultural world. (Adapted from Dixon, 1966)

which have evolved from biology, e.g., anatomy and physiology, the knowledge from which is used by biomedical engineers.

2. *The manufacturing and technology stem.* Anything we design must be buildable or manufacturable and is limited by the existing state of the art in technology. For example, what materials, manufacturing processes, software, and analytical techniques are available today? Of course, the state of the art advances as new devices and systems are created. So, tomorrow we may do more than today.
3. *The humanities and social sciences stem.* The needs that engineers aim to satisfy derive from society—the devices and systems we create have an impact on society. We need to know our cultural background through the study of literature, philosophy, etc. The social sciences are likewise necessary: economics, because many of our decisions are based on costs, and we need to understand the economic and business environment; psychology, which helps us understand how we think and learn, and thus how people might react to a new product, how to work effectively in teams, and what influence human factors have in industrial design; and political science, because what we do is influenced a great deal by the political process, and vice versa. A study of philosophy and religion helps us understand ethics, something engineers must be very aware of.

4. *The art, architecture, and industrial design stem.* Engineers must keep in mind aesthetics, style, form, and function in the devices and systems they create. Artists, architects, and industrial designers are creative people who concentrate on these aspects of design, and whose activities may overlap. For example, a coffeemaker may be designed by an artist or an architect or an industrial designer with regard to appearance. The services of these professionals should be sought during product design whenever appropriate.

Arrows shown in Figure 11 indicate that these fields have an influence on engineering design, and that engineering design influences these fields as well. There are a number of other disciplines in which an engineer must be proficient that have thus far not been mentioned. Some of these are shown in the upper left corner of Figure 11. Mathematics, statistics, and computer science may be regarded as tools (very important, nevertheless) that we and nearly everyone else need in order to apply and develop our knowledge. As an example, physics tells us conceptually what the acceleration of a car is; to know how fast the car is going and how far it has gone, we need mathematics. Another example: Electromagnetic waves are a physical phenomenon; their application, as in the design of a Stealth aircraft, would be impossible without mathematics.

A mastery of our language is important because we must communicate with others: colleagues, other professionals, and customers.

SUMMARY

The product realization process spans the time from recognition of the need for a product to the time it is in the hands of the customer. The different phases of the process may carry different names, but in general they may be described as recognition of the need, market studies, concept development, feasibility assessment, preliminary design, prototype development, process planning, and manufacture. The emphasis on each of these steps varies with the type of product. In this chapter we have also discussed company organizational structures and development resource requirements. The six trade-off parameters involved in the product development decision have been described and illustrated by an example. The relationship of engineering design to technological and cultural fields was discussed. In the following chapter we will look at the design process in detail.

BIBLIOGRAPHY

Dixon, J. R. *Design Engineering*. New York: McGraw-Hill, 1966.

Hayes, R. H., S. C. Wheelwright, and K. B. Clark. *Dynamic Manufacturing*. New York: Free Press, 1988.

Smith, P. G., and D. G. Reinertsen. *Developing Products in Half the Time*. New York: Van Nostrand Reinhold, 1991. The overriding message of this book is that it pays to develop products in as short a time as possible—that reduces risk, improves quality, and lowers cost.

Ulrich, K. T., and S. D. Eppinger. *Product Design and Development*. New York: McGraw-Hill, 1995.

The Design Process

Products are made of materials. The process for making them employs one or more of three basic building blocks: energy, material, and information. We can thus summarize the core aim of engineering design as follows: Determine the shapes and arrangement of material that will provide the optimum flow of *energy, materials,* and *information* to fulfill the desired *requirements*.

In this chapter we will describe a systematic method of design, which takes the problem from an expressed need to its final embodiment.

THE SYSTEMATIC DESIGN METHOD

The method of systematic design (or, methodical design) is based upon concepts and principles rather than on preexisting solutions and physical objects (Hubka et al., 1988 and 1992; Pahl and Beitz, 1988). The different stages of the design method are as follows:

- Clarify the task and develop its specifications.
- Determine the logical relationships and set up the function structure.
- Choose the best physical processes to fulfill the functions.
- Determine optimum shapes, motions, and materials, i.e., the embodiment.
- Develop the final design.

In the systematic design method we begin with the definition and clarification of the problem. Next we look at the function(s) of the product without giving consideration yet to solutions or hardware. Once the functions have been identified, we consider the physical effects and processes that can be used to satisfy them. Only then do we look for shapes and materials by which the physical phenomena can be utilized. The method therefore tackles problems from a very fundamental level.

Some of the important specific features of the method are:

- By abstraction of the application requirements, it reduces the problem to general, solution-neutral terms.

- It breaks up the problem into subproblems and looks for more than one solution for each part.
- By combining subsolutions it can produce a number of solutions (variants) for the problem.
- It places emphasis on selecting the best physical processes as the basis for the design.
- At each stage, a number of alternatives is generated—function structures, solutions, concepts, modules—among which a choice is made after an evaluation procedure.

It is during the design phase that most of the properties of a product are set, in particular, its functionality, manufacturability, and reliability. This is especially true of costs. Yet the design phase itself constitutes one of the smallest portions of the cost. The major document coming out of the task clarification step is a requirements or specifications list. The list includes the overall function of the device and any subfunctions that may be foreseen by the designer. The specific requirements are classified according to (1) life phases of the product (design and development, manufacturing, marketing, use, and disposal) and (2) types of requirements (technical, economic, ergonomic, and legal). The most important are the technical requirements for use of the product.

Conceptual design is the most important phase of design; it has the single largest influence on costs. As stated earlier, 70 to 80% of the costs of a product have been committed after only a small portion of the development resources have been expended in this phase. As an example, consider the design of a heat exchanger. When the designer chooses the concept—parallel flow, counterflow, cross flow—he or she has already made the major decision. The concept determines the arrangement of elements through a function structure and thus the flows of energy, material, and signals in the system. Functional analysis is the basis of value analysis, which is an accepted technique for product improvement (see Chapter V and Fowler, 1990, and Miles, 1992). The physical effects used determine how subfunctions are realized. The solution principles used determine how complex the product will be. Major advances in technology are the results of new concepts rather than improvements in embodiment. Examples are the internal combustion engine replacing the steam engine, the ballpoint pen replacing the fountain pen, and fiber-optic cable replacing copper wire.

The next step is to use the various effects to obtain solution possibilities in a sketch form, i.e., without considering the actual shapes the parts might have. By considering different solutions for the various subfunctions and using systematic combination, a number of different solution concepts can be generated. The concept that best satisfies the specifications is chosen.

The preliminary or embodiment design phase consists of determining (1) shapes or surfaces, (2) motion, if any, and (3) principal material properties. The final or detail design phase leads to production drawings. The final decisions on dimensions, arrangement, and shapes of individual components and materials are made. The design proceeds from a more abstract level at task clarification and takes on a more concrete form as it approaches this phase. In the case of the heat exchanger, these are the phases in which the sizes of the tubes and cavities are optimized and materials and manufacturing methods are chosen.

CLARIFICATION OF THE TASK

In the clarification phase, information about the problem is collected. Some sources of information are the customer, the marketing department, and past experience with similar products. The specific requirements of the task (also called conditions, constraints, criteria) are tabulated in a requirements list. The various specifications are identified by their importance, broadly, as "demands" or "wishes." Demands are those properties that the product must (or must not) have; wishes indicate additional properties desirable to improve its performance, lower its cost, etc. Such a list is updated as more information becomes available. It ensures that no aspect of the problem is overlooked and reduces chances of misunderstandings between the customer and the company and the various groups within the company.

Solution-Neutral Statement of the Problem

During systematic design, importance is placed on dispelling prejudice on the part of the designer, thus removing any unneeded constraints that might preclude unconventional solutions. A useful step toward this end is the abstraction of the problem from specific to more general, solution-neutral terms. The following steps show how it is done.

AIM: State a problem in solution-neutral terms.
HOW: Use a process of abstraction.

STEPS:

Have no personal preferences.
Omit requirements with no bearing on function.
Change from quantitative to qualitative.
Generalize and express in solution-neutral terms.

As examples, consider the sample problem statements given in Table 2. In the first column they are stated in the concrete form in which a problem is generally presented to a designer. These statements are already focused on a particular solution. However, if the designer wants to look at all possible solutions to the generic problem, its statement must be cast in a form that does not preclude other solutions. Thus, following the steps given above, we can state the respective problems as given in the second column.

An example of the broadening of the problem statement from specific to general, solution-neutral terms is given by Krick (1969). The problem

TABLE 2 Prejudicial and Solution-Neutral Statements of Problems

Prejudicial	Solution-Neutral
Design a keyed shaft/hub connection	Provide means to secure shaft to hub
Design a labyrinth seal	Keep lubricant inside and contaminants outside
Design a packing machine	Find a method to pack material
Measure height of a person	Measure length
Put a boom in a boat	Move objects from inside to outside of a boat
Install a garage door	Protect car from weather and intruders
Use a lever	Change line of action of a force
Clean a spark plug by sandblasting	Separate oil and carbon from the surfaces

involves filling, weighing, stitching, and loading bags of animal feed. The bags were stored in a warehouse, loaded on a truck, and shipped to consumers. In order to improve the process, the author analyzes the problem and eliminates unnecessary steps and consolidates some existing steps. Some of the steps in reformulating the problem are:

- Take feed from mixing bin; fill, weigh, stitch, and load the bags and transfer to warehouse.
- Transfer feed from mixing bin to stacked bags in warehouse.
- Transfer feed from mixing bin to stacked bags on delivery truck.
- Transfer feed from mixing bin to consumer's storage bin.
- Take feed ingredients at the source and transfer to consumer's storage bin.
- The last statement gets to the heart of the problem: fill the consumer's storage facilities.

Define the Real Problem

Before starting a design task, find out what the real problem is!

This seemingly obvious statement is often overlooked when starting on a design task, particularly when redesigning an existing product. In such cases the existing hardware can produce a mindset which can tend to preclude new and unusual solutions. Kamen (1995) cited two examples:

- The holes in a circuit board assembly were a problem during manufacture. Perhaps a sharper drill was called for. Then a radically different solution was found—use surface mount technology and do away with the holes altogether.
- A home dialysis unit was to be redesigned. The electromechanical valves in the unit were a critical part of the system. Each valve consisted of 67 parts. Rather than redesign the unit, the designers took a fresh look at the design, starting from the basic problem statement: "Move the fluid from the bag into the patient." By simplifying the design, using multifunctional parts, a cheaper and more reliable unit was designed.

Munro (1995) lists some typical reasons why people do not want to change their design; here are just a few:

- "We have always done it this way."
- "This is the boss's idea."
- "We have too much invested in this idea."

He cites some typical mindsets in different industries:

- Use of iron and steel in automobiles
- Use of threaded inserts by the military
- Use of rivets in aircraft frames
- Apparent adherence to UL standards in the appliance industry
- Continued marketing of a product by the medical industry "because it works and passed the tests"

In order to break away from preconceived ideas, Munro suggests the following:

- Challenge the existing designs
- Get outside perspectives
- Look outside your own industry for new ideas
- Use other industries to benchmark your operation

Benchmarking is used to compare a company's operations with other companies. It identifies one's strengths and weaknesses. Such comparison is not limited to similar companies: a significant example is Xerox Corporation's studying the operations of the mail-order retailer L.L. Bean, Inc., regarding the efficient storage and movement of material (Hayes et al., 1988).

THE REQUIREMENTS LIST

The requirements list is a most crucial document, prepared at the start of the development cycle. It should be prepared by design and manufacturing (at the minimum) working together. It should define for the yet nonexistent product its functions and the constraints on it—not how the functions will be realized, and not its physical features. The constraints should be real constraints, not those perceived by engineering or marketing. The requirements should be dictated by the customer's needs, expectations, and problems, benefits to the customer, and expected improvements over existing products. The requirements list is not an engineering document per se, nor is it the manufacturing plan. This is not the list of specifications that will appear with the product when it is marketed. It should be as abstract as possible, as far as physical features are concerned, and not rule out any options by containing false constraints. It should pay particular attention to the environment in which the product will operate. For example, the design requirements on ovens for commercial and domestic use are different. Commercial units are designed for heavier duty cycles, greater reliability, and use by professionals and must conform to professional standards.

The preparation of the requirements list can be expedited by dividing the requirements into categories. As shown in Appendix C, the requirements may be categorized in various ways, e.g., according to life phases of the product and according to type (Matousek, 1963; Pahl and Beitz, 1988; Slocum, 1991). All requirements lists include items such as geometry, kinematics, forces, energy, material, signals, safety, ergonomics, production, quality, assembly, transport, operation, maintenance, costs, recycling, and schedules. The brief checklist given in Table 3 is helpful in developing the requirements list.

In order to ensure that customer wishes are incorporated into the requirements, the method of Quality Function Deployment (QFD) has proved to be effective. This is described later in Chapter V.

PHASES OF DESIGN

The systematic design method can be summarized by the flowchart shown in Figure 12. It shows the principal steps and the results produced at each step. As we proceed from the conceptual toward the final design stage, the design goes from the abstract to the more concrete. The three phases in design partially overlap each other; in fact, it is generally difficult to say for a given project when one phase ends and the next begins. At the end of each step a number of alternatives are produced—function structures, solutions, etc.—which are evaluated. Depending on the evaluation, there are feedbacks from later stages to earlier steps (not shown in the figure). However, going back from detail design to, say, the conceptual

TABLE 3 Checklist for Developing the Requirements List

Main Item	Items to Be Checked
Environment	Physical environment
	Other products to interface
	Power and other facilities
Manufacture	Production equipment available
	Sourcing for parts
	Assembly
Safety	Protection systems
	Liability
Ergonomics	Use pattern
	User fatigue
	Ease of use
	User abilities
Quality	Quality control
	Reliability
Marketing	Transport
	Distribution
Legal	Standards
	Patents

stage can be costly. What is even costlier is making design changes after production has started, or worse, after the product has been put into use. As a general rule, the later any changes are made, the more expensive they get. This is expressed by the so-called rule of 10, i.e., changes at each successive stage are 10 times costlier than at the previous stage, as shown in Figure 13 (Ehrlenspiel, 1985). A similar set of ratios is given for software development by Boehm (1981).

Conceptual Design

The conceptual phase is by far the most important phase of design for a new product, since the concept determines to the largest extent the cost of the product. We first look at the functional requirements of the product, i.e., what it is supposed to do, its functions. This brings us to a more abstract level, away from concrete shapes and solutions. We consider first

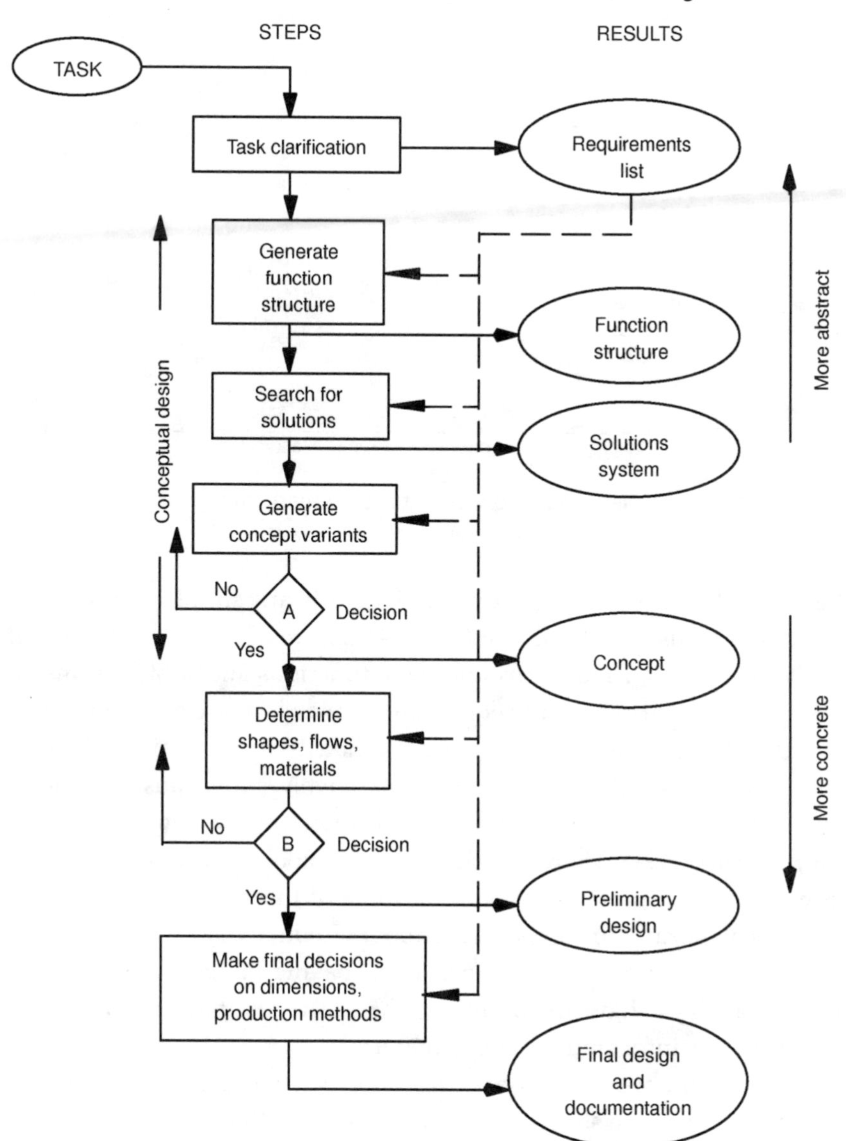

FIGURE 12. Cost of changes at different stages. (Ehrlenspiel, 1985)

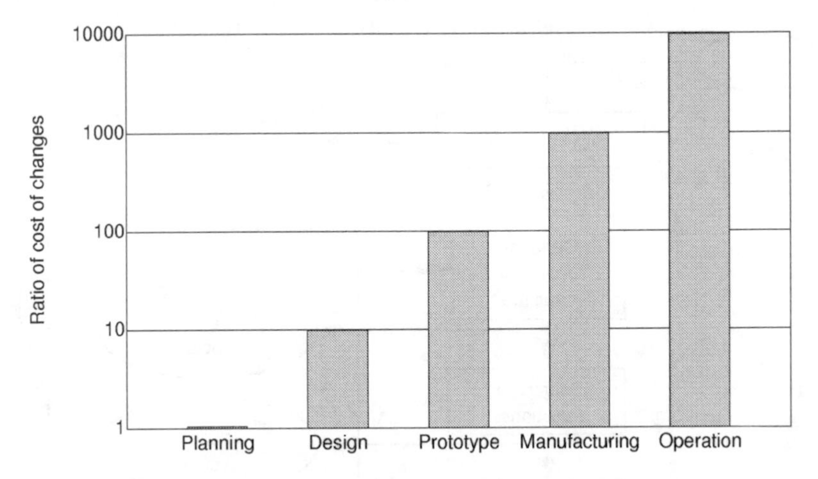

FIGURE 13. Flowchart for systematic design.

the overall function of the product and identify any simpler subfunctions that might be used to fulfill the former—i.e., we divide the problem into smaller problems. In most instances the functions and subfunctions of a product can be arranged in sequential form as a block diagram. One of the contributions of the systematic method is to recognize that every technical process embodied in a machine involves the transformation of one or more of three fundamental quantities: energy, signals (information), and material. Therefore the function blocks will have inputs and outputs consisting of one, two, or all three of these quantities.

Let us elaborate further. We talk about a "coffee machine," a "copying machine," a "computing machine," or a "paving machine." What do all these "machines" have in common? The answer is that they all process energy and/or information and/or material.

Aristotle said over 2500 years ago, "In order to understand something (a natural object) we should look at its function" (Rodenacker, 1984). His words are of great significance to us today. To design something new we begin with its functional requirements. In order to improve an existing design we must look at the functions of its parts.

In this first stage no consideration is given to the solutions, shapes or hardware that might be involved. But the next step in conceptual design is to look for solutions for each subfunction. By considering different physical (and chemical) processes it is possible to obtain a number of solutions for each function. For example, to remove oil and carbon deposits from a spark plug (see Table 2), the following processes offer potential solutions:

- Mechanical—scraping, sandblasting
- Chemical solution
- Heating—flame or electrical

The subsolutions thus found are then combined in a systematic and rational way to obtain a number of solution concepts (concept variants) for the task. At this stage, after an initial screening on the basis of the "demands" of the specification, some of the concepts are rejected. The rest are judged by looking at the "wishes" in the specifications list. Systematic methods that assign weights to different requirements are used to arrive at the final solution concept. The best concept is selected from those that meet the requirements, shown at "A" in Figure 12.

It should be mentioned that the term *conceptual design* means different things to different people. For instance, Michaels and Wood (1989) consider it synonymous with "preproposal and proposal phases," and market and product exploration.

Preliminary or Embodiment Design

Starting from the concept, the embodiment phase develops a definitive layout for the project in the form of general arrangement drawings. The chief considerations during embodiment design are shapes, motion, and principal material properties. Some preliminary computations are carried out at this stage to help in making decisions regarding general shapes, sizes, motion, and spatial arrangements. Several designs are developed and evaluated on the basis of technical and economic requirements. Another decision point in the process is at the end of this phase, shown as "B" in Figure 12. Technical and cost characteristics of the design are compared with the required values. If these requirements are not met, alternative designs must be tried. Methods of evaluating design alternatives are discussed in the rest of this chapter and in the examples that end it. At the end of the embodiment phase all the important decisions regarding the following will have been made:

- Materials
- Manufacturing methods
- Dimensions, tolerances
- Standard, purchased, and repeat parts

An important aspect of the embodiment design phase is the building of a mockup or prototype, as appropriate. A mockup is usually made of substitute, easier-to-work-with materials and is usually built at the full scale of the device. The mockup is not a working model, but is built to ensure that parts fit properly, that there are sufficient clearances, etc. A prototype, on the other hand, is an actual working preliminary model of the device. It is designed and built so that it may be easily changed and altered to try out the embodiment of different concepts. Some authors use the term *prototype* in a wider sense to include not only mockups but also computer models. Prototype development is further discussed later in this book.

Final or Detail Design

The detail design phase leads to production drawings, a bill of materials, manufacturing and assembly instructions, and operation and maintenance manuals. Final decisions on dimensions, arrangement, and shapes of individual components and materials are made. The traditional engineering "design" courses strongly emphasize this portion of the design process. In reality, design analysis forms only a small part of the overall product realization process.

Most of the detail design work, being routine in nature, is done on the computer—analyses, drafting, preparing bills of material, and project documentation. (In contrast, the activities during conceptual design tend to be more creative in nature, and the computer has found the least use there, at least until now.) The detail design work is being addressed by developing programs that use databases of physical effects and solution catalogs and incorporate expert systems.

TYPES OF DESIGN

Not every project needs to be carried through all the phases—conceptual, embodiment, and final design. The phase at which we begin depends on the type of design—the degree of originality involved. In this respect designs may be classified as (1) original, (2) adaptive, or (3) variant designs (Pahl and Beitz, 1988).

Original Designs

Original designs are those that involve developing new solutions. The systematic approach finds its most useful application in these types of designs, since new functions and physical principles may have to be used and new solutions may have to be developed. In such cases we begin with conceptual design. About 25% of all designs fall into this category.

Adaptive Designs

In adaptive designs, a known solution is adapted to a new task. The general structure of the product is better known in this case. Only new shapes, types of motion, and perhaps materials need to be investigated. The majority of designs produced are of this type. Two of the many instances of this type of design are:

- Friction clutches of various forms, perhaps using different materials, which are nevertheless variants on the same basic friction clutch concept
- Improvisation, such as using a ski as a leaf spring in an experimental vehicle

Variant Designs

In variant designs, sizes, materials, throughputs, and arrangements in an existing solution are varied. Development of modular products falls into this category. In these cases only new detail designing is necessary. About 20% of all designs are variant designs. Instances of this type of design are:

- Pumps or motors of sizes different from those existing
- Buildings and plants based on existing designs

With the development of new materials (and manufacturing methods), existing devices are continually being redesigned to achieve higher performance, lower cost, or lower weight. The evolution of the bicycle is a case in point.

The types of design, and how they relate to the different design phases, are illustrated in Figure 14. The categories given above are not rigid. For certain systems it may happen that one part requires completely new design while another needs only minor changes.

Ullman (1992) proposes a classification of mechanical designs as follows:

- Selection design, where a component is selected from a manufacturer's catalog
- Configuration design, which involves arrangement of existing components
- Parametric design, where system parameters are determined by solving mathematical equations
- Original design

The correspondence of these categories to those of the systematic design method and its phases can be readily seen.

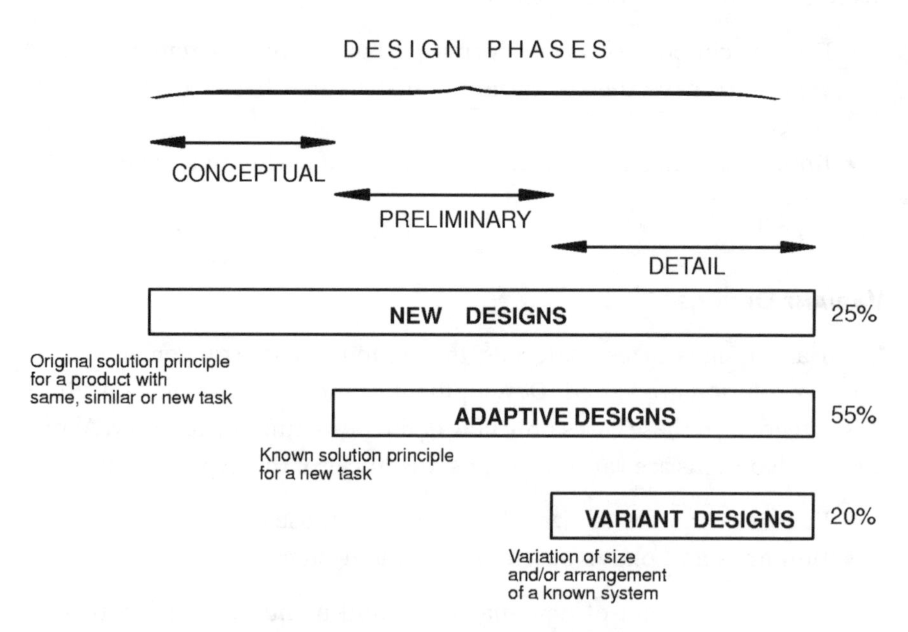

FIGURE 14. Types of design and the design phases.

INDUSTRIAL DESIGN

Industrial design deals primarily with two aspects of a product: (1) aesthetic appeal of the product, as regards its look, feel, sound, etc.; and (2) its interfaces with the human user—ergonomics. Industrial designers can provide services to engineers where these aspects are important. Application of industrial design can help in the following:

- *The product's interfaces*. The controls, dials, handles, and switches should be easy to use, unambiguous, intuitive, and clearly communicative to the user of their purpose.
- *Cost*. The product's features should allow for low-cost manufacture and easy maintenance.
- *Looks*. The product should be pleasing to the eye in its shape, color, materials, and texture. It should offer pride of ownership. It can communicate corporate identity to the public through its appearance and use of logos.

Industrial design got its start in Europe in the early 1900s as a joint effort of architects and engineers. One of the early industrial designers in the United States was Raymond Loewy (Loewy, 1990), who, among other products, designed the covers for locomotives of the Pennsylvania Railroad during the 1930s, to give them a streamlined look. The appearance of these locomotives conveyed the impression of modern design, speed, and power. However, the covers were installed on locomotives that were already built; there was no integration of mechanical and industrial design. Examples of successful application of industrial design among present-day products are handheld instruments, power tools, and portable telephones and computers.

Requirements that affect ergonomic design are discussed in Appendix C.

METHODS FOR GENERATING SOLUTIONS

During designing we are constantly looking for solutions. This happens at any stage of the design, but particularly during concept development.

There are many ways of finding solutions, as described below. They are broadly classified as conventional, intuitive, and systematic procedures.

Conventional Methods

From our own memories and experiences, we know of solutions that apply in similar situations. This is perhaps the most common method we use. A beginner, however, is at a disadvantage here, having a limited repertoire of available answers. There is also a possible trap in using our experience: we might be prejudicial and overlook unusual solutions.

We look for possible solutions for our use by searching the published literature. This includes looking at textbooks, professional and reference books, professional and trade journals, manufacturers' catalogs, and our own company's and competitors' literature. Patent search is a well-known means of idea survey. Many of these sources are now available in the form of computerized databases.

We find help in nature. Natural systems have been undergoing evolutionary optimization for millions of years. Examples of natural systems giving rise to technical solutions are many. A study of beehives led to honeycomb structures; thistles gave the inventor the idea to develop the hook-and-loop fastener. Many other lightweight structures have evolved from the study of plant stems and bird bones and skeletons. The long bones in our bodies, with hollow, round cross sections, are models for structural members that are subject to multiple load types (axial, bending, torsion) but use a minimum of material. The shapes and communication and propulsion methods of fish and other marine animals hold promise of improvements in submarine design.

Intuitive Methods

Solutions come to us in a flash when we are engaged in a totally unrelated activity or are observing a natural or human-caused phenomenon. Creative thinkers are known to synthesize knowledge from other fields and apply it in new situations. There are methods for improving one's creativity (von Oech, 1983).

There are a number of methods in which a group of people generate ideas and modify each other's ideas in what is generally known as "brainstorming." Brainstorming is an essential part of many solution generation methods, including value analysis (see Chapter V). There are other, similar procedures: a derivative of brainstorming called the 635 method; the Delphi method; and synectics. There are several do's and don'ts for successful brainstorming sessions (Pahl and Beitz, 1988):

- The group shall be 5 to 15 people and should include generalists as well as specialists.
- All members should be treated as equals.
- Individuals should shed their prejudices and not reject ideas for reasons such as lack of soundness, practicality, or outlandishness.
- Ideas should be put down on paper, to be built upon by others to produce new ideas.

The results of a session, which should last no longer than 30 to 45 minutes, should be evaluated by experts for their feasibility.

Creativity requires that we break out of our normal, ordinary set ways of thinking. The idea of using provocation has been suggested as a means toward this end (DeBono, 1992). Methods of generating a provocation include:

- *Escape*. Write down your ideas and methods. Then generate ideas totally different from these.
- *Reversal*. Look at possibilities of doing things in a completely opposite manner.
- *Exaggerate*. Change sizes and numbers to outlandish values.
- *Distort*. Think of changing time and space sequences.
- *Word association*. Think of a (random) word. Develop other words, each related to the previous in some manner.
- *Aim for the sky*. Use wishful thinking to generate (seemingly) impractical ideas.

Systematic Methods

There are a number of procedures that can be used by an individual to find solutions in a step-by-step fashion.

Study of System Equations

If a physical device is found suitable for fulfilling a function, we look at the mathematical equation describing its operation. By recognizing the part played by different physical effects and/or parameters in the equation, it is possible to consider variations in these; each variation can lead to a different solution.

Rodenacker (1984) shows how different concepts for a capillary viscometer can be developed using this method. The expression for fluid viscosity μ is:

$$\mu = \frac{\Delta p\, R^4}{QL} \tag{1}$$

where Δp is the pressure drop, Q the volume flow rate, L the length, and R the radius. The experimental setup is shown in Figure 15. Since there are four parameters: Δp, Q, L, and R, four possible concepts can be developed by choosing one of these at a time as the variable.

Search With Classification Schemes

Classification schemes in the form of two-dimensional matrices have been developed to help in the search for solutions. Typically, for a given function a table is prepared with one parameter designating the rows and the other, the columns.

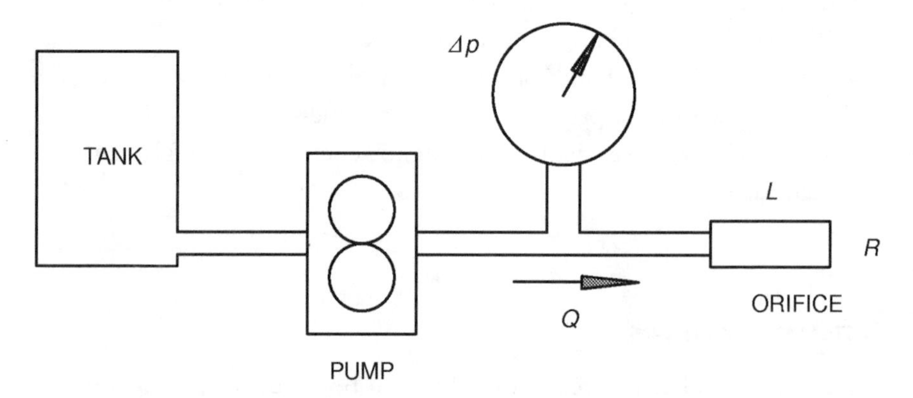

FIGURE 15. Viscosity measurement system.

Appendix F shows a collection of solutions for functions, an excerpt from which is shown in Table 4. This is an example of using a classification scheme for generating solutions. Table 4 shows the example of solutions to the the basic function "convert" (see Appendix D for classification of functions). In the context of converting mechanical energy to other forms of energy, it shows the descriptors "displacement" and "mechanical" for the inputs. The solutions that yield the different forms of energy and other kinds of output are shown in the middle column.

Use of Design Catalogs

Somewhat similar to classification schemes, design catalogs offer more concrete solutions to fulfill given functions. They contain the type of information found in handbooks, manufacturers' catalogs, and standards. Design catalogs provide tables containing classifying criteria, available solutions, and, importantly, selection characteristics. The solutions may be descriptive or in graphical form and in some cases include governing equations for the solutions. Roth (1982, 1994) gives a collection of 39 design catalogs that span the whole mechanical design field. The catalogs contain information on design entities such as force transmission types, sliding systems, logic conjunction and disjunction linkages, force genera-

TABLE 4 Excerpt from Solution Table

Function	Input and Descriptors	Solutions	Output
Convert	Energy		Energy
	Mechanical		
	Displacement		Force
		Elastic element	
	Displacement		Electrical
		LVDT	
		Piezoelectric element	
	Displacement		Heat
		Friction	
	Force		Displacement
		Elastic element	

tion, one-way motion, shaft-hub connections, permanent and separable joints, bearings, and variation of shape.

Table E2 in Appendix E gives an example of a solution catalog that applies many different physical effects for the commonly occurring function: "generate force." Table E3 shows some other physical effects. Figure 16 shows an excerpt from such a solution catalog.

Generating Concept Variants From Subsolutions

Each of the subfunctions and their corresponding solutions are arranged in matrix form, as shown in Figure 17. The first column shows the subfunctions in verbal and/or symbolic form. Each function has one or more solutions; these are shown in the corresponding rows. In the figure, the functions are indicated by symbols F_i. The solutions for function F_i are indicated by the symbols S_{ij}. The rows will generally be of different lengths.

If we take one of the solutions for each function and combine them, a concept variant is created. A vast number of concept variants can thus be created. Naturally, due to incompatibility and other reasons, not all such variants are feasible.

Figure 17 shows the creation of two concept variants. The first variant is created by combining the subsolutions S_{11}, S_{22}, S_{33}, S_{41}, and S_{i3}. The

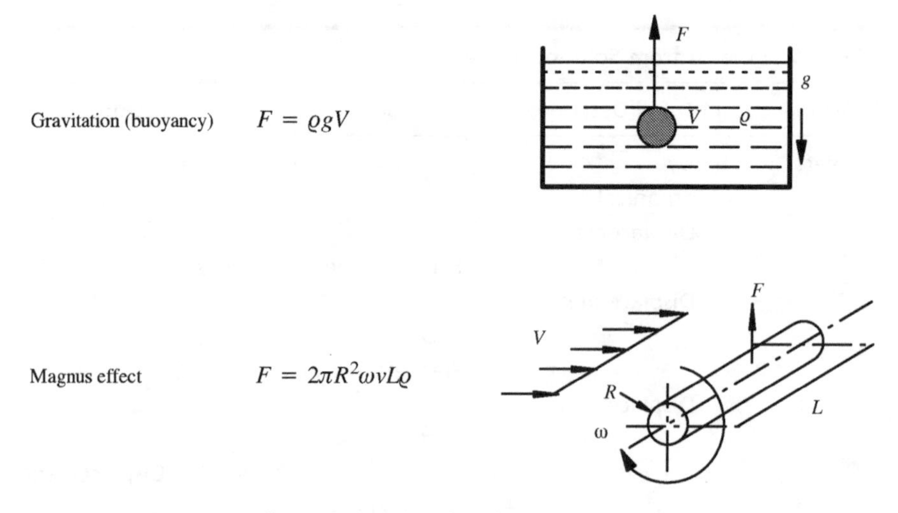

Figure 16. Excerpt from design catalog for the function "generate force."

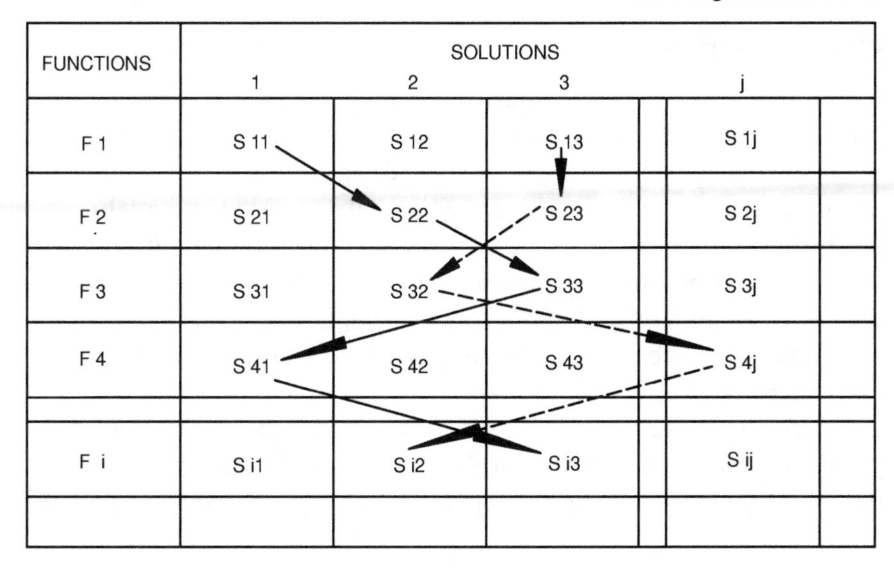

FUNCTIONS
SOLUTIONS
1
2
3
j
F 1
F 2
F 3
F 4
F i
S 11
S 12
S 13
S 1j
S 21
S 22
S 23
S 2j
S 31
S 32
S 33
S 3j
S 41
S 42
S 43
S 4j
S i1
S i2
S i3
S ij

FIGURE 17. Concept variants from the solutions matrix.

second variant is created by combining the subsolutions S_{13}, S_{23}, S_{32}, S_{4j}, and S_{i2}. We indicate these as follows:

$$\text{Variant 1: } S_{11} + S_{22} + S_{33} + S_{41} + S_{i3}$$
$$\text{Variant 2: } S_{13} + S_{23} + S_{32} + S_{4j} + S_{i2}$$

EVALUATION OF DESIGN ALTERNATIVES

The systematic method of design allows us to generate alternative solutions at each stage of the design process. These alternatives must be evaluated and the best one selected before proceeding to the next stage. The earlier the design stage, the less the information available to help with the decision making. Yet, a decision made on a rational basis, even with the small amount of information, is better than one made spontaneously. In the published literature a number of decision-making methods are given (Johnson, 1978; Pahl and Beitz, 1988; Roth, 1994). We will present some of these methods here.

Initial Concept Selection

Figure 18 shows a scheme for selecting among concepts at early stages. This initial selection process requires a minimum of numerical data; rather, a judgment on the part of the designer leads to rejecting the unsuitable concepts. The criteria used are fairly general and might seem so obvious that one might be tempted to say that those solutions should not have been brought this far. But this runs counter to the discursive process of systematic design. All concepts ought to be considered, evaluated, and then rejected in a rational procedure, if they violate these general criteria.

The evaluation criteria to be used in the above matrix would vary from problem to problem and the purpose for which the matrix is used. For the initial selection among concepts, Pahl and Beitz (1988) suggest the following:

- Compatibility with the rest of the system or process
- Fulfills the "demands" in the requirements list
- Is basically realizable
- Can be accomplished within available resources
- Its safety is assured
- Is preferred in own company or field

ANALYSIS & EVALUATION	DECISION
Evaluation according to criteria: + yes – no ? need more information ! check requirements list ← CRITERIA →	Mark decisions: + proceed further – solution unacceptable ? supply information ! check requirements list

A	B	C	D	E	F	CONCEPT/Remarks	
+	+	+	+	+	+	Concept # 1	+
+	+	+	+	+	+	Concept # 2	+
+	+	+	+	–	+	Concept # 3	–
+	+	+	?	+	–	Concept # 4	?
						....	

FIGURE 18. Initial concept selection matrix.

A similar selection matrix can be used for evaluating subsolutions in a concept.

Other Methods

The examples that follow show the method of weighted values, also called the decision matrix technique (Johnson, 1978), for a more detailed evaluation. Table 5 shows the decision matrix scheme for evaluating concept variants.

The decision matrix can be prepared by using technical criteria or economic criteria, or a combination of both. Since not all criteria are equally important, they are assigned weights. For each concept variant, each criterion is assigned a value (on a scale of 1 to 10, say). The weighted value is then found by multiplying the value and the corresponding weight. The sums of the values and the weighted values are then found for each concept variant. The sums of the weighted values are generally normalized to 1.0 or 10.

While the actual criteria would vary from case to case, there are certain generic terms which are pertinent to most problems. A typical list might be:

TABLE 5 Decision Matrix

		Concept Variants							
		CV 1		CV 2		CV 3		. . .	
Criteria	Weights	Value	WV	Value	WV	Value	WV		
Criterion 1									
Criterion 2									
Criterion 3									
Criterion 4									
. . .	. . .	. . .	. . .	. . .	. . .	. . .	. . .	. . .	. . .
Sum: Weighted sum:									

Technical Criteria	Economic Criteria
Sturdiness	Material costs
Simple operation	Number of parts
Low wear	Assembly costs
Safety	Operating costs
Robustness	Testing

Figure 19 shows the values of the weighted sums of four concept variants plotted on an evaluation chart. In this case the criteria have been normalized to 10. The ideal design is one for which the normalized weighted sums for both technical and economic criteria are 10, that is, the point at the top right of the chart. The arcs drawn in Figure 19, with the ideal point as the center and different radii, are the lines of equal "goodness" of the concept variants. Of the four concept variants plotted, we

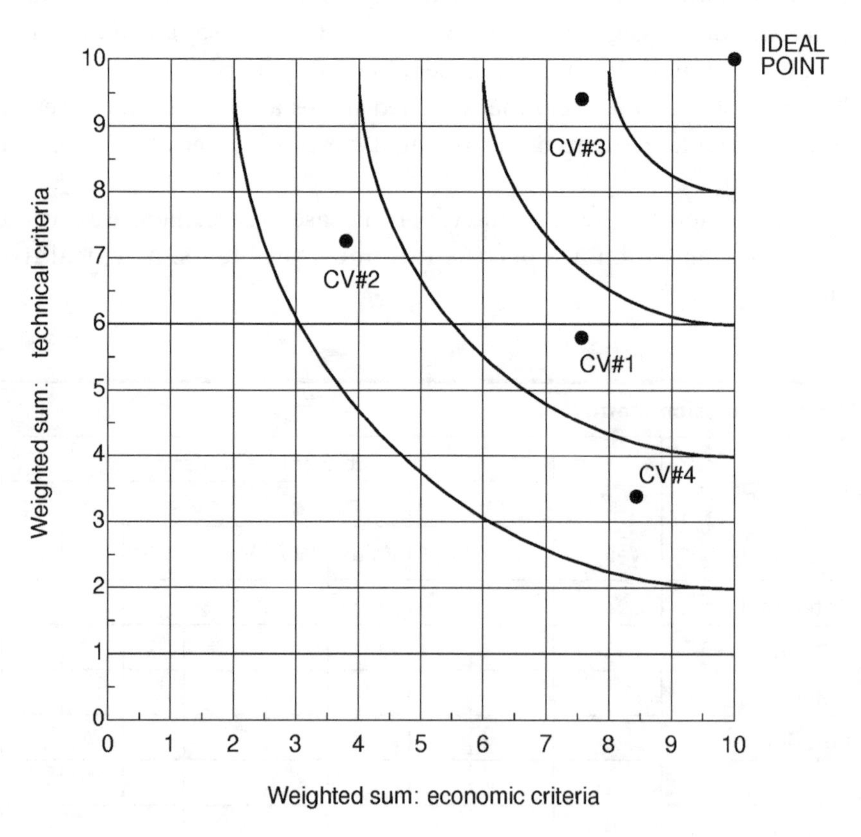

FIGURE 19. Evaluation of concepts on technical and economic criteria.

see that variant 3 is the best; next comes variant 1. Variants 2 and 4 are tied for the third place.

The objectives tree method is shown in the variable-speed transmission example at the end of this chapter (see Figure 32) to illustrate how weights for different criteria are obtained.

SUMMARY

An introduction to the systematic method of design has been described. It is a scientific method (i.e., it proceeds in a systematic manner and applies systematized knowledge) that can be used in a variety of fields and is easily learned and taught. The method should supplement the usual procedures of design which depend on intuition, existing solutions, and past experience. It is most fruitful for developing new products. New designs for existing products can be developed by analyzing their function and that of their parts to uncover insights and new ideas. While a gifted designer is irreplaceable, an "average" designer can improve his or her productivity by applying these principles.

The steps involved in systematic design are:

- Clarification of the task; preparation of a requirements list
- Conceptual design: function structure, physical effects, solution
- Embodiment design: shapes, materials, manufacturing processes
- Prototype development
- Detail design

One of the strengths of the systematic design method is the number of alternatives that can be generated at each step, as shown in Figure 20. Starting at the top level we see that for a given purpose a number of function structures are possible; for each function structure a number of different physical effects can be called upon, which can give rise to a variety of geometries and resulting concepts. The concepts can further be embodied by using different shapes, materials, and flows. The extent of the field of concrete solutions is larger the earlier we start in the process.

At each step of the way in this process, since a number of alternatives become available, an evaluation must be made to choose the best alterna-

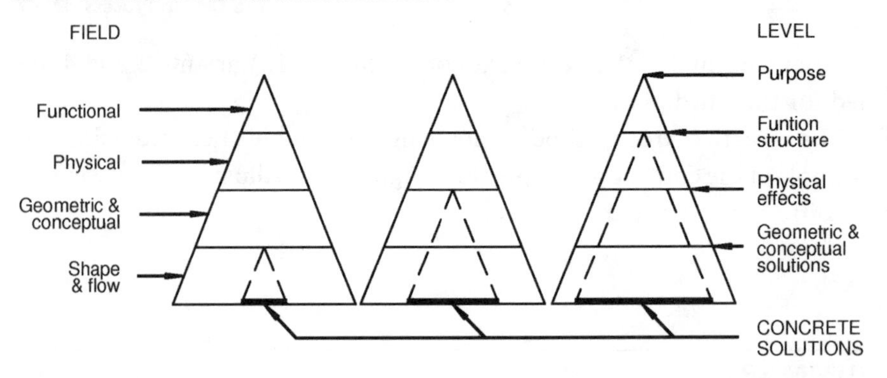

FIGURE 20. The multiplicity of alternatives generated in systematic design.

tive with which to proceed to the next step. Otherwise the solution field would become too large. The choice needs to be narrowed at each step. This is illustrated by Figure 21. We select one or possibly two of the function structures with which to proceed to the physical field. For each of the functions, one or more physical effects might be utilized. Thus,

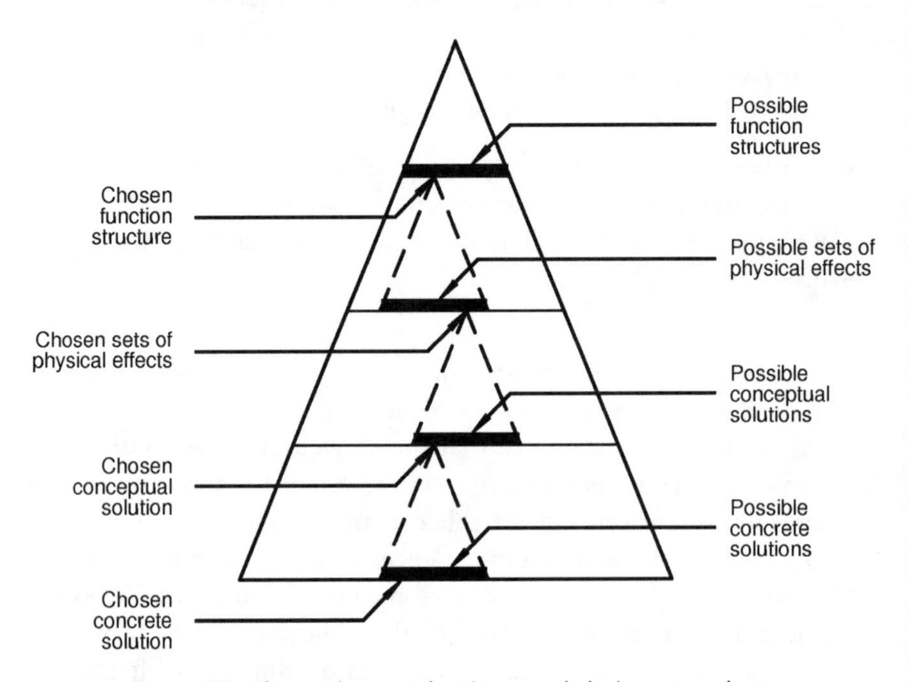

FIGURE 21. Alternatives, evaluations, and choices at each step.

more than one set of physical effects may be used to develop the concepts. After evaluating the concepts, we proceed with the best concept and develop a number of embodiments, by considering variations of shapes, materials, and flows. An evaluation of the embodiments then leads to the optimum concrete design.

Figure 22 shows more of the details of the steps in conceptual design that we have discussed here. This is an expanded form of Figure 21.

More details on the systematic method of design are given in the appendices. Appendix C provides details on developing requirements lists: requirements are classified according to importance, type, and the product life phase. In Appendix D we discuss function structures further: how variants may be obtained by relocating, subdividing, combining, or eliminating functions, and the simplification of function structures by function reduction and integration. Appendix E shows tables of physical effects and elaborates on some of those in sketches and equations. Appendix F shows solutions for the six basic functions in tabular form. An example of a design catalog is also given.

Two examples that show the application of systematic design conclude this chapter. The examples stress the conceptual design stage and to some extent the embodiment design. Detail design is not covered here, since that is adequately dealt with in traditional design texts. These examples show how a number of alternative designs can be developed by considering various physical effects, solution principles, and geometries. Examples that involve the application of embodiment design principles are given later when we discuss materials and manufacturing processes.

BIBLIOGRAPHY

Boehm, B. W. *Software Engineering Economics.* Englewood Cliffs, N.J.: Prentice Hall, 1981.

Clausing, D. *Total Quality Development.* New York: ASME Press, 1994.

DeBono, E. *Serious Creativity.* New York: Harper Collins, 1992.

Ehrlenspiel, K. *Kostengünstig Konstruieren (Cost-Effective Designing).* Berlin and New York: Springer-Verlag, 1985.

Flurscheim, C. *Industrial Design in Engineering: A Marriage of Techniques.* London: Design Council; Berlin, New York: Springer-Verlag, 1983.

Fowler, T. C. *Value Analysis in Design.* New York: Van Nostrand Reinhold, 1990.

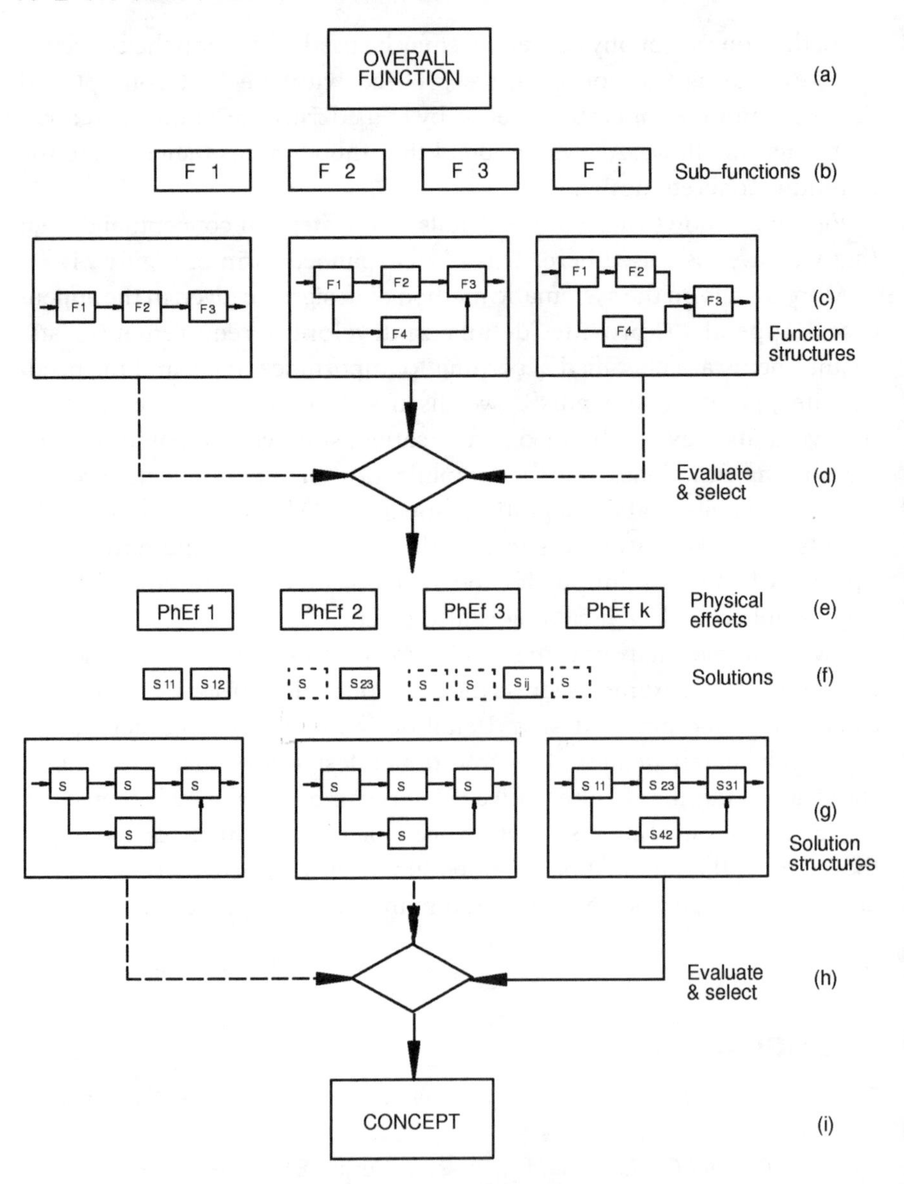

FIGURE 22. Steps in conceptual design.

Hayes, R. H., S. C. Wheelwright, and K. B. Clark. *Dynamic Manufacturing.* New York: Free Press, 1988.

Hubka, V., M. M. Andreasen, and W. E. Eder. *Practical Studies in Engineering Design.* London: Butterworths, 1988.

Hubka, V., and W. E. Eder. *Engineering Design.* Zurich: Heurista, 1992.

Hundal, M. S. "A Systematic Method for Developing Function Structures, Solutions and Concept Variants." *Mechanisms and Machine Theory* 25(3): 243–256 (1990).

Johnson, R. C. *Mechanical Design Synthesis.* Huntington, N.Y.: Krieger Publishing, 1978.

Kamen, D. Keynote address, National Design Engineering Conference, Chicago, 1995.

Krick, E. V. *An Introduction to Engineering and Engineering Design.* New York: Wiley, 1969.

Loewy, R. *Raymond Loewy: Pioneer of American Industrial Design.* Munich: Prestel-Verlag, 1990.

Matousek, R. *Engineering Design—A Systematic Approach.* London: Blackie, 1963.

Michaels, J. V., and W. P. Wood. *Design to Cost.* New York: Wiley, 1989.

Miles, L. D. *Techniques of Value Analysis and Engineering.* New York: McGraw-Hill, 1961; also 2nd ed., 1992.

Munro, A. S. "Let's Roast Engineering's Sacred Cows." *Machine Design*, February 9, 1995, 41–46.

Pahl, G., and W. Beitz. *Engineering Design—A Systematic Approach.* Berlin and New York: Springer-Verlag, 1988.

Rodenacker, W. G. *Methodisches Konstruieren (Methodical Designing).* 3rd ed. Berlin and New York: Springer-Verlag, 1984

Roth, K. *Konstruieren mit Konstruktionskatalogen (Designing with Design Catalogs).* Berlin and New York: Springer-Verlag, 1982; also 2nd ed. (2 vols.), 1994.

Slocum, A. H. *Precision Machine Design.* Englewood Cliffs, N.J.: Prentice Hall, 1991.

Ullman, D. G. *The Mechanical Design Process.* New York: McGraw-Hill, 1992.

Von Oech, R. *A Whack on the Side of the Head.* New York: Warner Books, 1983.

• • • • •

EXAMPLE
A TEMPERATURE MEASUREMENT DEVICE

This example will show the design of a device to measure temperature, such as a household thermometer for outdoor or indoor use. The process has been described in Hundal (1992). The technique will show how a problem can be attacked at a fundamental level, the different solution possibilities that are produced, and how one proceeds through the different phases. The emphasis in this example will be on conceptual and preliminary design.

Task Clarification

A thermometer can take various forms. Let us remove ourselves from existing shapes and solutions and look at the problem at an abstract level. The basic function which it should serve is:

Give a visual indication of the ambient temperature.

The problem definition in this case is quite simple: design a device to fulfill this function.

Requirements List

A list is prepared showing the requirements for the device. For a household thermometer it might look as shown in Table 6; for other temperature measurement devices, the specifications could be more extensive and stringent. The requirements on the device are listed as demands (D) or wishes (W), with numerical values given where pertinent. The list includes the overall function of the device and any subfunctions that may be foreseen by the designer. Appendix C gives more details on classifying different types of requirements.

Conceptual Design

We will discuss the various stages of developing the concept.

Functional Requirements

The overall function of the device is stated in the task clarification. It is to convert a signal (ambient temperature) into another signal (visual display). Figure 23a shows the overall function structure (block diagram). This function may be broken up into two parts: sensing the temperature, and displaying it, as was shown in the requirements list and is depicted in Figure 23b. We incorporate a further possibility—modifying the measured signal, perhaps ampli-

TABLE 6 Requirements List for the Temperature Measurement Device

D vs. W	Requirement	Value
	Overall function: Provide a visual indication of temperature	
	Subfunction: Convert temperature to signal	
	Subfunction: Convert signal to visual display	
D	Temperature range	−50 to 120°F
D	Provide a visual indication	
D	Require no auxiliary energy	
D	Accuracy	±3°F
W	Accuracy	±1°F
D	Cost per unit	< $5
W	Production quantity	10,000
D	Size	6-in maximum dimension
D	Weight	< 2 oz
W	Reading distance	10 ft maximum

*D stands for demand, W stands for wish.

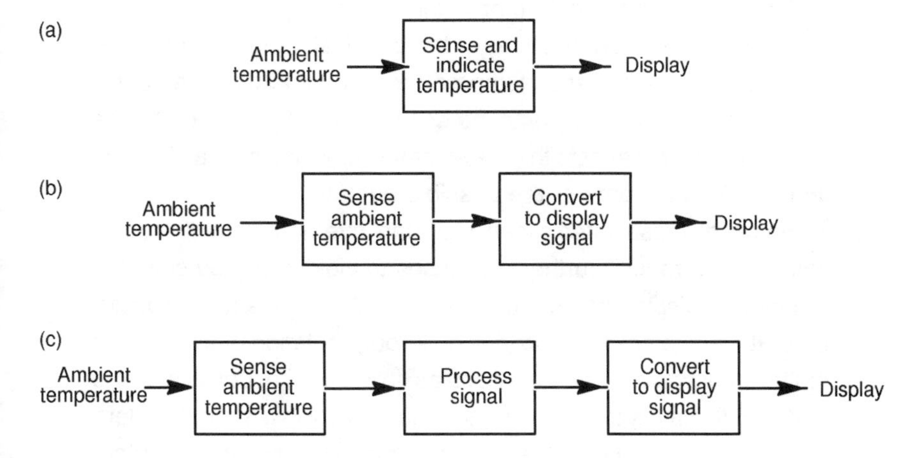

FIGURE 23. Function structure development.

fying it and/or correcting for nonlinearities. This is shown in Figure 23c. After we have proceeded further along the design process and have gained more insight into the problem, we may return and modify the block diagram. Appendix D shows more details on generating function structures and their variants.

Search for Solutions

After the required functions and subfunctions have been established, we look for solutions to satisfy them. There are two possible approaches in this regard: find solutions directly, or proceed at a more fundamental level and consider the physical principles that may be applicable. Then, for each of these, look for possible solutions.

Physical Effects. All processes in machines which are of interest in design are physical processes. This is true even for chemical or biological equipment. These processes are tied to the transformation of energy, signals, or materials. Physical processes (effects) are governed by laws of physics and expressible in the form of equations. It is fruitful, therefore, to study which effects might apply to the pertinent functions of a device. Appendix E provides more possibilities for choosing different physical effects.

The input to a thermometer is thermal energy, represented in the form of a signal (or energy), the physical variable "temperature." The function "sense temperature" Figure 23b has temperature as the input and some other variable as the output. The principal physical effects that have temperature as input are shown in Table 7, with some of the possible outputs.

Some of these effects can fulfill the overall function directly, while others require further processes to yield a display signal. If the type of display has not been specified, it needs to be considered. It may be analog mechanical, through rotation or translation; digital; or of some other form, such as color change. Let us assume that the display is to be analog mechanical (displacement). (Here we are looping back to the requirements list. This type of output could have been specified at the start.) Some of the pertinent

TABLE 7 Physical Effects with Temperature as the Input and Various Outputs

Input	Physical Effect	Output
Temperature	Thermal expansion of solid	Displacement, force
Temperature	Thermal expansion of fluid (change of density)	Volume change, pressure change
Temperature	Thermoresistance	Resistance change
Temperature	Thermoelectric	Voltage
Temperature	Quartz oscillator frequency	Frequency change

physical effects with displacement as the output are given in Table 8.

Extensive tables of physical effects and laws are available in the literature. While we may have in the backs of our minds devices that utilize specific physical effects to fulfill various functions, we should keep this stage solution-neutral (more abstract, less concrete).

Solution Principles. The next step is to use the various effects to obtain solution possibilities in a sketch form, i.e., without considering the actual shapes the parts might have. Some of the physical effects are ruled out by requirements such as permitting no external energy source. Let us look at the first two effects—thermal expansion of solids and fluids, the latter of which results in a significant change in density. Figures 24 and 25, corresponding to Tables 7 and 8, show these physical effects and some of the ways they may be applied.

TABLE 8 Physical Effects with Various Inputs and Displacement as the Output

Input	Physical Effect	Output
Displacement	Lever	Displacement
Density change	Buoyancy	Displacement
Electric current	Electromagnetic	Displacement

There are several other possibilities than those shown here. For example, expansion of a gas can be used. The density effect can be used with a column of liquid; with one or more solid objects of different densities floating in it, the solids will rise or fall as liquid density changes. As an aid in the process of finding solutions, catalogs of solutions for common functions, so-called design concept catalogs, have been published. We need to choose one (or more, but as few as possible) before proceeding to preliminary design.

The Solution Matrix. Having found one or more solutions to each of the functions, we arrange them in the form of a morphological matrix. The left column of the matrix lists the functions. The remaining columns contain the corresponding solutions. This matrix contains the complete solution field for the design task.

The input to the device is thermal energy and its output is visual display, which might take different forms—i.e., a mechanical displacement, or optical analog or digital display. At the input is the function "sense" temperature, which is of the basic type "convert." It changes heat, manifested as temperature, into a signal of another form. Some of the solutions for this function are shown in Figure 24. The function "convert" to display signal is at the output. Figure 25 shows some of the solutions that would produce an output in the form of mechanical displacement. The inputs for these solutions are of various kinds.

The solutions for the three functions are compiled in matrix form; some of them are shown in Table 9. It should be noted that the solutions at this stage are given only in a rudimentary form. The determination of the various shapes in which they may be utilized is the domain of preliminary design.

Concept Variants

If we take one solution for each of the functions from the solutions matrix, we create a concept variant, i.e., a potential solution. Thus, a large number of solutions are possible, depending on the number of subfunctions and the number of solutions for each. A systematic method for implementing this process is called for that ensures

FUNCTION: 'SENSE' TEMPERATURE
INPUT = HEAT

PHYSICAL EFFECT	OUTPUT (CHANGE)	SOLUTION SKETCH
EXPANSION OF SOLID	DISPLACEMENT, FORCE	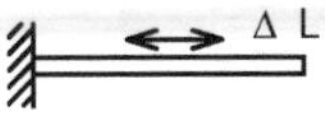 Δ L
EXPANSION OF FLUID	VOLUME, PRESSURE	 Δ L Δ P
THERMO-ELECTRICITY	VOLTAGE	 V
THERMO-RESISTANCE	RESISTANCE	 Δ R

FIGURE 24. Physical effects and solutions for the function "sense."

the compatibility of the subsolutions in any of these combinations. Once the concept variants have been established, they must be evaluated to determine the best solution concept. These steps are described below.

Forming Concept Variants. We begin with the first function in the matrix and choose one of the given solutions. We then look at the adjoining function(s) and pick one of the solutions. This process is repeated until all functions have been covered. This random approach can be improved upon by applying the following criteria: (1) starting with the most important rather than the first matrix function, (2) considering the geometric as well as physical compatibilities of the adjoining solutions, and (3) using a value factor with each solution in a row to indicate the order of preference.

FUNCTION: 'CONVERT' TO DISPLACEMENT

OUTPUT = MECHANICAL DISPLACEMENT

INPUT (CHANGE)	PHYSICAL EFFECT	SOLUTION SKETCH
DISPLACEMENT	LEVER	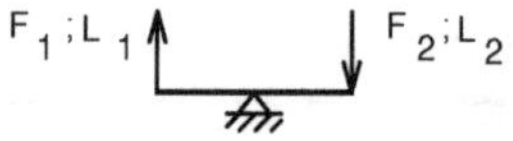 $F_1; L_1$ $F_2; L_2$
DENSITY	BUOYANCY	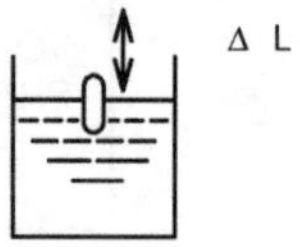 Δ L
MAGNETIC FIELD	MAGNETIC FORCE	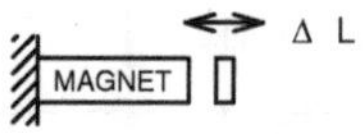 MAGNET Δ L

FIGURE 25. Physical effects and solutions for the function "convert."

These criteria can reduce the number of possible combinations considerably.

Evaluating Concept Variants. We evaluate the concept variants by using a decision matrix technique as described by, for example, Johnson (1978). A list of criteria for evaluation is developed, based upon the specifications. Each criterion is assigned a weight depending on its importance. We now assign values to each of the concept variants depending on how well it satisfies the specifications. These values may be obtained by calculation—e.g., life of a part—or are based upon our judgment. The latter is more often the case at the conceptual design stage, since accurate information is not yet available. The products of the two corresponding sets of numbers are the weighted values for each criterion. The sum of these products for each of the variants is an overall numerical value that ranks them in an order suitable for choosing the winning combination—i.e., the solution concept.

TABLE 9 Solutions Matrix

	Subsolutions*			
Functions	$j = 1$	$j = 2$	$j = 3$	$j = 4$
$i = 1$ Sense temperature	I = temperature Solid expansion O = displacement	I = temperature Liquid expansion O = density change	I = temperature Liquid expansion O = displacement	I = temperature Thermoelectric O = voltage change
$i = 2$ Convert signal	I = voltage Amplification O = voltage	I = displacement Piezoelectricity O = voltage		
$i = 3$ Convert to display signal	I = displacement Lever O = displacement	I = density Buoyancy O = displacement	I = voltage Biot-Savart O = displacement	I = voltage Liquid crystal O = optical

*Physical effect is shown for each solution; I = input; O = output.

The solutions given in Table 9 produce a large number of concept variants. For the purpose of illustration, four of the variants are chosen for evaluation, as given below. In this list (i-j) refers to subsolution j for function i.

- Variant 1: (1-1), (3-1); solid expansion, lever
- Variant 2: (1-2), (3-2); liquid density change, float
- Variant 3: (1-3); liquid in tube
- Variant 4: (1-4), (2-1), (3-3); thermocouple, amplifier, voltage indicator

For evaluating these concept variants, the following criteria are chosen:

1. Number of parts used in the design
2. Estimated cost
3. Accuracy of the device
4. Whether external energy is required for its operation
5. Sturdiness of the product

Each of these criteria i is assigned a weight w_i (with $\Sigma\ w_i = 1.0$). Each of the concept variants j is assigned a value v_{ij} for each criterion on a scale of 1 to 10; i.e., the more favorably the criterion is fulfilled, the higher is the value. For example, the greater the number of parts in a design, the lower is the corresponding v_{1j}. These data and the calculated results are shown in Table 10. The weights can be chosen based upon the designer's judgment. Later examples show the methods of the objectives tree (traction drive example) and binary weighting (material selection example) to arrive at these numbers.

Two types of overall results are shown, the unweighted sums $\Sigma\ v_{ij}$ and the weighted sums $\Sigma\ w_i v_{ij}$. These sums rank the concept variants in the order of their value. Variant 1 is ranked the highest by both of these sums under the assumed evaluation criteria.

Preliminary and Detail Design

Let us, for the sake of this example, choose the solutions produced by using the thermal expansion effect (Figure 24). In the preliminary design stage the chief criteria are (1) shapes or surfaces, (2) motion, if any, and (3) principal material properties. We con-

TABLE 10 Evaluation of Concept Variants

Criterion	Weight w_i	Concept Variant							
		$j = 1$		$j = 2$		$j = 3$		$j = 4$	
		v_{i1}	$w_i v_{i1}$	v_{i2}	$w_i v_{i2}$	v_{i3}	$w_i v_{i3}$	v_{i4}	$w_i v_{i4}$
$i = 1$ Parts	0.20	7	1.4	5	1.0	8	1.6	3	0.6
$i = 2$ Cost	0.15	8	1.2	5	0.75	7	1.05	2	0.3
$i = 3$ Accuracy	0.30	6	1.8	2	0.6	7	2.1	10	3.0
$i = 4$ Energy	0.15	10	1.5	10	1.5	10	1.5	5	0.75
$i = 5$ Sturdiness	0.15	8	1.2	1	0.15	1	0.15	5	0.75
Overall results	Σv_{ij}	39		23		33		25	
	$\Sigma w_i v_{ij}$		7.1		4.0		6.4		5.4

sider variations in these to develop several designs. Figure 26 shows some of the possibilities, where physical effects have also been indicated. Further calculations still need to be made to determine actual size of the components.

The principal material property in this case is the coefficient of thermal expansion, on the basis of which the material will be selected. Volume expansion of a liquid can be used directly, as

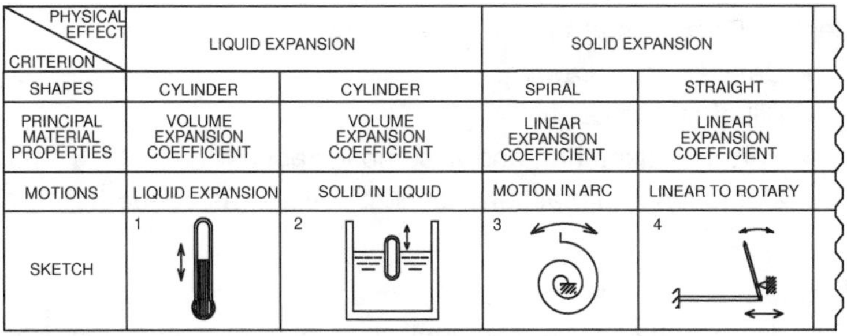

PHYSICAL EFFECT / CRITERION	LIQUID EXPANSION		SOLID EXPANSION	
SHAPES	CYLINDER	CYLINDER	SPIRAL	STRAIGHT
PRINCIPAL MATERIAL PROPERTIES	VOLUME EXPANSION COEFFICIENT	VOLUME EXPANSION COEFFICIENT	LINEAR EXPANSION COEFFICIENT	LINEAR EXPANSION COEFFICIENT
MOTIONS	LIQUID EXPANSION	SOLID IN LIQUID	MOTION IN ARC	LINEAR TO ROTARY
SKETCH	1	2	3	4

FIGURE 26. Preliminary design examples.

in design 1, or a solid can float in the liquid at different heights as the density of the latter changes, as in design 2. For a solid the expansion element can have many possible shapes, among which are straight and spiral. The corresponding motions are straight line and circular. The former can be converted to rotary motion, which provides amplification for better display. In both cases the motion is being amplified.

We note that designs 1 and 2 correspond to the function block diagram shown in Figure 23a, whereas in designs 2 and 4, sensing and display functions have been separated, as in Figure 23b.

A commercially produced variation of design 2 from Figure 26 is shown in Figure 27. The so-called Galileo thermometer consists of a number of spheres, each of a slightly different density, immersed in a clear liquid. Each sphere has a tag attached to it, which indicates the corresponding temperature value. As the temperature increases, the spheres rise up one by one in the column. The tag value of the last sphere to rise is nearest the room temperature. Certainly, if aesthetics were weighted high and sturdiness low, this design would rank high.

Getting back to the solid–expansion-based concept, further work in embodiment involves decisions regarding items such as the following:

- Size of the expansion element
- Materials
- Fastening methods
- Manufacturing methods

In addition, the housing design needs to be carried out. The detail design is developed based upon the chosen preliminary design. Decisions at this stage relate to:

- Detailed part drawings
- Bill of materials
- Part manufacturing and assembly steps
- Operation and maintenance instructions (as needed)

Summary

This example has shown the application of the systematic design procedure, with emphasis on the conceptual design phase. We

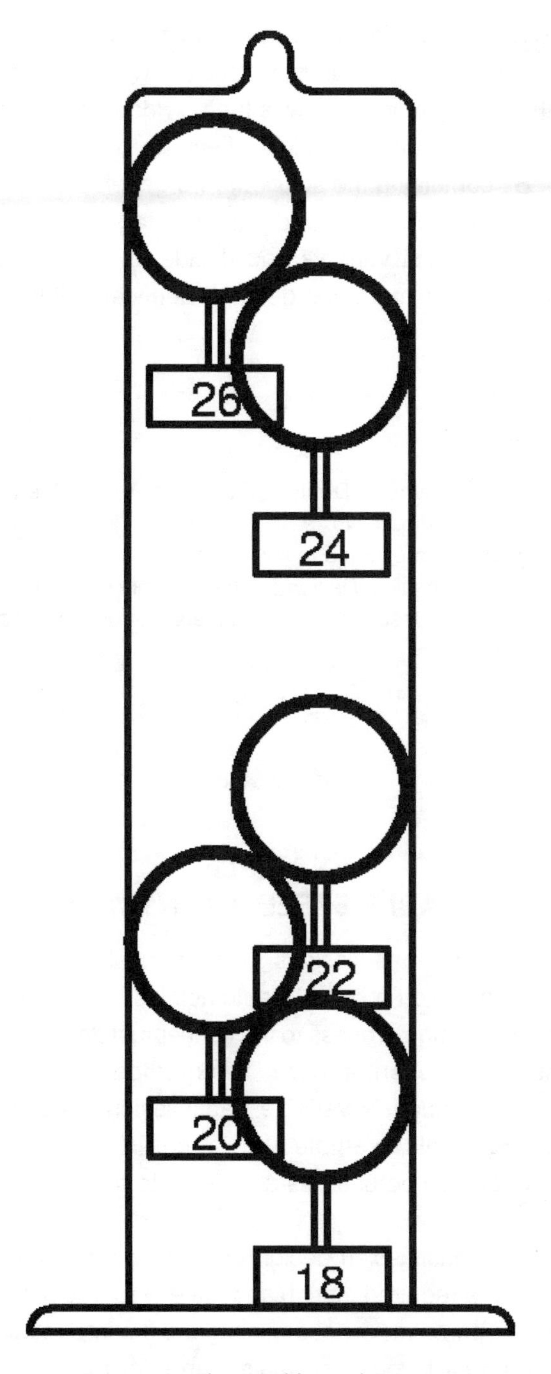

FIGURE 27. The "Galileo" thermometer.

started with task clarification and followed it with preparation of the requirements list. The function structure was developed, starting with the overall function; it was then divided into subfunctions. We investigated various physical effects in order to search for solutions. By combining the solutions, we generated a number of concept variants. We then used the weighted values method to evaluate the concept variants and decide on the final concept. The steps in embodiment and detail design were alluded to here only briefly.

BIBLIOGRAPHY

Hundal, M. S."Systematic Design of a Temperature Measurement Device." ASME, Paper No. 92-DE-1, New York, 1992. The above example was first presented in this paper.

Johnson, R. C. *Mechanical Design Synthesis.* Huntington, N.Y.: Krieger Publishing, 1978. Describes, among others, the decision matrix technique.

● ● ● ● ●

EXAMPLE
A VARIABLE SPEED TRANSMISSION

Mechanical power transmissions form the most crucial part of all machines—from food mixers to automobiles. The application of the systematic design process to the conceptual design of a power transmission is presented here. The method is applied to the design of a continuously variable transmission, as described in Hundal (1987). This example shows that in most mechanical designs, shape can be used as a variable to generate a number of concepts.

The power transmission design problem can be stated briefly as follows: For a required motion and force or torque at the output, determine the prime mover at the input and the intermediate components and their layout. The main step prior to the conceptual design is the preparation of a requirements list. This list will contain the specifications under the following general headings:

- Geometry: space availability; mounting locations
- Kinematics: types and directions of motion
- Kinetics: forces and torques at input and output; weight; natural frequency constraints
- Energy: output; efficiency
- Signals: measurement; control
- Safety: protection systems
- Environmental conditions: temperature; humidity; chemical actions
- Production: simple parts; ease of assembly
- Maintenance requirements: lubrication, wear, and replacement of parts
- Cost

Abstracting to Identify the Essential Problem

This step permits us to conduct the widest possible search for solutions, by reducing the problem to its most essential form. The functional relationships in the specification are formulated explicitly and arranged in order of importance. The most important specifications for a power transmission are the torques and speeds at input and output, followed by space requirements and environmental conditions. Upon abstraction of the power transmission problem—i.e., omitting unimportant details, transforming from quantitative to qualitative terms, and stating the problem in solution-neutral terms—we obtain the statement:

Transmit mechanical power from one point to another at various speeds and torques.

This brief statement is free of all unnecessary detail, most notably quantitative facts. The most important specifications stated in solution-neutral terms are the following:

- Output speed continuously variable within certain limits
- Speed adjustable by a mechanical signal
- Output speed to be indicated

ESTABLISHING FUNCTION STRUCTURES

In this step we consider the overall function and subfunctions. The overall function of the device is to transmit power from one point to another. This is broken down into subfunctions to facilitate the search for solutions and to combine these to produce a function structure. In the function structure for a power transmission the primary flow is that of energy, and the second is of signals for control. The transmission will be composed of some known subassemblies which can be treated as complex subfunctions. Conceptually new subassemblies require further subdivision into simpler subfunctions.

Development of the function structure of the device is shown in Figure 28:

- The basic function is "change speed ratio," shown in the top of the figure.

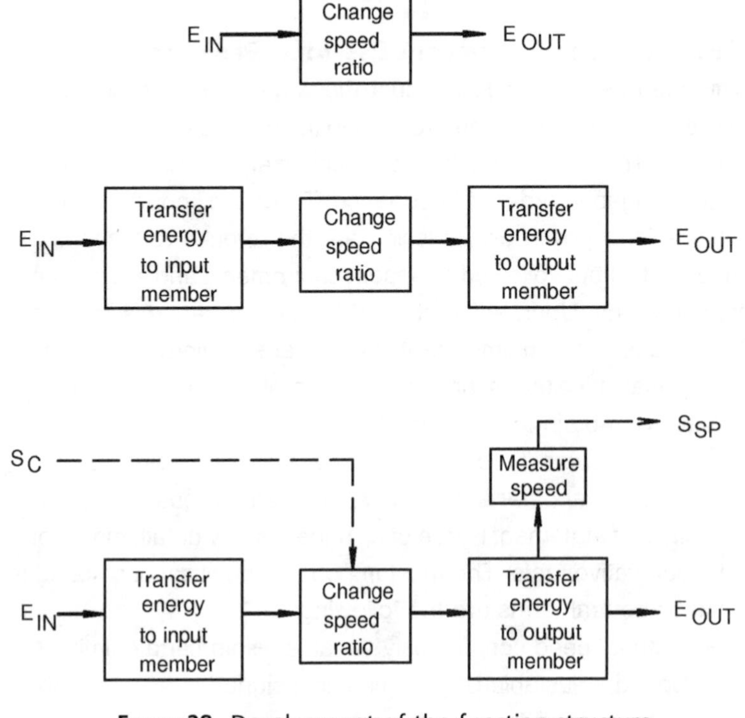

FIGURE 28. Development of the function structure.

- To this we add the functions indicating the transfer of energy to the device from the input and from it to the output.
- Finally, the bottom diagram shows the addition of (1) the signal to change the speed ratio and (2) the function "measure speed" and the signal for speed display.

Searching for Solution Principles to Fulfill the Subfunctions

A solution principle reflects the physical effect needed to fulfill a given function. It is generally, though not necessarily, closely tied to a physical shape. For instance, for the function "transfer torque," one possible physical effect is friction. The physical shape the function may take is a hub on a shaft. Thus, many solution principles may generally be found, which form a solution field. There are a number of established methods of finding solution principles.

Systematic methods of generating solutions were presented earlier. We will look at the application of those to this problem.

Study of System Equations

If we develop equations describing the operation of a device, we can observe how varying certain parameters in the equations can lead to new solutions. In the case of a power transmission, it is possible to generate variants for certain types of elements, such as speed-varying mechanisms.

Search With Classification Schemes

As an example in the present case, the subfunction "transfer torque by friction" gives rise to a number of possibilities. The scheme in Table 11 considers the "driving surface" and the "transmitting element" as the two parameters. The solution principles shown in the table are those used in some currently available traction drives.

The spaces left blank in the table offer possibilities of other types of drives. Some combinations may be incompatible; however, the scheme does indicate how the search for solution principles may

TABLE 11 Classification Scheme for Solution Principles

Transmitting	Driving Surface			
Element	Flat	Ring Cylinder	Toroidal	Spherical
Sphere	Ball disk		Ball variator	Offset sphere
Cylindrical		Ring roller		
Conical		Roller variator		
Flat			Toroidal drive	
Spool			Spool drive	

be systematically expanded. Further use of this scheme is illustrated below.

Commercially available traction drives (see Khol, 1994) include the ball variator, Beier disk, fixed-ratio, friction disk, planetary, ring cone, ring roller, and toroidal drives.

Combining Solutions to Fulfill the Overall Function

Once a solution field has been generated, the solutions are combined to form concept variants that fulfill the overall function. The governing role in this is played by the already established function structures. There are some obvious principles to be observed in forming the combinations:

- A solution must be physically compatible with those adjacent to it in the structure.
- Another requirement is to select an economic and technically feasible combination.

A convenient method of combining solutions is to use the classification table prepared for the search of solutions. Here functions are designated by rows and solutions by columns. Table 12 shows some of the functions involved in a mechanical power transmission and various solutions for them. Combinations from this table can be used to generate some of the common types of transmissions. Two examples are:

1. Gear drive: (1, 1)-(2, 1)-(3, 1)-(4, 2)-(3, 1)-(2, 1)-(1, 1)
2. Rack and pinion: (1, 1)-(2, 2)-(3, 1)-(4, 1)

TABLE 12 Functions and Corresponding Solution Principles

Function	Solution Principles 1	2	3	4
Transfer torque—axial	Shaft	Coupling		
Transfer torque—radial	Keyed hub	Shrunk hub		
Convert force or torque	Wheel	Rotating link	Cam and follower	
Transfer force	Axial member	Contacting members	Belt or chain	Friction

The key subfunction is "speed ratio change," which in effect requires a transfer of energy, for which we next look for solution principles. Energy can be transferred by mechanical effects from the input member to the output member by a solid or a fluid medium. The physical effects involved are:

- Contact (solid to solid)
- Friction (solid or fluid)
- Fluid (static or dynamic)

Using these alone, the search for solution principles can be expanded and will indeed produce a large solution field. Let us restrict ourselves for the sake of this example to the solid medium and solid friction effect. The basic principle to be used in a continuously variable transmission is the change in operating radius at the driving and driven members. Solid transmitting media may be broadly divided into two categories:

1. Flexible—including belts and chains
2. Rigid—basis of most traction drives

We now look further at a solution search using a rigid transmitting element with solid friction between surfaces. It is recognized, however, that in practice elastohydrodynamic lubrication would exist.

We now expand further on the idea of using classification schemes in the search for solution principles. Whereas Table 12 listed only some of the currently available traction drives, the search will now be at a more basic level. From the most promising candidate geometries for the driving, transmitting, and driven elements, the following are selected in this example:

1. Flat disk
2. Cylindrical roller
3. Cylindrical ring
4. Spherical roller
5. Toroidal ring
6. Conical roller
7. Conical ring

Figure 29 shows the search process from the function to the solution principles stage for the subfunction "speed ratio change." There are theoretically ($7^3 =$) 343 combinations of driving, transmitting, and driven members of these geometries. However, there are other considerations involved in generating a solution field, as discussed below.

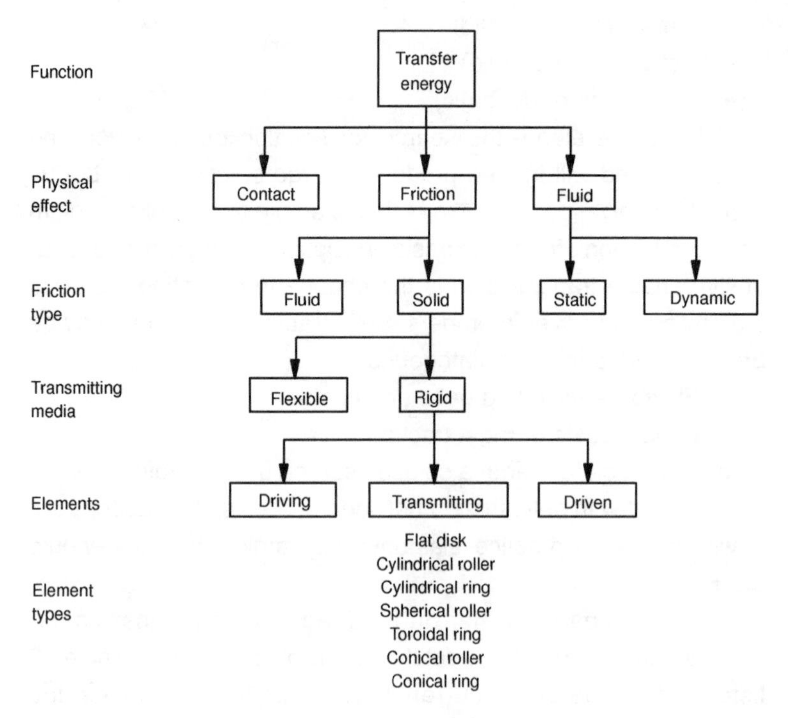

FIGURE 29. Developing solution principles for speed ratio change.

We next consider the following points in generating the solution field and the subsequent evaluation of concept variants in the design of this transmission:

- Geometric compatibility of contacting surfaces
- Type of contact—point, line, or surface
- Extent of sliding between mating surfaces, for wear
- Type of motion of members for speed change

Figure 30 shows a tabulation of the first two items for each combination of contacting surfaces. The different types of contact—surface, line, or point contact—are indicated, as well as the instances where the mating surfaces are incompatible.

Figure 31 shows some of the possible solutions that can be obtained by applying this classification scheme using five of the surfaces mentioned above. The combinations that are incompatible (*) and those which do not allow speed variation by radius change (+) are indicated. We note that in some cases the transmitting element itself could be the driven element. In this figure, one can recognize some of the solutions as the basic components of commercially available traction drives. The scheme, however, points out other possibilities.

	Flat disk	Cylindr. roller	Cylindr. ring	Spherical	Toroidal	Conical roller	Conical ring
Flat disk	□	▭	✳	○	✳	▭	✳
Cylindr. roller		▭	▭	○	✳	▭	▭
Cylindr. ring			✳	○	✳	▭	✳
Spherical				○	▭	○	○
Toroidal					✳	✳	✳
Conical roller						▭	▭
Conical ring							✳

□ Surface ▭ Line ○ Point ✳ Incompatible

FIGURE 30. Types of contact between mating surfaces.

Driving element	Transmitting element				
	Flat disk	Cylindrical	Spherical	Toroidal	Conical
Flat disk			(1) (2)	*	+
Cylindrical	+	+		*	
Spherical					
Toroidal	*	*	(1) (2)	*	*
Conical	+			*	

FIGURE 31. Solution matrix for traction drive.

EVALUATING THE CONCEPT VARIANTS AGAINST PROBLEM SPECIFICATIONS

After the concept variants have been generated, we reject those that do not satisfy the requirements and decide on the most suitable concept based on economic and technical criteria. For the evaluation of concept variants, evaluation criteria including weighting factors must be established.

The important part remaining is the evaluation of the solutions against the specifications list, and on the basis of technical feasibility and economic considerations. The evaluation is carried out in the following steps:

1. The evaluation criteria are established from the specifications. Use of an objectives tree is helpful in this regard, in which the objectives are arranged in hierarchical order (Ertas and Jones, 1993).
2. Weights are assigned to the evaluation criteria; thus, the relative importance of the criteria plays a deciding role in this process.
3. The final step is the actual evaluation.

Evaluation criteria and the determination of their respective weights are shown in Figure 32 in the form of an objectives tree. The criteria shown here apply to most products. We proceed as follows:

- The total weight of 1.0 is given to the primary criterion "reliable and inexpensive device."
- This is divided into fractions for the subcriteria "long service life" (which is weighted somewhat higher, 0.6) and "easy manufacture" (0.4).
- Service life is deemed to depend on two factors, "low wear" and "easy maintenance," which are assigned the relative weights of 0.6 and 0.4, respectively.
- The absolute weights of these two criteria are found by multiplying the relative weights and the weight of the higher-level criterion ("long service life," 0.6).
- The other criterion, "easy manufacture," is likewise dependent on the ease of component production and assembly.
- Finally, the production of components is governed by their number and shapes.

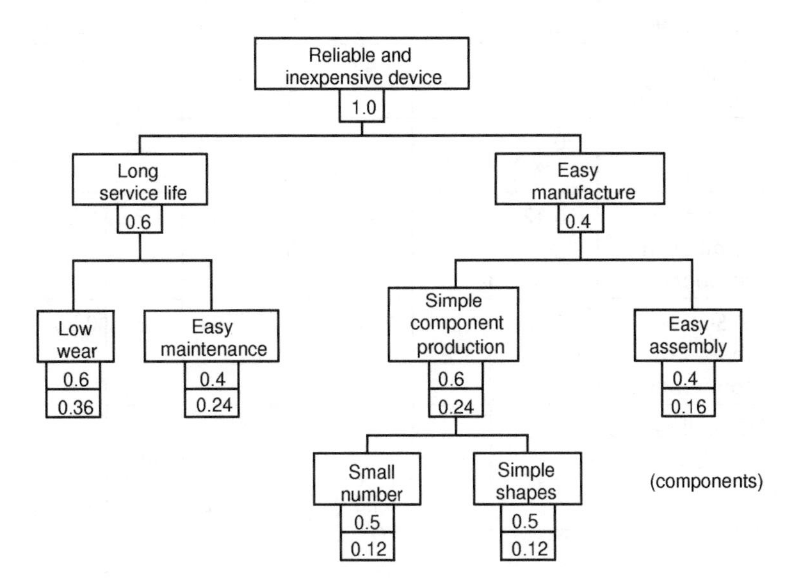

FIGURE 32. Objectives tree for traction drive.

The lowest-level criteria and their weights are carried into the concept evaluation table, Table 13. For the assigned weights and the values, we see that the flat-disk drive has the highest rating. It is, of course, obvious that with different criteria and weights the choice might be different.

The designer assigns the weights using his or her judgment. At this stage the decisions are made on the basis of approximations and not detailed and precise calculations. The aim is to choose a concept that promises the optimum final design.

TABLE 13 Evaluation of Concept Variants

		Concept Variant					
Criterion	Weight	 $j = 1$		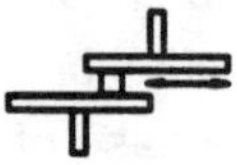 $j = 2$		 $j = 3$	
$i =$	w_i	v_{i1}	$w_i v_{i1}$	v_{i2}	$w_i v_{i2}$	v_{i3}	$w_i v_{i3}$
1. Low wear	0.36	3	1.08	5	1.80	8	2.88
2. Easy maintenance	0.24	8	1.92	5	1.20	3	0.72
3. Easy assembly	0.16	8	1.28	3	0.48	2	0.32
4. Small number of components	0.12	2	0.24	5	0.60	3	0.36
5. Simple shapes of components	0.12	8	0.96	7	0.84	3	0.36
Overall results Σv_{ij}		29		25		19	
$\Sigma w_i v_{ij}$			5.28		4.92		4.64

Summary

We have discussed the application of the systematic design process to the design of power transmissions. Conceptual design is the most important part of the design process from both economical and technical viewpoints. The conceptual design steps of establishing a function structure and the use of a classification scheme to generate a solution field for the key subfunction—"change speed ratio"—have been considered. It was shown how a variety of solutions may be generated even under such narrow confines. Solution fields for transmitting energy to the driving element and from the driven element, as well as for speed measurement and indication, also need to be developed. This example shows that the degree of embodiment is fairly high at the concept stage. This is typical of mechanical designs.

Bibliography

Ertas, A., and J. C. Jones. *The Engineering Design Process.* New York: Wiley, 1993.

Hundal, M. S. "Conceptual Design of Mechanical Power Transmissions." *Proc. ICED 87* I: 246–253. New York: ASME, 1987. The above example was first presented in this paper.

Khol, R., ed., *Machine Design 1994/5: Basics of Design Engineering.* Cleveland: Penton Publishing, 1994. Discusses all aspects of design engineering, including power transmissions. Provides a list of commercially available traction drives.

• • • • •

Role of Management: Time and Cost

Manufacturing has become more globally competitive in recent years. It has been emphasized (Smith and Reinertsen, 1991) that the key ingredients in a manufacturing company's economic viability are (1) product innovation and (2) quickly developing products and bringing them to market. The product development cycle time—from the moment a company realizes that a product should be developed to the time it is in the customer's hands—can be reduced by applying certain techniques of *time-driven development*. This is a procedure used by management to focus on compressing the time taken to develop a product. Other terms used for this technique are *rapid product development* and *accelerated product development*. Companies that have applied the technique of time-driven product development have found time reductions of 50% or more in the planning-design-manufacturing cycle. Some of these companies, with their products in parentheses, are listed in Figure 33.

The highly competitive nature of the marketplace requires that manufacturers be able to deliver products to the customer while satisfying each of the following:

- High quality
- Short time
- Low cost

CONCURRENT ENGINEERING

A shorter time to market does not necessarily mean higher costs. A properly managed program can yield high-quality products at low cost and under time constraints. These three properties depend not only on the physical features of the product—i.e., geometry, materials, tolerances, etc.—but also upon how the development process is carried out. This will become apparent from the following discussion.

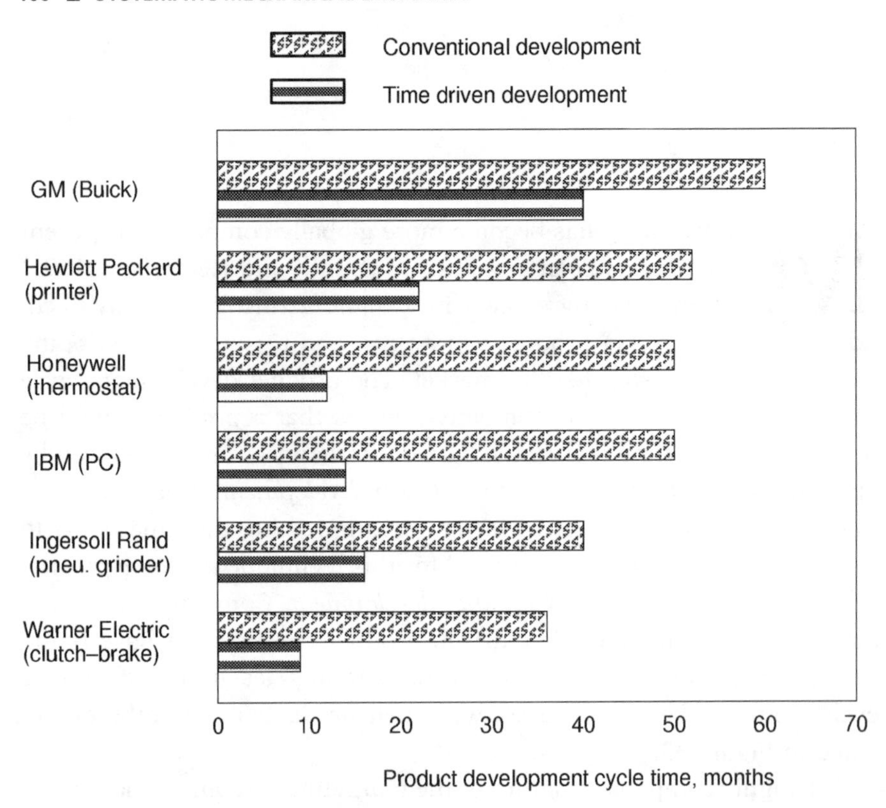

FIGURE 33. Examples of accelerated product development. (Smith and Reinertsen, 1991)

From the time the development of a product begins, its cost starts to grow because of the resources used for its realization—personnel, facilities, equipment, etc. In a free market economy, however, the price of competitors' products already on the market will be dropping with time. There are several reasons for the latter phenomenon:

- Continuing rationalization of the manufacturing process
- Cost-driven improvements in existing designs
- Increased knowledge about the product—the "learning" process

These two variables—competitive price and development cost—are shown in simplified form in Figure 34. The reasons for accelerating the development of products are obvious from this figure.

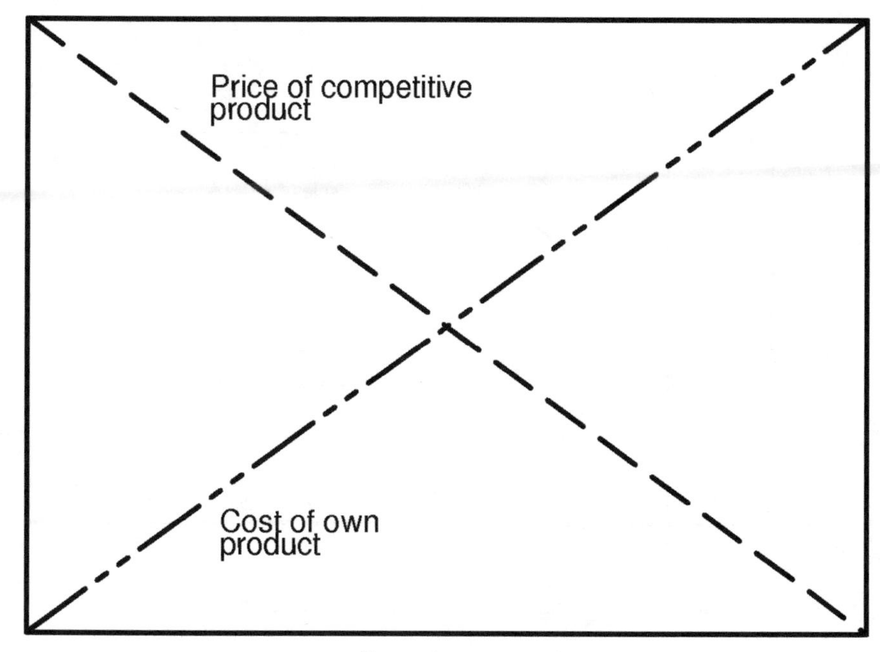

FIGURE 34. Cost and price change with time.

BENEFITS OF RAPID PRODUCT DEVELOPMENT

Time-driven product development, as opposed to the conventional development process, benefits a company in many ways:

1. It extends the product's sales life.
2. As a result of early introduction, the product has a marketplace advantage by gaining early customers who lock on to it, develop loyalty, and are less likely to switch.
3. The company gains a pricing advantage and gets on the learning curve ahead of the competition.

These points are illustrated in Figure 35. Sales curves of two products, A and B, are shown. Product A is developed faster than B and therefore launched earlier than B. A therefore has that much longer sales life. It also shows higher sales than B due to its earlier introduction.

Suri (1995) mentions another advantage:

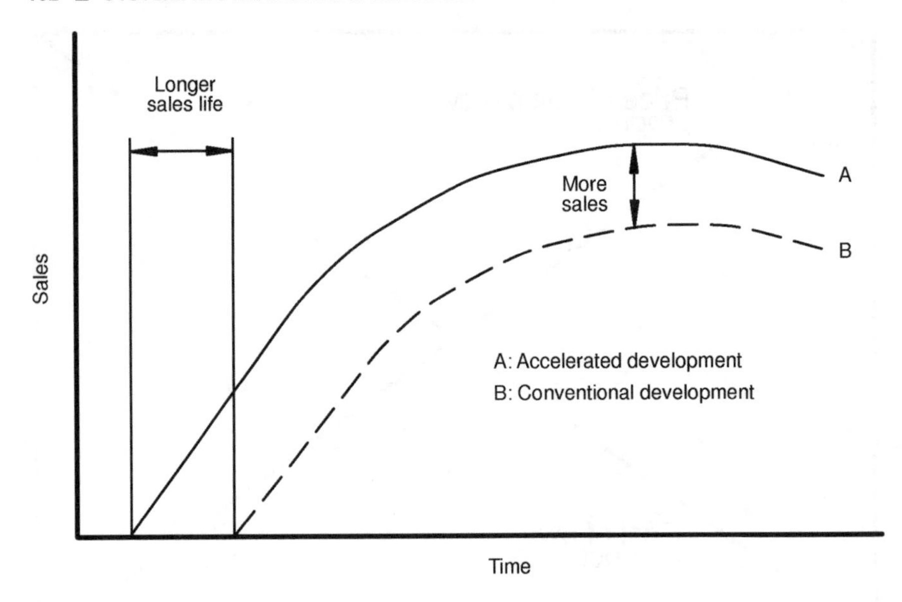

FIGURE 35. Advantages of time-driven product development.

4. A company that applies rapid development methods and starts on a product later than a traditional company will use more up-to-date technology in the product.

This point is illustrated in Figure 36. The line marked "State of the technology" indicates how the technology is advancing with time. Company B, carrying out the traditional development process, foresees a 6-year development cycle. When it starts its product development, it locks on to the technology of that point in time. Company A applies rapid development methods and foresees a 3-year development cycle. It thus starts with a higher state of technology and an advantage over company B. Both products reach the market at the same time (time 0).

Figure 36 also illustrates another point. The line marked "Risk" shows the manner in which the risk that a company takes increases the earlier it starts product development in relation to the expected launch date. Since company A begins its product development only 3 years before the launch date, it is taking much less risk than company B, which must look ahead 6 years when it begins its product development.

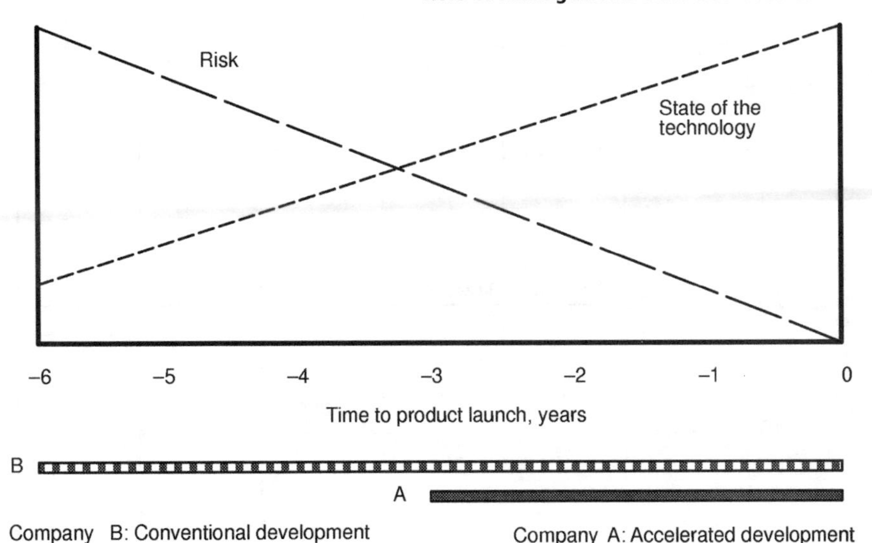

FIGURE 36. State of technology, starting time, and risk.

The effect on profits (Wolfram and Ehrlenspiel, 1993) of deviations in development time, production costs, and development costs is shown in Figure 37. For example, an increase of 50% in development costs can decrease profits by 10%, but an increase of 10% in development time can reduce profits by up to 30%.

EARLY STAGES OF A PROJECT

A lack of management of a product in its earliest stages can account for the greatest loss of time. A project can go dormant without people realizing it, as there is no one responsible to keep track of it. There may be several reasons for this. There might be a perceived need for the product but not enough importance assigned to it. Or the company may not be sure it has the technology for all parts of the product. Meanwhile, expenses on the product start to increase sharply just as its importance is being recognized, while the market window of opportunity is closing. There *is* a management scheme that can shorten product development, and that is to carry out product planning and design concurrently rather than sequentially. The planning and design are broken up into small steps and

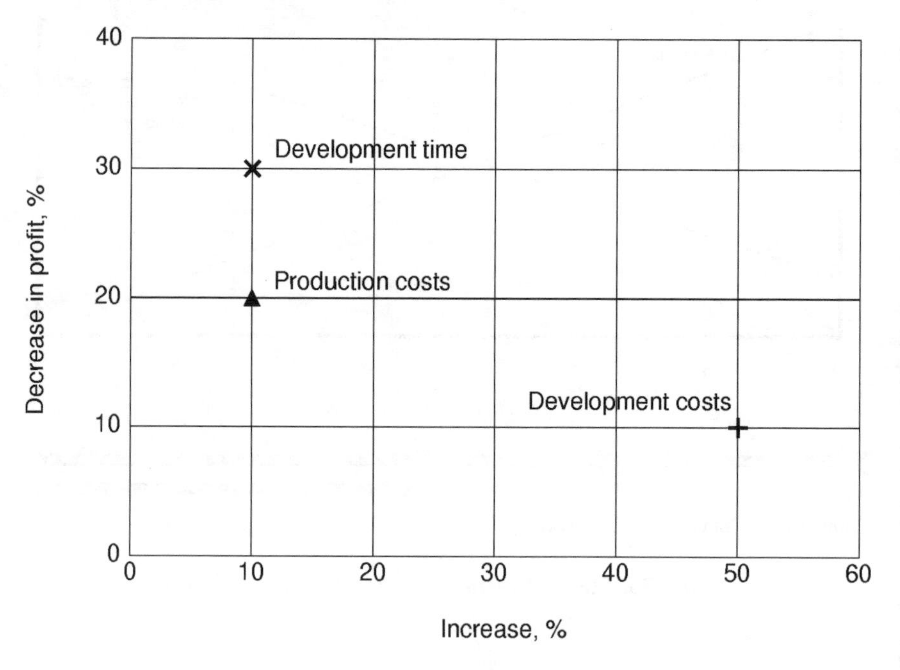

FIGURE 37. Effect on profits. (Wolfram and Ehrlenspiel, 1993)

meshed together. Indeed, manufacturing process planning and certain early steps in manufacturing can also be carried out concurrently with design.

PRODUCT INNOVATION BY STAGES

All new products and new parts thereof entail learning on the part of personnel. The greater the degree of novelty, the more are the unknowns and therefore the risk attached to development. Rather than innovate on the whole product, it pays to use existing technologies as far as possible and to improve only parts of the product at a time.

• • • • •

EXAMPLE
MITSUBISHI HEAT PUMP

An example of such incremental innovation is the development of a heat pump by Mitsubishi, as described by Stalk and Hout (1990).

Over several years the product was improved in a step-by-step fashion as shown in Figure 38. Mitsubishi started out with a system that had a sensor of contemporary design, a controller with wired circuits, and an AC motor driving a reciprocating compressor. At the end of the development cycle the heat pump had a fiber-optic sensor, a control system with integrated circuits and a microprocessor, remote control and learning circuitry, and a motor supplied by an inverter for better speed control to drive a rotary compressor. At each step the company used existing technology, rather than strive for dramatic breakthroughs. At the end of a nine-year period, the company had an entirely new unit, but at no time did it expose itself to excessive risk.

● ● ● ● ●

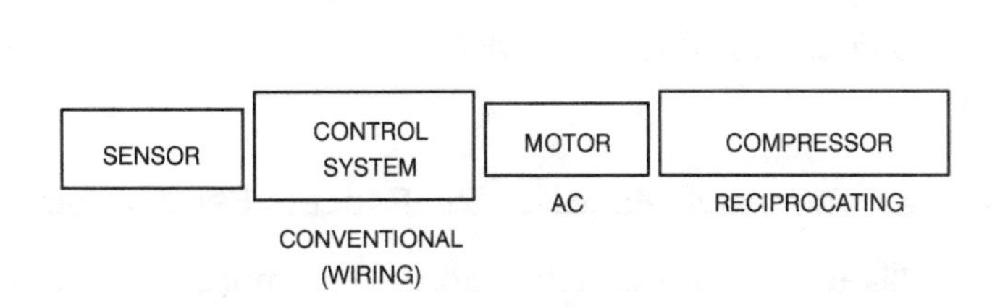

YEAR	FEATURE INTRODUCED	BENEFITS ACHIEVED
1980	Integrated circuits	Improved EER; Reliability
1981	Microprocessor	Easier to install & service; Increased market share
	'Quick – connect' Freon lines	Sold through hardware stores
1982	Rotary compressor	Higher efficiency
	Louvered fins; inner–fin tubes	Better heat transfer
	New electronics	Improved Energy Efficiency Ratio (EER)
1983	New sensor	Better control of cycle
	More computing power	
1984	Inverter	Better speed control; Higher EER
1985	Shape–memory alloys	Control of air louvers
1986	Fiber optic sensor	Cycle adjustment; Higher efficiency
1987	Remote control	Personalized control
1988	Learning circuitry	Optimum cycle; Timely defrosting

Figure 38. Incremental improvement of the Mitsubishi heat pump.

Other well-known examples of innovating in small steps are:

IBM PC (and compatibles):

- Started with 8088 CPU. Upgraded in steps to 286, 386, 486, and Pentium.
- Put in larger memory chips as CPU was upgraded.

Sony Walkman:

- Introduced in 1979, it was an unattractive design.
- Many small changes were made sequentially.
- Manufacturing improvements were made, which reduced cost, and improved performance.
- Mechanical components were gradually replaced with electrical components.

Chrysler Minivan:

- Planning was started in 1978, when gasoline prices were high.
- Introduced in 1984, with a 4-cylinder engine (gasoline prices had dropped).
- V-6 engine introduced in 1987.

Need for Concurrent Product and Process Realization

Present-day technical products tend to be complex in order to fulfill the functional and other requirements imposed on them. Product complexity has led to the emergence of complex organizations, with individual departments addressing the different functions—e.g., design and manufacturing. In most companies there are "walls" between these different departments; i.e., the departments are physically separated and/or there is poor communication among the departments. As a result of such separation:

- The product idea is given out from marketing to design and development.
- Design department works on the design.
- After the design is complete, the documents are "thrown over the wall" to manufacturing.

This can be understood from Figure 39a, which shows the product realization phases of the total product life cycle that was depicted in Figure 3 in Chapter II.

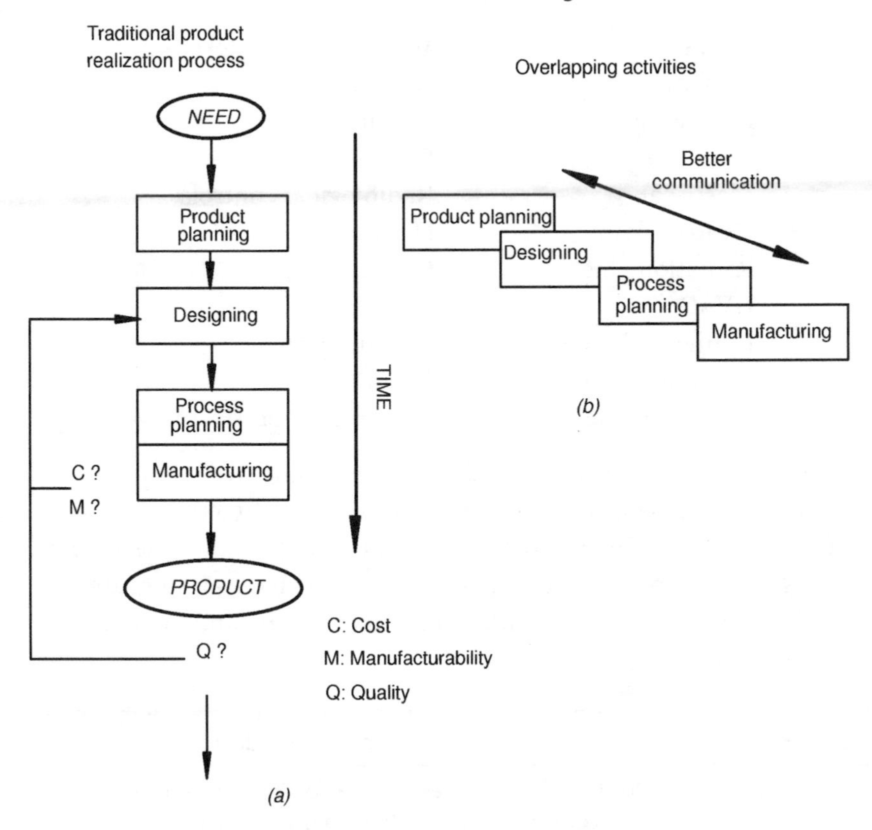

FIGURE 39. (a) Traditional and (b) concurrent engineering.

This method results in several problems, chiefly due to most designers' insufficient familiarity with the manufacturing processes. A misunderstanding on the part of designers of customers' wishes may also play a significant role. Often, last-minute changes must be made at the manufacturing stage; a consequence is poor product quality as well as higher life cycle costs.

The design process, rather than being isolated, must obtain inputs from all parties involved in the product's life cycle: planning, marketing, manufacturing, and the customer. The parties involved and their number will vary, depending upon the product and the company. As an example, for the development of a nuclear attack submarine, the following departments at General Dynamics Electric Boat Division are quoted as being involved (see Ashley, 1995):

Piping	Welding and materials
Structures	Nuclear
Hull design	Facilities
Acoustics	Chemical
Electrical	Radiological controls
Hydrodynamics	Human factors
Radiation shielding	Quality control
HVAC	Computers
Lifting and handling	Construction
Automated processes	Planning
Environmental	Customer—the U.S. Navy

Customer input is crucial in product development. We will discuss this aspect further in the sections of Chapter V on quality function deployment.

There are yet other reasons for delays in the product development process. The different departments' perspectives on the product are different. The following list shows how marketing, design, and manufacturing view the process.

Department	Views on the Product
Marketing	Product should be as good as or better than the competitors'; it should have more features.
Design	Use the latest technology; make it "sexy."
Manufacturing	Have a stable production line; processes should be easy to set up.

It is apparent that these views are in conflict. It is incumbent upon management to develop a sense of common purpose among the various groups.

IMPLEMENTING CONCURRENT ENGINEERING

"Concurrent" involvement of different functions and overlapping of their activities, along with better communication between departments, has been proven to reduce development time, reduce costs, and increase quality, as shown schematically in Figure 39b. The modern product realization paradigm is that *the process must be designed at the same time as the product.* More details on concurrent engineering are given in several excellent books on the subject, e.g., Parsaei and Sullivan (1993). This brief mention is not intended to minimize its role in product management. We shall discuss here only how concurrent engineering and other management initiatives may be applied in lowering costs.

• • • • •

EXAMPLE
CONCURRENT ENGINEERING IN THE "SKUNK WORKS"

Lack of communication between departments is a problem in large organizations. This problem can be addressed by formation of cross-functional teams that operate with a high degree of autonomy. One of the most famous examples of such a unit within a large company is the "Skunk Works" of Lockheed Aircraft Co., formed during the Second World War under Kelly Johnson. This unit was responsible for the quick development and manufacture of many superb airplanes such as the P-80 Shooting Star, the F-104 Starfighter, the U-2, the SR-71 Blackbird, and the F-117 stealth fighter, among others. Kelly Johnson had rules (Rich and Janos, 1994) for operating the Skunk Works, including:

- "There will be only one object: to get a good airplane built on time."
- "Engineers shall always work within a stone's throw of the airplane being built."
- "Special parts or materials will be avoided whenever possible. Parts from stock shall be used even at the expense of added weight."
- "Everything possible will be done to save time."

The program manager on the development of F-117 stealth aircraft pointed out how concurrent engineering was implemented on this project: "Our engineers were expected on the shop floor the moment their blueprints were approved. Designers lived with their designs through fabrication, assembly and testing. Engineers could not just throw their drawings at the shop people on a take-it-or-leave-it basis. . . . Our designers spent at least a third of their day right on the shop floor; at the same time, there were usually two or three shop workers up in the design room conferring on a particular problem" (Rich and Janos, 1994). It may be argued that perhaps these people were motivated by national security concerns, but they were also proud to be building excellent products under severe time constraints.

● ● ● ● ●

Munro (1995) suggests making the product designers spend a day on the production line making and assembling their designs.

Overlapping Activities

Fully concurrent engineering is an ideal that is seldom achieved. A frequent compromise is a partial overlapping of each phase of the product realization process with the next phase (Hayes, Wheelwright, and Clark 1988). The advantage of overlapping over sequential activities is shown with the help of Figure 40 (adapted from Hayes et al., 1988). Let us say that activity i is designing and $i + 1$ is manufacturing. The results of activity i are delivered upon its completion to $i + 1$. The group pursuing activity $i + 1$ needs some time to digest the information (designs, documents) given to it. It then begins with its task and proceeds to complete it in a certain length of time as shown at (a). However, such an ideal, trouble-free process does not often take place in real life. There are glitches—for example, it may be found that parts cannot be made economically as designed, or the assembly of parts may run into problems. Thus, quick fixes and changes must be made, leading to mistakes that in turn require more fixes. The time taken for activity $i + 1$ is stretched out, as shown at (b). At the same time, because of the unplanned changes and hurried decisions product costs increase and quality suffers.

At (c), we see an improved mode of operation. There is a continual sharing of information between groups engaged in consecutive activities,

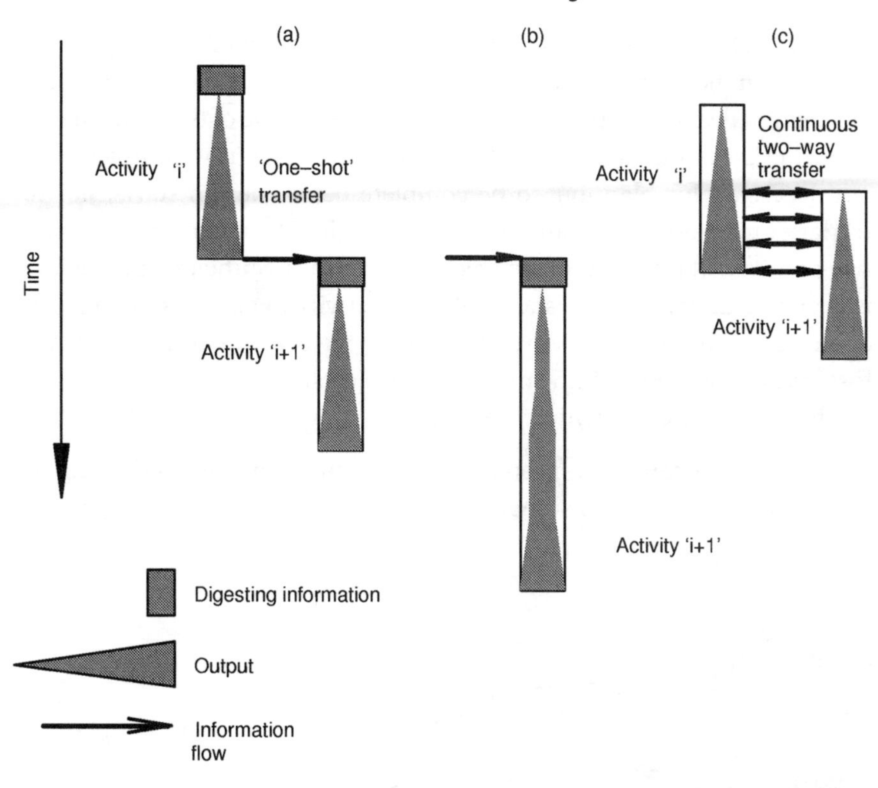

FIGURE 40. Advantages of overlapping activities. (Hundal, 1995)

as well as an early resolution of conflicts. Between product planning and designing, such information might be about the state of technology, which affects the feasibility of a design. Sharing information between process planning and the design groups can uncover manufacturability concerns earlier than would otherwise happen.

The most obvious question that comes to mind is, How can the group prerforming activity $i + 1$ begin its activity on the basis of incomplete information? The key to resolving this paradox is that the downstream group should not accept the information as final. It should also be able to make quick changes in its processes in response to revised information from the upstream group. A high degree of trust is required between the two groups as regards each other's motives and competence.

Figure 41 shows a high degree of interplay among all three of the key activities in the product development cycle. This process goes well beyond concurrent design and process planning. Indeed, some of the actual manufacturing steps may be started before all detail design is complete. It is not necessary for planning to be complete before design starts, nor for the latter to be done for manufacturing to begin. There can be continuous two-way flow of information between the activities. Sufficient information is available at appropriate stages of each activity to enable starting some operation in the next activity. For a large project, this interplay can be fostered by dividing it into smaller subprojects.

The sequence of activities is as follows:

- The group can begin preparing the requirements list and doing the technology survey as soon as planning has identified the market

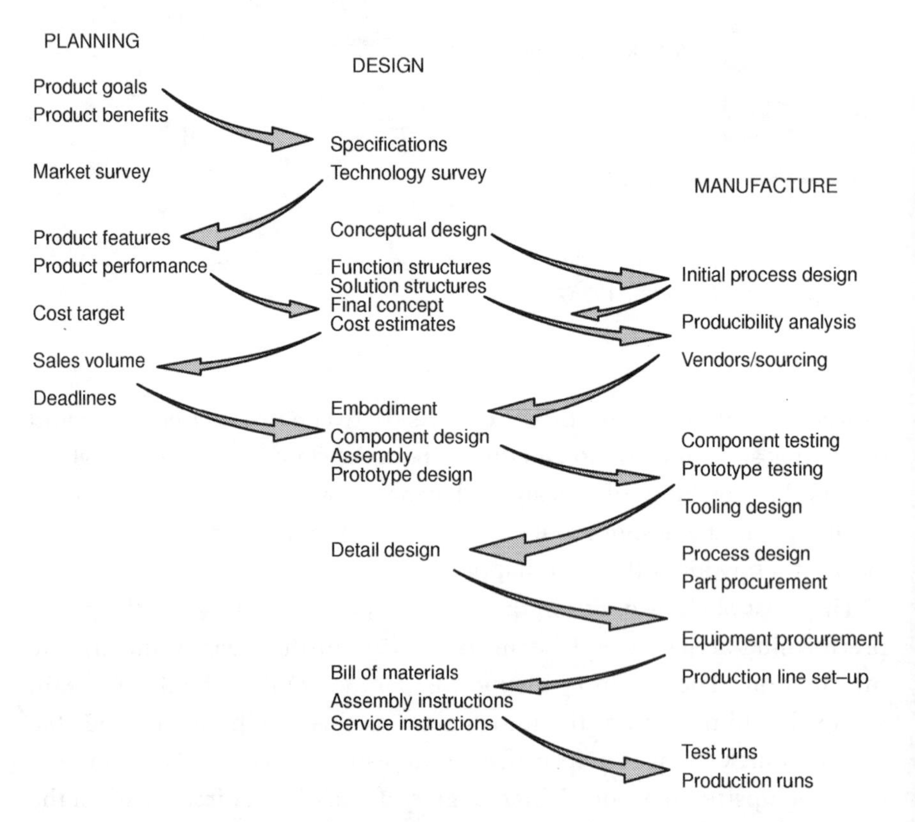

FIGURE 41. Overlapping of activities in planning, design, and manufacture.

position of the product, its price range, new features, and its relation to other company products.

- When specific desired features and performance have been identified and a cost target set, design can use this input to work on the concept.
- The last stages of planning—sales forecasting and setting deadlines for subsequent activities—are carried out concurrently with, or follow, the concept development.
- During conceptual design, as the product structure is finalized, manufacturing begins to look at the production process design, analyzes producibility, and begins selecting vendors.
- At this time, embodiment has started, leading to specific shapes of components and assembly and then to the initial prototype design.
- Test component production or procurement is succeeded by the building and testing of the prototype.
- Following development of the prototype, the details of the design are finalized while final tooling and process design are implemented.
- The final activities are the procurement of parts and any special production equipment, and setup of the production line.
- Design prepares the final documents on bill of materials and assembly and service instructions, while manufacturing sets up the test run, followed by the production.

● ● ● ● ●

EXAMPLE
TIME TO MARKET WITH TRADITIONAL AND CONCURRENT ENGINEERING

Figure 42 shows the differences in time required for bringing a product to market through traditional methods and through a time-driven approach. The example is that of a company making specialty mechanical power transmissions (Smith and Reinertsen, 1991). The company had branches in the United States and in Japan. The Japanese branch, using cross-functional teams and

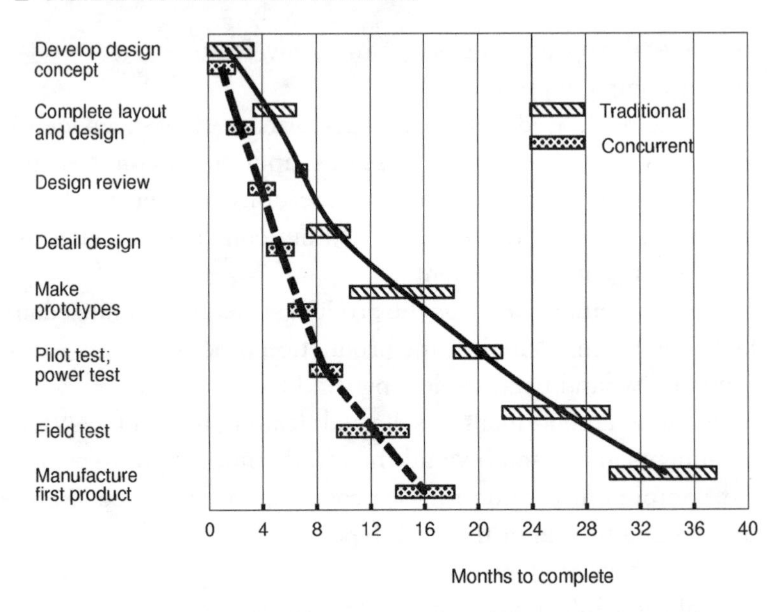

FIGURE 42. Time to market for a mechanical transmission.

overlapping of the various product realization activities, was able to finish the product in less than half the time.

• • • • •

METHODS FOR LOWERING MANUFACTURING COSTS

At the design stage, the following items have the largest influence on manufacturing:

- *Overall design arrangement* relates to the manufacturing process as affected by the nature and number of subassemblies and components.
- *Shape design* of components determines the production methods used and the quality of the components.
- *Materials* selected for components determine the production methods, storage and handling, and quality.

- Number and type of *standard and purchased parts* relate to production capacity and storage.

Improvement of a Product by Redesign

When a new product is put into service all of its parts are, as a rule, not optimally designed. They may need to be strengthened as failures occur. What does not show up in this manner, however, are parts that may be too costly to begin with, due to overdesign. Figure 43 shows the progress of costs in an industrial vacuum cleaner brush assembly as individual parts are redesigned. Beginning with the initial design, the motor exhibited failures and was replaced with a larger motor. This increased the total cost by 10%. Then the belt drive had to be redesigned, increasing the cost by another 10%. This 20% increase in the total cost could be

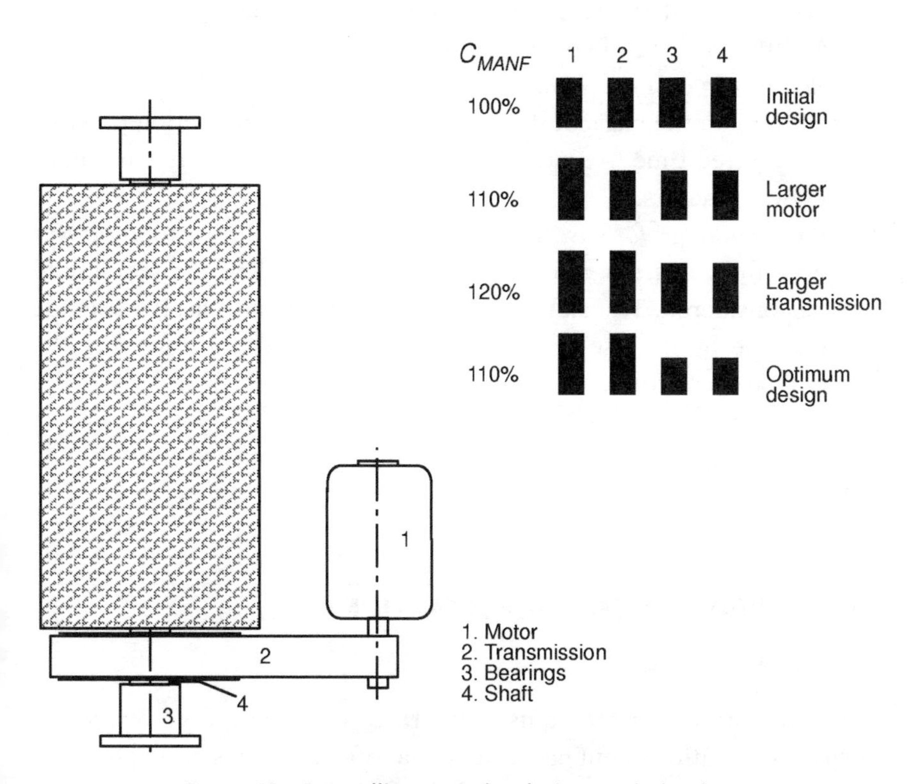

Figure 43. Controlling costs by design optimization.

lowered to 10% by reducing the size of other components that had initially been overdesigned.

Sources for Cost Information

The goal of lowering manufacturing costs cannot be realized by the design department alone. Other departments' expertise also needs to be utilized for cost reduction—at the minimum, purchasing, process planning, and cost engineering, as shown in Figure 44. It is management's job to facilitate communication among these groups.

Needs in Designing for Cost (DFC)

1. *Providing cost information.* Current costs of parts and assemblies are needed, along with information on cost-influencing parameters and the makeup of cost structures.
2. *Reaching a cost decision.* The most cost-favorable solution must be chosen, on the basis of the materials used, the production process, the desired time to market and other deadlines, quality, and other relevant factors.
3. *Product design for cost.* DFC needs to be viewed in its totality. Not only the costs but also the deadlines and technical risks should be considered. Translating these requirements into design is made possible by the use of the systematic method and observing the embodiment rules.

HOW TO PROVIDE COST INFORMATION

Tasks of Purchasing

Cost information is a prerequisite for DFC. It is necessary not only to provide information about part prices, manufacturing costs, etc., but also to establish the cost structures and the interaction of the costs of the

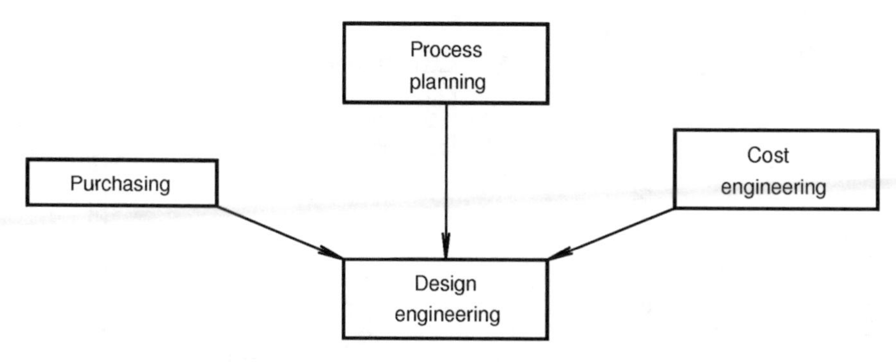

FIGURE 44. Design for cost: interaction of design with other departments.

different products under consideration. Manufacturing costs can vary widely, as shown by Ehrlenspiel and Fischer in their article on relative costs of spur gears in a study of 12 gear manufacturing firms (1982, 1983). The results of four of these firms are shown in Figure 45. The manufacturing costs of individual firms are seen to be as much as 60% above and 40% below the average costs of the 12 firms.

Purchasing can help designers by:

1. Providing specialized knowledge and information about typical purchased items such as semifinished parts and housings.
2. Providing the designers with suppliers' cost structures.
3. Helping to build up a user-friendly cost database that can be called up by design.

Design can help purchasing in outsourcing for parts. When providing specifications to vendors, only the most essential requirements should be given, and the vendor should be allowed the freedom to come up with the optimum design and product. This follows the philosophy of the systematic design method described in this book.

TASKS OF PROCESS PLANNING

Process planning brings together the knowledge about the production processes and facilities to be used. Process planning plays an important role in the company's make-or-buy decisions. It decides which of the parts, assemblies, or whole products will be made in-house or purchased.

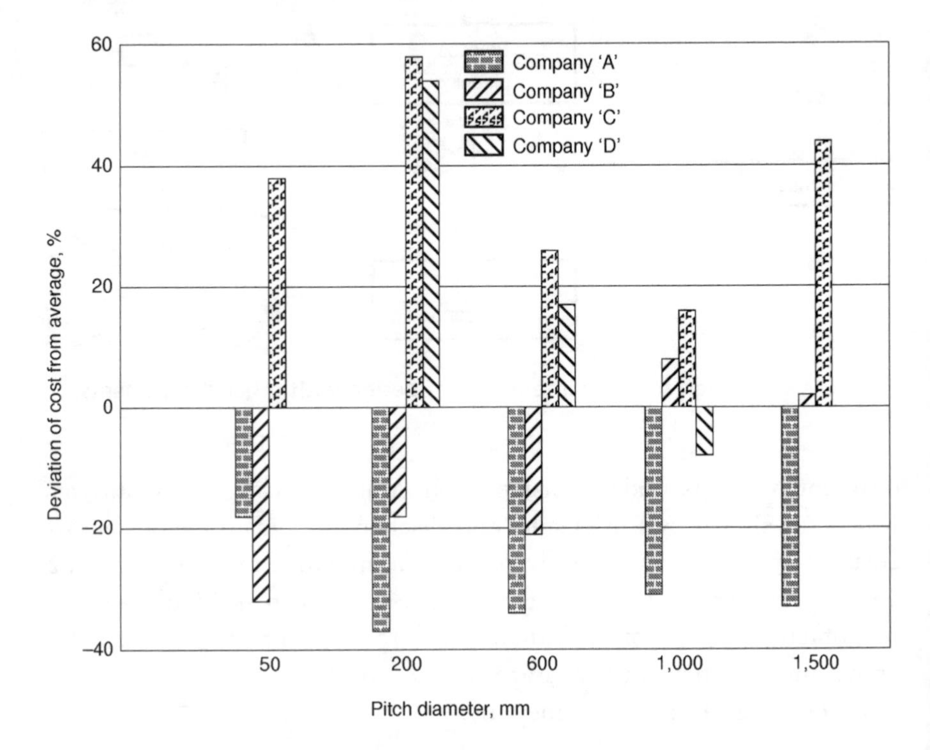

FIGURE 45. Manufacturing costs for gears from four firms. (Hundal, 1995)

For example, if a part is to be made in-house, what manufacturing facilities are available? If the company has facilities and expertise in welding, then the appropriate parts ought to be designed as weldments.

Concerning the cost information, process planning has the following tasks:

1. Provide cost information related to the different manufacturing processes used.
2. Develop an easy-to-use strategy for making make-or-buy decisions.
3. Provide information about in-house and external production facilities.

Tasks of Cost Engineering

Cost engineering has the most expertise and knowledge about costs. Cost engineering calculates the factory cost, which includes design and development costs.

Cost engineering can help designers as follows:

1. Provide cost comparison of similar parts, assemblies, and products so that early design decisions can be made at the rough layout stage. For example, provide information for making material and process decisions, e.g., choosing between sheet metal and a rolled section assembly for a chassis.
2. Provide relative cost information data with examples arranged according to part groups, materials, manufacturing processes, etc.

HELP IN MAKING COST DECISIONS

The design department must produce several concepts and, after choosing the best concept, produce several possibilities for embodiment. Only then can an evaluation be made that leads to an optimum design. The choices and evaluations must be made under both technical and economic criteria. It is not proper design procedure to lock on to an idea early on in the process and pursue only that idea.

When all necessary cost information is available, a decision can be made about which design is to be taken to completion. The decision should be made after viewing the product in its totality in the form of a life cycle analysis.

Aids for the Designer

1. Use a systematic method so as to generate a number of concepts and to produce a number of unbiased designs.
2. Use the expert advice from purchasing, process planning, manufacturing, and cost engineering.
3. Use design improvement techniques such as value engineering, failure mode and effect analysis, and quality function deployment. These topics are presented in Chapter V.

It has been found (Kiewert, 1979) that in the machine industry, manufacturing costs can be reduced to 60 or 70% by applying value engineering. The average of 33% of the unnecessary costs was seen to divide up as follows: design, 65%; process planning, 19%; and purchasing, 15%.

CONTINUOUS COST CONTROL

Costs and cost structures of a product can change drastically during its evolution and development. The changing conditions and requirements in other departments of the company also play a role. Examples of product-related governing parameters might include changing tolerances, or a change in maximum operating temperature, which may require use of a new material and thus possibly also a different manufacturing process. A new manufacturing method introduced in the company's shop for one purpose may change the way another part is made. Purchasing might one day find that a part that was made all along by casting suddenly becomes more cost-attractive if made by forging. This is a type of information that purchasing must continually keep up to date.

• • • • •

EXAMPLE
MATERIAL AND PROCESS SUBSTITUTION

As suggested above, the design department, with the help of other groups, must be constantly on the lookout for more appropriate materials and manufacturing processes. Ulrich and Eppinger (1995) give the example of an automobile intake manifold which was redesigned from an aluminum alloy to glass-filled nylon, resulting in the cost savings shown in Table 14.

TABLE 14 Intake Manifold Cost

	Material	Cost
Original design	Al alloy	$18.56
After redesign	Glass-filled nylon	7.67

(Ulrich and Eppinger, 1995)

• • • • •

It is important that all departments work together in cost-oriented development, rather than leave cost entirely the responsibility of cost engineering. In this way the response time to a changing environment can be shortened.

SUMMARY

This chapter has stressed the importance of time in product development. Sequential development activities take more time and are more likely to lead to mistakes in the process than concurrent activities involving cross-functional teams. It is important for all departments—marketing, purchasing, design, manufacturing, sales, cost engineering, industrial engineering—to work together. Designers must on the one hand be aware of the customer's wishes, and on the other understand materials and manufacturing processes. Purchasing, process planning, and cost engineering can ensure that the designers have appropriate and on-time information to effectively design for cost under time constraints.

BIBLIOGRAPHY

Ashley, S. "Designing a Nuclear Attack Submarine." *Mechanical Engineering* 117(4): 66–69 (1995).

Ehrlenspiel, K., and D. Fischer. "Relativkosten von Stirnrädern in Einzel und Kleinserien Fertigung" (Relative Costs of Gears in Single and Short Run Production). *FVA Forschungsvorhaben* 61(116, 146). Frankfurt: FVA, 1982 and 1983.

Hayes, R. H., S. C. Wheelwright, and K. B. Clark. *Dynamic Manufacturing.* New York: Free Press, 1988.

Hundal, M. S. "Time and Cost-driven Design." *Design for Manufacturability 1995,* DE-Vol. 81: 9–20. New York: ASME, 1995.

Kiewert, A. "Systematische Erarbeitung von Hilfsmitteln zum Kostengünstigen Konstruieren" (Systematic Working on Help Media for Cost-Effective Designing). Dissertation, Munich Technical University, 1979.

Munro, A. S. "Let's Roast Engineering's Sacred Cows." *Machine Design,* February 9, 1995, pp. 41–46.

Parsaei, H. R., and W. G. Sullivan. *Handbook of Concurrent Engineering.* London: Chapman & Hall, 1993.

Rich, B. R., and L. Janos. *Skunk Works.* Boston: Little, Brown, 1994.

Smith, P. G., and D. G. Reinertsen. *Developing Products in Half the Time.* New York: Van Nostrand Reinhold, 1991.

Stalk, G., and T. Hout. *Competing against Time: How Time-Based Competition Is Reshaping Global Markets.* New York: Free Press, 1990.

Suri, R. N. "Common Misconceptions in Implementing Quick Response Manufacturing." *J. Applied Manufacturing Systems* 7(2): 9–20 (Spring 1995).

Ulrich, K. T., and S. D. Eppinger. *Product Design and Development.* New York: McGraw-Hill, 1995.

Wolfram, M., and K. Ehrlenspiel. "Design Concurrent Calculation in a CAD-System Environment." *Design for Manufacturability 1993,* DE 52: 63–67. New York: ASME, 1993.

Improving Product Design

In this chapter we look at methods of improving product design. These include value engineering, quality function deployment, failure mode and effect analysis, and Taguchi methods.

Customers' dissatisfaction with a product arises from its lacking the features and quality they expect for the price and from its having characteristics they don't like. Typically, in a manufacturing enterprise organized along functional lines, customer needs, as perceived by marketing, are passed on to development and design. The designs are then "thrown over the wall" to manufacturing. At each step where such information transfer takes place there is potential for misunderstandings and mistakes. The design department does not fully comprehend the customers' needs; manufacturing spends hours and days trying to fix problems which are really design problems to begin with. To make this process smoother, i.e., translate customer needs into quality products, the method of quality function deployment (QFD) was developed in Japan in the 1970s. It uses structure trees to understand customer requirements and the development process. It is closely related to the value engineering technique developed in the United States in the 1940s. Toyota reduced its auto body preproduction and startup costs by 60%, while achieving quality improvement, by applying QFD. One of the first papers to appear in the English language on QFD is by Sullivan (1986). QFD is presently being used by such major United States manufacturers as Ford, GM, AT&T, HP, and DEC. QFD is also used by the service industry to improve processes to better address customer needs. The method can be applied to a complete system—say, a car—or to subsystems, such as the car door.

Quality function deployment is a technique for understanding the design problem. By applying QFD, we convert customer requirements to technical requirements. It also allows customers to evaluate competitors' products and company personnel to undertake a competitive technical assessment. QFD can be applied not only in product planning, but also in process planning. QFD is an important tool in concurrent engineering. It should be noted that the systematic design method, discussed earlier, helps us generate altogether new designs; QFD and other methods discussed in this chapter, on the other hand, are tools for improving existing products. However, QFD has also been used to generate improved concepts based on existing product lines.

Hauser and Clausing (1988) provide a detailed discussion of QFD. They use the example of a car door design to satisfy customer requirements. They start with requirements such as "easy to close," "stays open on a hill," "doesn't leak," and "no road noise." Results of customer surveys show how the company's product compares with those of the competitors. The next step is to develop a set of technical requirements which are to be met in order to satisfy the customer requirements. This list includes items such as "energy to open door" and "door seal resistance." The correlation between the technical requirements and the customer requirements are then determined. This allows the engineers to set priorities on which items to address. Finally, they discuss how target values are set for each of the technical requirements and compare their results with competitors' products.

Steiner (1991) discusses the experience at GE of introducing and applying QFD. He compares it with earlier design tools—structured product analysis (SPA) and value analysis/value engineering (VA/VE) (see Fowler, 1990, and the discussion below)—which, like QFD, use structure trees to translate customer requirements to product and process details. Clausing (1994) shows how QFD can be used as the basis for competitive design and manufacturing.

VALUE ENGINEERING

Value engineering (also called value analysis, value control, value management, and VA/VE) aims to maximize the value of a product. The technique dates back to the 1940s, when it was developed at the General Electric Company by Miles (1972). A number of books have since appeared on the subject, including that by Fowler (1990). The value of a product is defined by the equation

$$\text{Value} = \frac{\text{Product function and performance}}{\text{Product cost}}$$

Value engineering aims to obtain maximum performance for minimum cost. VA/VE procedures employ cross-functional teams to evaluate each step in the product realization process—design, procurement, manufacturing—to achieve its aims.

VA/VE consists of a number of phases, listed here with the activities in each phase:

- *Preparation phase*: Selection of the product to be analyzed; selection of the team members; gathering technical, market, and cost data
- *Measurement of user acceptance*: Gathering user input by focus group and questionnaire
- *Information phase*: Function analysis, cost, acceptance; comparative value analysis
- *Analysis phase*: Identifying targets, problem areas
- *Creativity phase*: Developing ideas, innovation; brainstorming
- *Synthesis phase*: Converting ideas to solutions
- *Development phase*: Plan for implementation; Gantt chart; decision making
- *Presentation*: Preparing the presentation, use of visual aids, report
- *Follow-up*: Documents, policies, and procedures

Value engineering focuses particularly on the design process, materials used, and the manufacturing process to reduce costs and improve performance. The team asks questions pertaining to each of these areas:

Area	Questions
Design	Simplify the design?
	Look at alternative designs?
	Combine or eliminate features and parts?
	Make it smaller and lighter?
	Relax tolerances?
	Ease of assembly?
	Eliminate fasteners?
	Make in-house?
	Use standard parts?
	Outsource?

Materials	Satisfy minimum requirements?
	Replace by cheaper materials?
	Materials suitable for manufacturing?
	Materials available in the required form, finish, size, tolerance?
	Reliable supplier?
Manufacturing	Consider other processes?
	Best fit of materials, processes, quantities to be made?
	Can near net shapes be produced?
	Are finishing operations necessary?
	Is special tooling required?
	Scrap produced—how much, and of what value?
	Investigate automation, group technology?
	Quality control in place?

Fowler (1990) provides the results of several case studies of value analysis, among them:

- Cost reduction by the redesign of a heat pump which reduced the number of check valves from five to two
- Manufacturing cost reduction by a change in the configuration of a heat pump which eliminated angles other than 90°, for example, in mountings or structural parts
- Cost reduction and efficiency improvement of a heat exchanger by redesign of fins
- Enhancement of worth by redesign of heat pump side panels for easy disassembly and for minimizing injury to maintenance personnel (this change increased the cost)
- A switch from $7 to $0.89 spark plugs for use during engine checkout and run-in
- A switch from $11 push rod adjusting screws made in-house to parts costing $2 purchased from a specialty shop, taking advantage of the supplier's efficient production setup
- Casting the lubrication header in a milling machine as an integral part of the bed casting, instead of a separate part fabricated from pipe

The case studies as listed by Fowler come with the analysis.

CUSTOMER REQUIREMENTS

Structure tree techniques of design improvement, such as VA/VE and QFD, help companies to listen better to the customer (keep in mind VOC: voice of the customer). This is accomplished by using focus groups, questionnaires, interviews, and the like. (The ideal situation would be for the customers to be able to talk to the designers directly!) The customers' responses to the various questions may be graded (on a scale, of, say, 1–5, or poor to excellent), or may be descriptive, using colloquialisms and clichés. It is very important to preserve their real intent. The next task for the company is to translate customer requests to business or technical requirements that it needs to implement. The QFD method uses a matrix such as that shown in Figure 46 to reach this goal. The shape of the picture suggests a house, and therefore the QFD matrix is also referred to as the "House of Quality" (Hauser and Clausing, 1988). The matrix highlights the important issues in the planning of a new product or improving an existing product.

FROM CUSTOMER REQUIREMENTS TO ENGINEERING REQUIREMENTS

The needs and desires of the customer, which are generally stated in everyday language and are vague, must be translated into engineering or technical requirements, which are stated in an objective manner and in measurable terms. Only then can these requirements be addressed during design.

Here are two examples of developing engineering requirements from customer requirements:

Customer Requirements	Engineering Requirements
Car door "easy to close"	Force needed: less than 5 lb
Vehicle should not rust	Use rust-resistant material
	Use appropriate coatings
	Avoid pockets where corrosive material may accumulate
	Control clearances to avoid wear of finishes

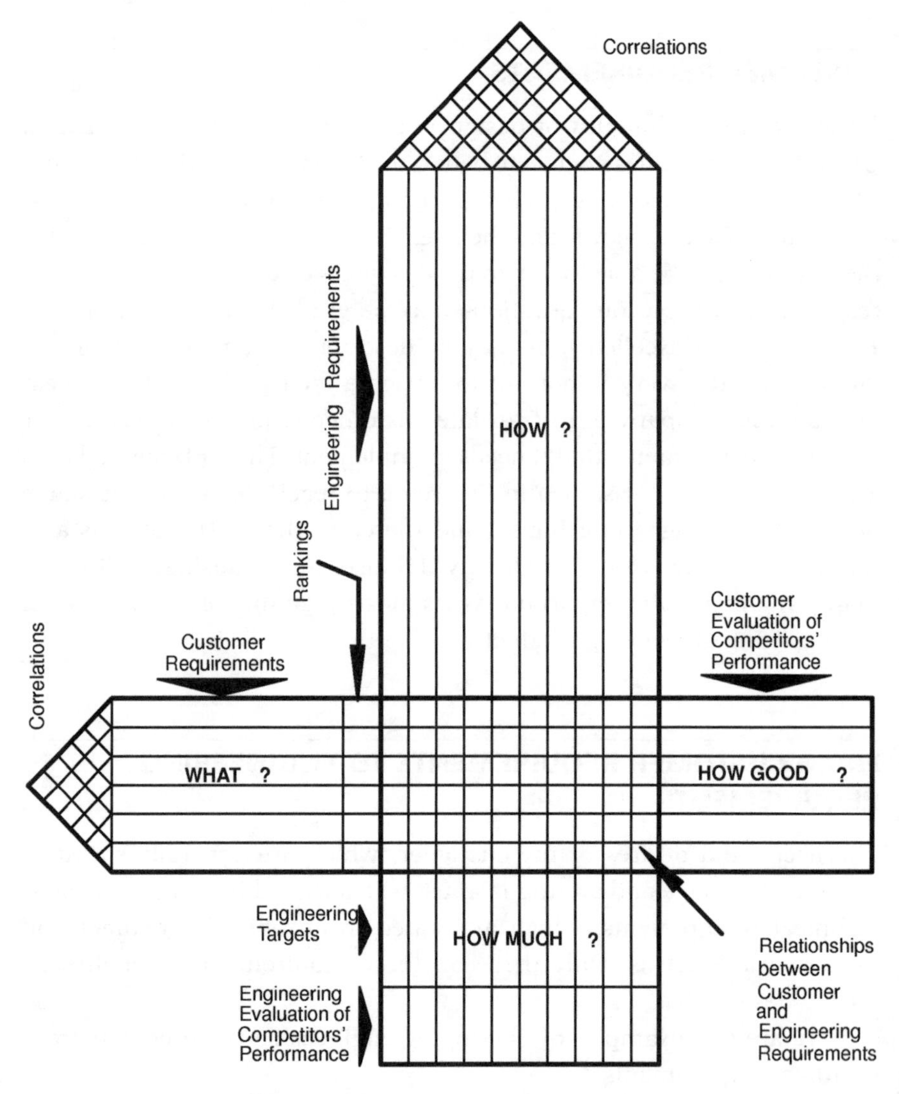

FIGURE 46. QFD matrix for product planning.

QFD must have the support of—be bought into by—top management. Management assigns people to the QFD team. It is important to have people from different departments on this team—as a minimum, from marketing, design, and manufacturing. Depending on the product, some companies also include customers and suppliers on the team.

●●●●●

EXAMPLE
QFD FOR OUTDOOR THERMOMETER

Customer Requests

As a first example let us consider a company that makes outdoor thermometers. It wants to increase its market share and therefore improve its product. To do so, the company should also compare its design with that of its competitors' products. Figure 47 shows the QFD matrix for this case.

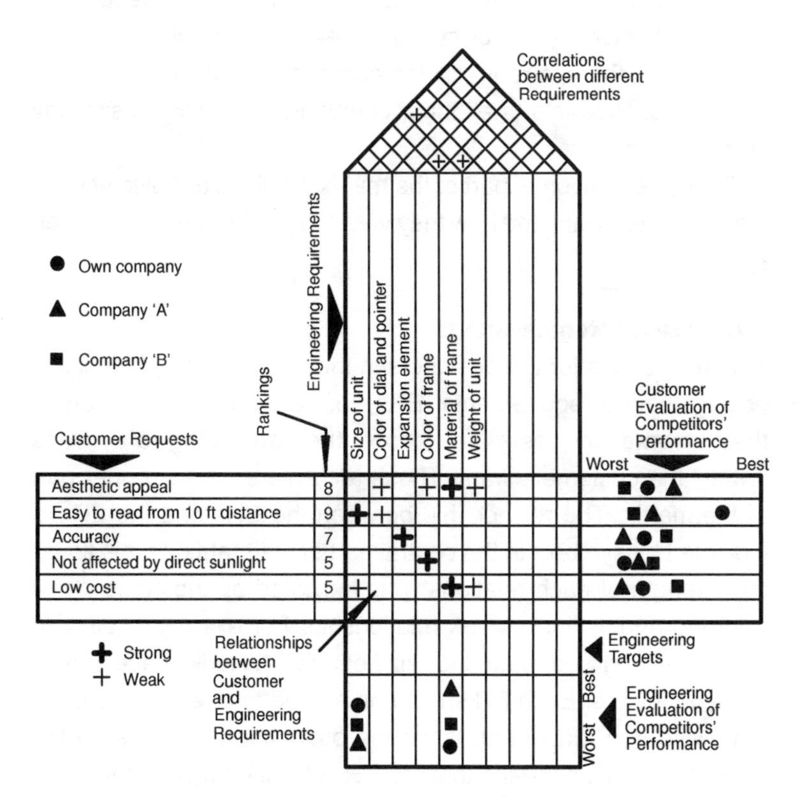

Figure 47. QFD matrix for the thermometer example.

The left side of the matrix shows the customer requests (CRs). Customers may be end users, retailers, or distributors, among others. There are also requirements pertaining to the legal aspects, recycling, and other peripheral constraints. We have shown a few of these CRs; there may be as many as 100 or more for some products.

Not all CRs are equally important. Customers may be asked to rank them, say, on a scale of 1 to 10. Often the company team developing the QFD matrix will assign such weighting to customer responses. These rankings are shown in the column next to the CRs.

On the right side of the matrix we show how customers view this company's product compared with those of two of its competitors. "Easy to read" shows that this company's product is doing very well, so no action is necessary. On the other hand, the responses to "not affected by direct sunlight," even though its ranking is lower, show there is room for the company to gain a competitive advantage. Products of all three companies are viewed similarly in this regard.

Thus, the horizontal part of the matrix provides a profile of what the customers want and how they view the products on the market.

Engineering Requirements

The next task is for the designers in the team to list the technical or engineering requirements (ERs) that are most likely to affect the customer requests, i.e., "how" they are going to address "what" the customers want. These are entered in the upper part of the matrix. The relationship between the CRs and the ERs is indicated by symbols in the central part of the matrix. In this figure the strength of each relationship is indicated by the signs ╋ for a strong and + for a weak relationship. For example, "easy to read" is strongly correlated to "size of unit." A blank space means no dependence. Each ER should be related to one or more CRs; otherwise it is redundant. If, on the other hand, there is no ER related to a CR, it means that the ER list is incomplete and there is potential for significant improvement in the product.

The technical members of the team (design and manufacturing) then evaluate how the competitors' products compare with their company's own product and enter the targets for each of the ERs in the bottom part of the matrix. This part of the matrix can point out the discrepancies between the customers' perceptions and the team's correlation of CRs and ERs. The "roof" of the house, the triangle at the top, shows how one ER may be affected by others. The vertical part of the matrix therefore shows how the company may respond to customer requests.

If a company uses the QFD method no further than has been described thus far, the method will more than pay for itself in identifying the weaknesses and strengths of the company's product in the marketplace and in planning a better product. The technique can, however, be extended further. Similar matrices can be developed for concept selection, part design, and planning the manufacturing process. This sequence is shown in Figure 48. Each phase starts with the "whats" and ends with the "hows." The next example shows how each of these matrices is used.

• • • • •

EXAMPLE
THEATER SEATING

Customer Requirements

Day (1993) gives the example of a theater seating problem to illustrate QFD. Customer requirements come from two sources: theater patrons and theater operators. Theater patrons' requirements are:

- Comfortable seat
- No need to stand to allow entrance
- Comfortable armrests

Theater operators' requirements are:

- Easy access for cleaning
- Maximum number of seats

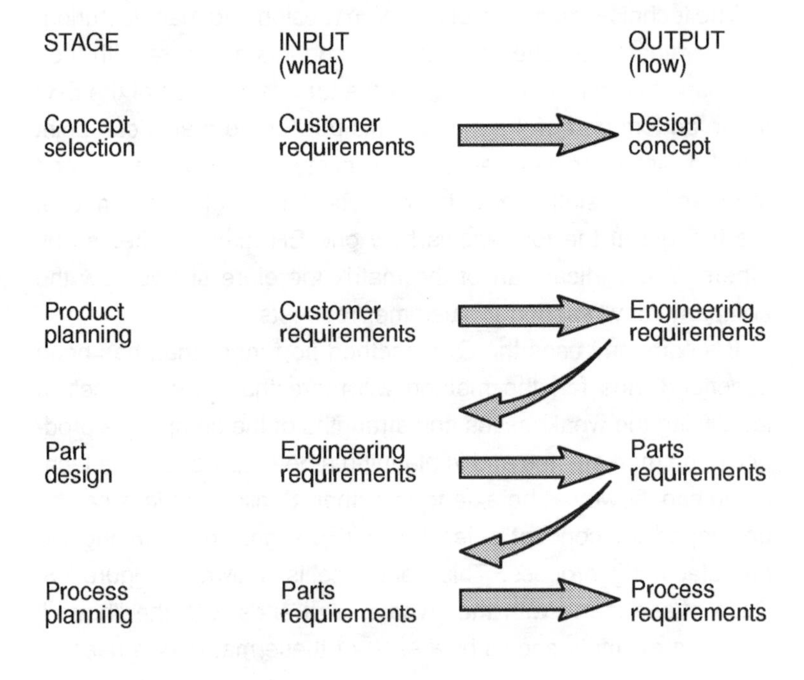

FIGURE 48. Applications of QFD in the various stages of product realization.

Customer competitive evaluation shows how the company's product is seen by customers when compared with a competitor's. This brings out the areas in which there is potential for gaining competitive advantage.

ENGINEERING REQUIREMENTS

Technical requirements that must be addressed in order to satisfy the customer requirements are listed as:

- Seat comfort index (on a scale of 1 to 10)
- Human accommodation range (say, for a 5 to 95 percentile male)
- Entrant clearance (in inches)
- Stiffness (inch/lb)
- Seat-up clearance (inches)
- Seats per 100 ft^2

Targets are set up for each of these, and company personnel perform a technical assessment of their own company's product relative to the competitor's. If the company's technical assessment of a certain item finds that its own product is superior to the competitor's while the customer competitive evaluation shows otherwise, it means that the company's test practices are at fault.

CONCEPT EVALUATION

A concept evaluation and selection matrix is then developed as shown in Figure 49. This is the step where the systematic design method can be used to advantage in generating concept variants. Notice the similarity to the concept selection matrix shown earlier in Figure 18 (Chapter III).

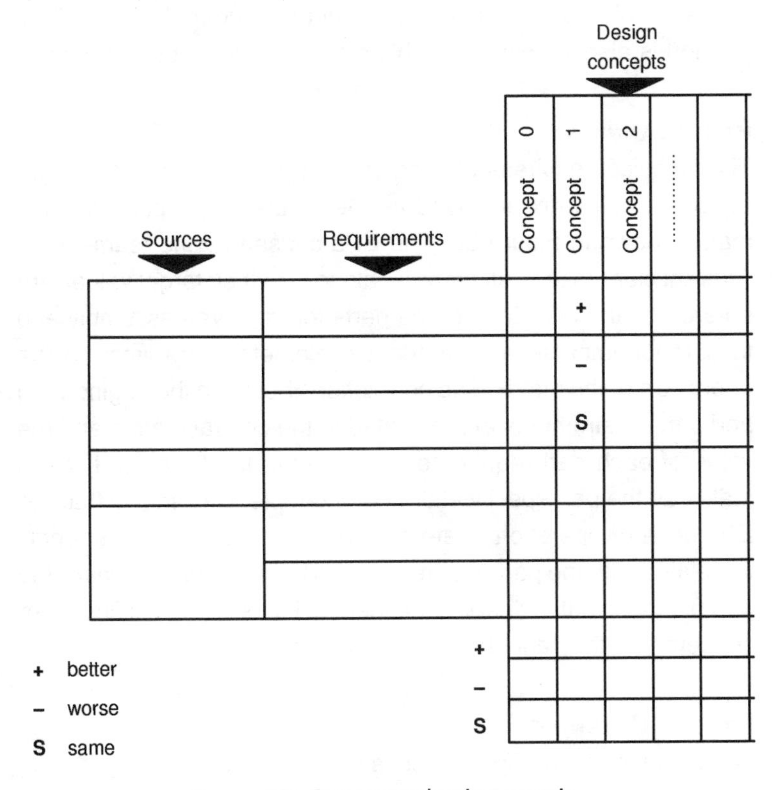

FIGURE 49. Concept selection matrix.

The requirements are listed on the left as before, with the sources of the requirements identified, perhaps as follows:

Source	Requirements
Product planning matrix	Clearance between rows, seat comfort index
Standards	Safety, ergonomics
Theater owners	Seats per 100 ft^2, cleaning access
Manufacturing	Parts, assembly, production equipment

Each of the new concepts is compared with the existing product (concept 0) on the basis of the various requirements. For each requirement, each concept is given a rating of +, −, or S, depending on whether the given concept is better than, worse than, or the same as concept 0. The totals in the lower part of the matrix indicate which of the concepts should be pursued further. This method is also known as Pugh's method of concept selection.

Part Planning

The technical requirements and their target values from the product planning matrix are used as the inputs to the part planning matrix shown in Figure 50. The ERs chosen by the team—e.g., "entrant clearance," "stiffness"—along with their target values are entered on the left. Each of the parts for the given assembly and their requirements—e.g., hardness, diameter—are listed in the upper part of the matrix. The correlations between the engineering and part requirements are indicated in the central matrix, and the value of each part requirement is noted in the lower part of the matrix as the part specification. For example, Day shows that the ER "ease of operation" (referring to seat movement) is strongly correlated with the part requirement "roller diameter." In short, the part planning matrix enables the team to focus on critical customer concerns as they affect part specifications.

Process Planning

The next step in the QFD process is the process planning matrix, shown in Figure 51. With the help of this matrix the team can

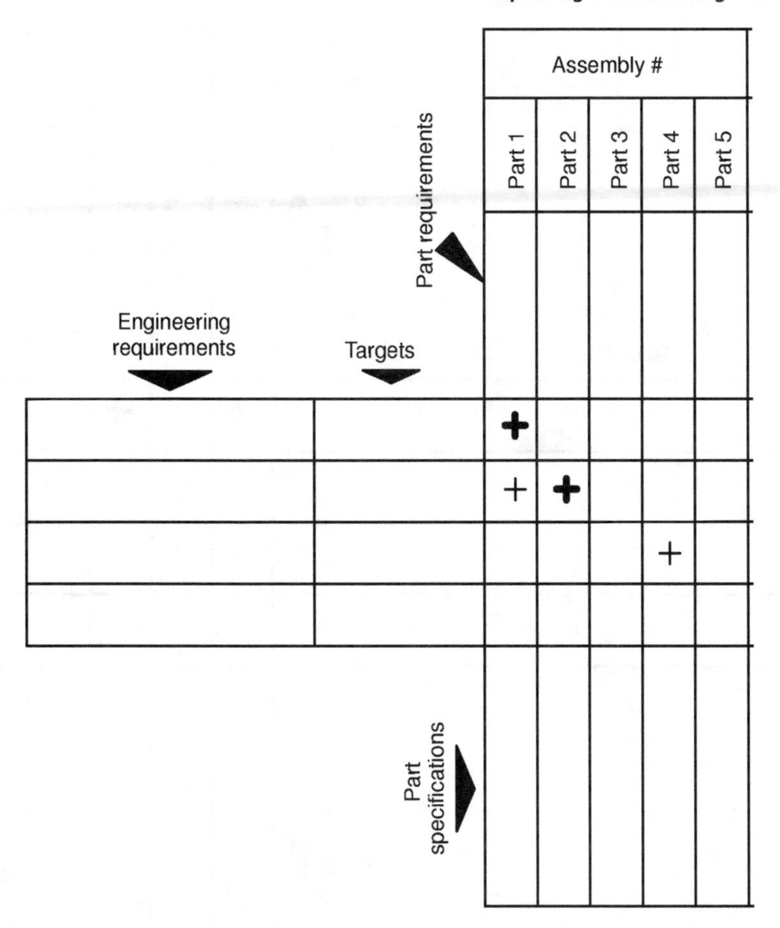

FIGURE 50. Part planning matrix.

examine the critical processes and controls which ensure that the parts produced will conform to the requirements. For example, for one of the operations, "assemble and weld pin to frame," the process requirements and corresponding specifications are:

- Weld current = . . . A
- Weld pressure = . . . lb
- Weld time = . . . s

• • • • •

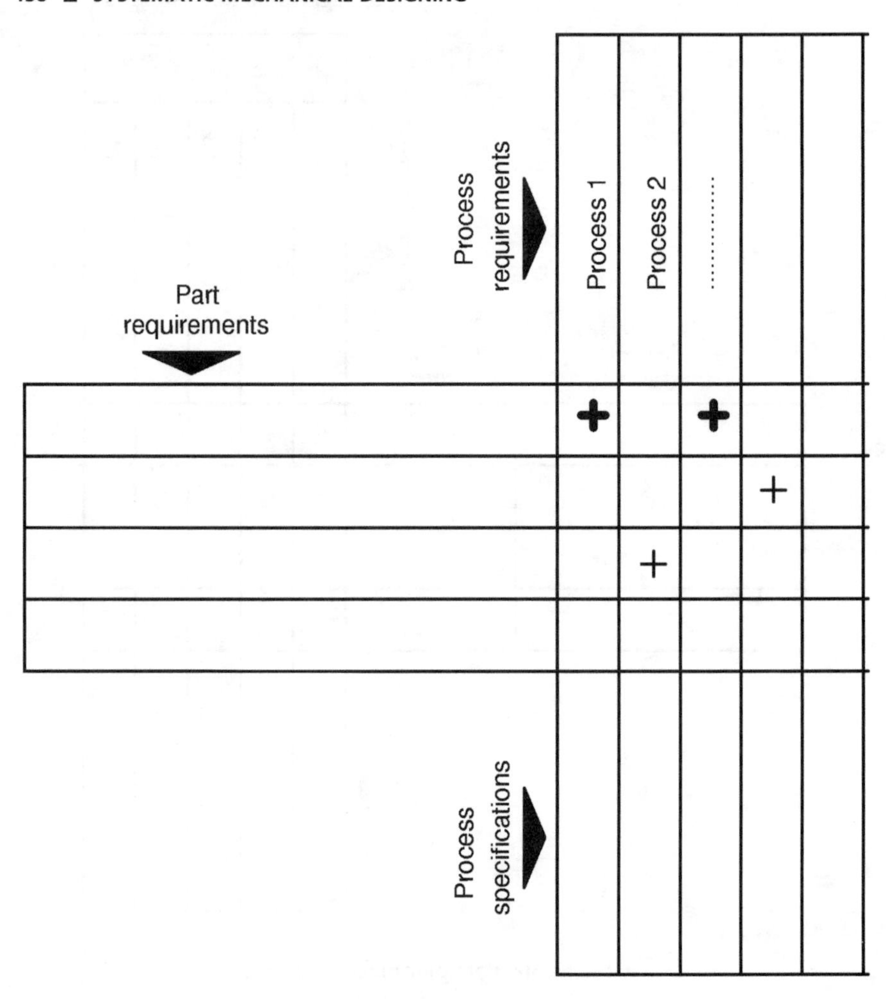

FIGURE 51. Process planning matrix.

FAILURE MODE AND EFFECT ANALYSIS

Failure mode and effect analysis (FMEA) is a means of determining how a product is most likely to fail and thus how it should be improved (Pugh, 1990). It can be carried out at the part or the product level. The attributes of the product—which may be either functions or parts, for example—are listed in the first column of a matrix. The corresponding failure modes are listed in the succeeding columns, along with their correlations with the attributes. The sum of the correlations, shown in the last column, provides an indication of the relative importance of the attributes.

Table 15 shows the application of FMEA to a table fan. The first column lists the various functions and other attributes expected of the fan: it should move air; the user should be able to vary the rate of airflow; the fan should be able to blow air in different directions, i.e., swivel; and it should be safe to handle, i.e., there should be no danger of electric shock. The columns to the right list the failure modes: failure of control and switch, insulation failures, blade damage, and failure of the swivel drive. The columns marked "Value" indicate the correlation between functions and failure modes. The numbers used are as follows:

9 indicates strong correlation.
3 indicates medium correlation.
1 indicates low correlation.

If no value is shown, it indicates that there is no correlation between a function and a failure mode. The second in each pair of columns shows the relative importance of the function. The numbers range from 1 (low importance) to 16 (high importance). The correlation values are multiplied by the function weights and listed in the adjacent columns marked "Weighed Value." These weighted values for each function are added and shown in the column marked "Gross weight." These numbers are shown in the last column as percentages. The results show that moving the air has the highest priority in this instance, followed by the other functions.

TAGUCHI APPROACH TO ROBUST DESIGN

Products and processes are always subject to factors that influence how well they perform their functions. These factors may be internal to the system (e.g., size variation, material property change) or external (manufacturing equipment, operators, environmental factors, usage). Deviation in product performance may lead to customer dissatisfaction. The essence of the Taguchi approach to design is to minimize the effects of these factors by adjusting appropriate system parameters.

Dr. Genichi Taguchi introduced his methods in the late 1940s at Japan's Electrical Communication Laboratories with the aim of improving the

TABLE 15 FMEA Matrix for a Table Fan

		Failure Mode									
		Control Failure		Insulation Breakdown		Blade Failure		Drive Failure			
Functions	Function Weight	Value	Weighted Value	Value	Weighted Value	Value	Weighted Value	Value	Weighted Value	Gross Weight	%
Move air	16	3	48	1	16	9	144		0	208	52
Vary airflow rate	9	9	81		0	3	27		0	108	27
Change flow direction	4	3	12		0		0	9	36	48	12
Safe to handle	4		0	9	36		0		0	36	9
Sums										400	100

product quality of Japanese telecommunication equipment. His methods are discussed in detail in a number of books (Ross, 1988; Roy, 1990).

Taguchi's philosophy of robust design may be summarized by the following:

- Design quality into a product. Do not use inspection to weed out poor-quality products.
- Set a target. The cost of quality is the deviation from the target value.
- Make the product insensitive to uncontrollable external factors.

Let us consider as an example the force required to close a car door. If the force required is too low, the door might close too readily if the car is parked on sloping ground, or slam shut due to wind gusts. If the door closing force (DCF) is too high, it causes inconvenience to users, particularly to those physically impaired.

THE GO/NO–GO LIMITS

Traditionally, since the beinning of the modern industrial age, variability in product parameters has been controlled by setting upper and lower limits on parameters. These are the so-called go/no-go or pass/fail limits. Parts or systems are accepted or rejected at the inspection stage depending on whether the parameters lie within the limits or not. Figure 52 shows how such limits might appear for the DCF if the lower limit is 5 lb and the upper limit is 15 lb.

There is, however, a fallacy in this go/no-go approach to design, which was, after all, developed originally to ensure the fit of mating parts. Let us say the DCF is 14.5 lb. According to the go/no-go philosophy and Figure 52, this value is acceptable, but a slightly higher value, say, 15.5 lb is unacceptable. Most people would not be able to feel the difference; in fact, both are nearly as inconvenient. Taguchi proposes that the design should aim for a target value. Suppose, in this case, it is 10 lb. Any deviation from the target value should be penalized, since it represents a cost, or "loss" in Taguchi's terms. Thus the parameters should aim toward their respective target values and not be allowed to move freely between upper and lower limits.

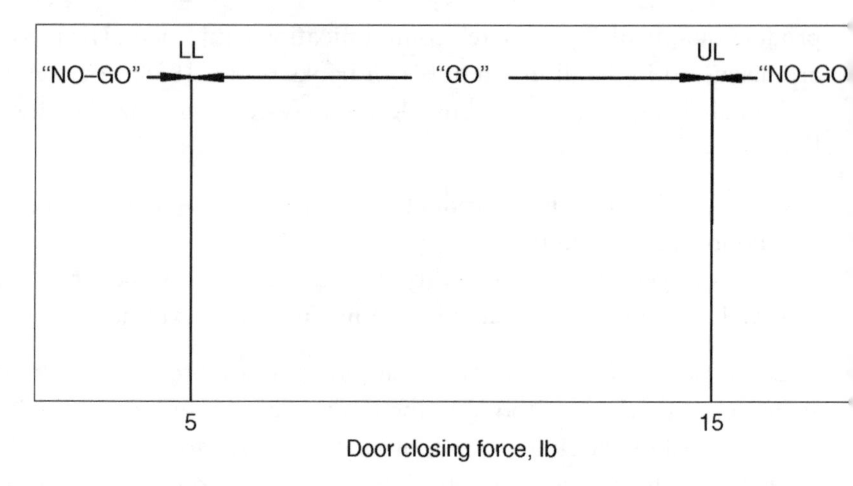

FIGURE 52. Upper and lower limits.

A quadratic function is a simple and effective way to define this loss. The loss function is given by

$$L = k(F - T)^2 \tag{2}$$

where

F = force
T = target value
k = cost coefficient

We shall see that the use of this function yields the most cost-effective limits, which should be used during inspection. There are, of course, other shapes of loss functions for specific problems.

Two cases of the "stay within specs" versus "aim for the target" problem are described by Taguchi and Clausing (1990). One is the Sony television color density problem. The density was to be of a nominal value of 10, with specification limits of 7 and 13. The company produced the same models at its San Diego and Tokyo plants. The San Diego factory produced sets with color density uniformly distributed between the limits. Any sets with density outside the limits were rejected and not shipped. Thus 100% of the sets shipped were "within specs." The Tokyo plant produced sets with color density variation clustered around the target value of 10. These sets showed uniformly high customer satisfaction, whereas those made in the San Diego plant produced a large number of complaints from customers. The other case is of Ford Motor Company, which was building a transmission for one of its cars. It asked Mazda to build an identical transmission (presumably to boost car production). When put in service, the Ford-built transmissions produced far more customer complaints and warranty costs than those from Mazda. Ford engineers disassembled both makes of transmissions and checked the tolerances on the parts. They found that while the parts in Ford-built transmissions were all within limits, those in the Mazda-built units were almost entirely "on target." In mechanical parts, small deviations from target tend to add up, causing greater friction, wear, and other undesirable effects.

THE "BEST" PRODUCT

According to the Taguchi approach the best product is one that shows the minimum variation in its performance from the target value. Let us say that four different manufacturers produce springs (a component of the car door mechanism) of a certain stiffness. The nominal stiffness of the springs is 1000 lb/in., with the upper limit at 1100 lb/in. and the lower limit at 990 lb/in. The probability distributions for the products of the four manufacturers are shown in Figure 53.

Even if all manufacturers carry out strict inspection of their products and ship out only those springs whose stiffness lies between the upper and lower limits, it is obvious that manufacturer B produces the best springs. Its springs show the smallest standard deviation. (We assume here that these statistical data are available to the purchasing company, or that the latter does its own testing to check the distribution.) Manufacturer C's product is also satisfactory, but its process is not as closely controlled as B's. Let us look next at manufacturers A and D. Manufacturer A has 10% of its product outside the limits, whereas D has 50% of its product outside the limits. Yet D has less variability in its process. By shifting its mean value to the target value (say, by changing a dimension), its product can be as good as that of manufacturer B.

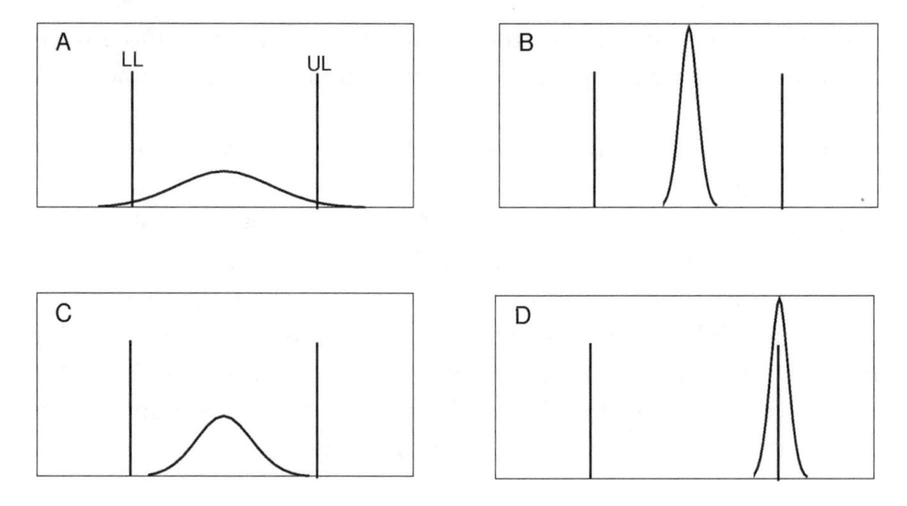

FIGURE 53. Probability distributions of springs from four manufacturers.

SYSTEM, PARAMETER, AND TOLERANCE DESIGN

A product or a process is designed so that its performance achieves its nominal (mean) value. This is the basic design, which is obtained by applying one's experience, or using the systematic design method elaborated on in this book, or flashes of brilliance, or any combination of these. Taguchi calls this the *system* or *primary design*, the first step in designing. If the designer then finds that the system performance needs further improvement, he or she can change parameter values or use closer tolerances.

Parameter, or *secondary, design* involves adjusting the system parameters to improve quality without increasing cost. The aim is to find parameter values and operating levels that make the system least sensitive to environmental conditions. Parameter design uses experimental design and other statistical techniques. This approach will be demonstrated in the tile kiln example to come. In a paper on motor vehicle quality, Taguchi (1988) describes the choice of a device in the ignition system. The characteristics of two such devices are shown in Figure 54. The device is to produce an output voltage of 20 kV. The expensive model can provide this output within close tolerances. The cheap device is very sensitive to

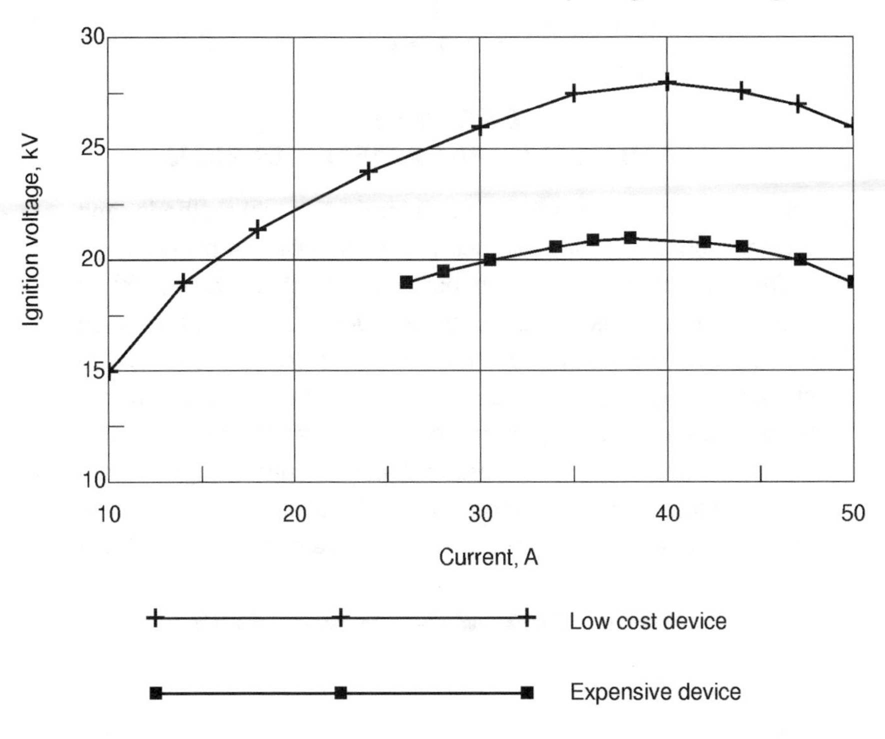

FIGURE 54. Characteristics of ignition devices.

the input when the output is near 20 kV. However it has a fairly stable output in the vicinity of 27 kV. An inexpensive solution to the problem is to use the lower-cost device, in conjunction with a voltage divider, which is also a low-cost and stable device.

Tolerance, or *tertiary, design* means not only tightening the geometric tolerances, but also including steps such as using higher-quality and higher-cost materials and using better processes and machines. All these steps increase costs. Tolerance design steps should be undertaken only after parameter design methods have been used to the utmost advantage. This approach will be illustrated in the door closing force example later on.

● ● ● ● ●

EXAMPLE
THE TILE KILN: PARAMETER DESIGN

What was perhaps the first application of Taguchi methods, and a classic one, was for a kiln used for baking tiles. A tile producer in Japan purchased a new kiln, but found that there was too much variability in the baked product quality, depending on the location of the tiles in the stack; see Figure 55. Tiles on the outside of the stack were being overbaked, while those on the inside were not being heated to the proper temperature. It was the resulting size variation of the tiles that rendered a significant percentage of them unacceptable. Figure 56 shows the distributions of sizes of tiles at the center and outside of the stack. As we can see, almost half

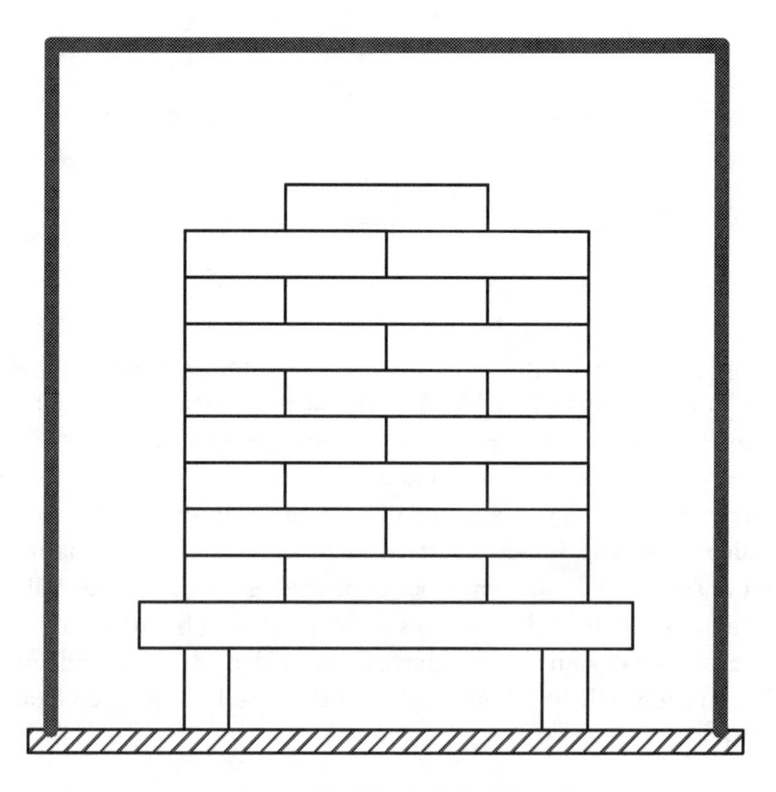

FIGURE 55. Tile stack in kiln.

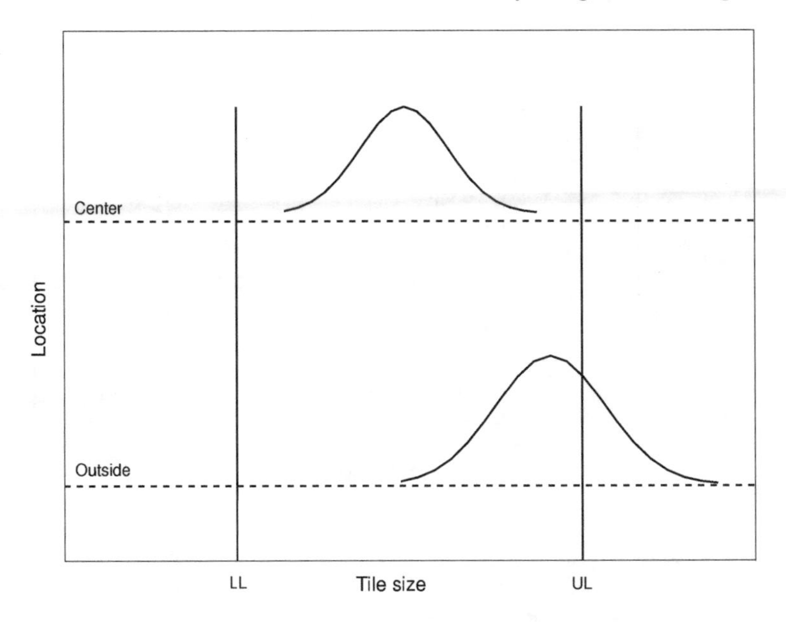

FIGURE 56. Tile size distribution before corrective actions.

of the tiles on the outside were of unacceptable size. Redesigning and rebuilding the kiln was deemed to be too expensive.

The plant scientists and engineers looked at the product itself and came up with the following determining factors for product quality:

- Limestone percentage
- Additive fineness
- Agalmatolite percentage
- Agalmatolite type
- Quantity of raw material
- Waste return content
- Feldspar percentage

The effects of these factors were tested by using experimental design and orthogonal arrays. Each of the parameters was varied over an appropriate range. It was found that the product quality was most sensitive to the limestone content. Figure 57 shows qualitatively the variation in the sensitivity of the final size of the tiles to their limestone and agalmatolite content. By increasing the

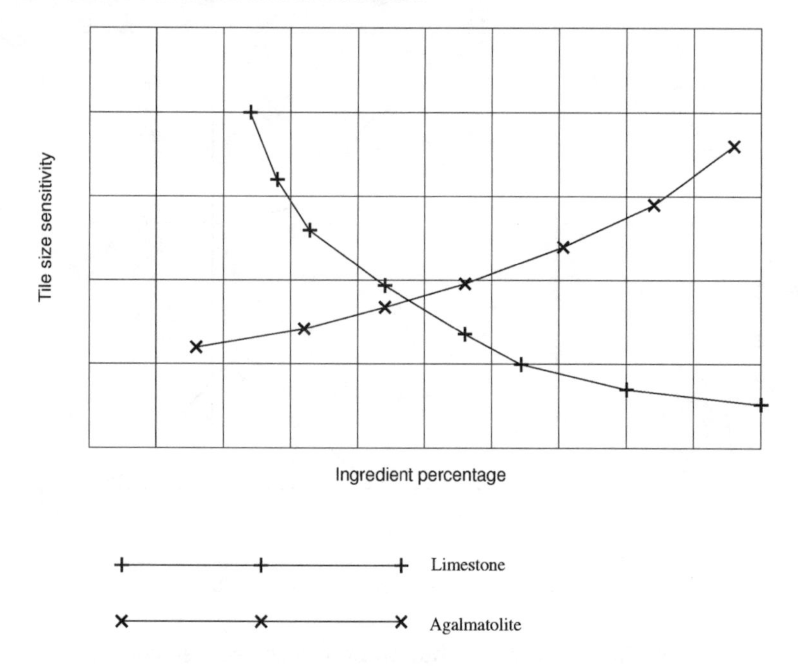

FIGURE 57. Sensitivity of tile size to ingredient content.

percentage of limestone (the cheapest of the constituents) from 1% to 5% and reducing the precentage of agalmatolite (the most expensive material), the variation in tile size from the outside to the center of the stack was brought within acceptable limits. At the same time (since limestone was the cheapest ingredient), the product cost was reduced.

Figure 58 shows the tile size distribution at the center and on the outside of the stack before and after the mix change. When the original mix was used, there was a large variation in the tile size at all locations in the stack, although the tiles near the center were all within the limits. With the revised mix, all tiles were well within the size limits, but equally significant is the fact that their size variation was much less.

This example disproves the popular belief that higher quality implies greater cost.

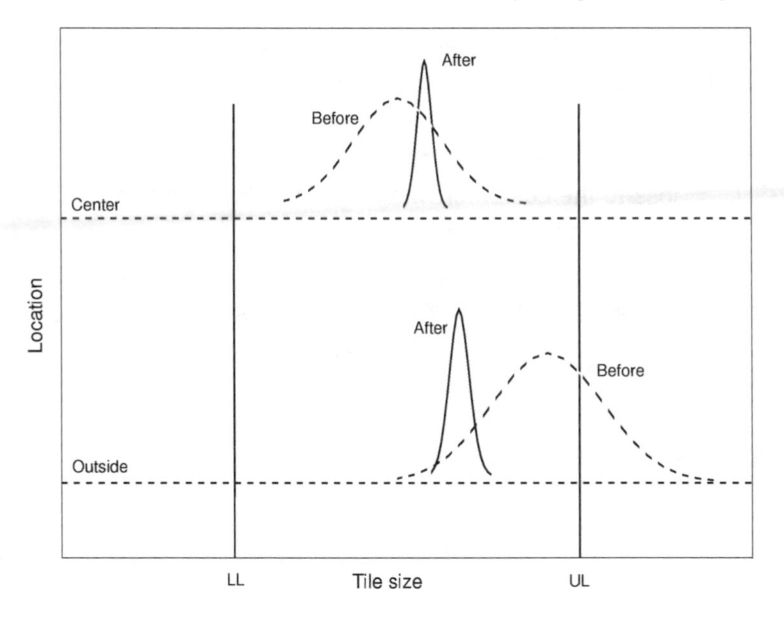

FIGURE 58. Tile size distribution after corrective actions.

• • • • •

EXAMPLE
CAR DOOR CLOSING FORCE: TOLERANCE DESIGN

We return to the example of finding the proper values (limits) for the door closing force (DCF). The target value T is 10 lb. At this value the customer is satisfied; there is no need for any adjustments or repairs by the car dealer. When the force F required to close the door deviates from T, there are costs associated with it—the warranty cost, time lost by the customer, etc. These costs are denoted as the "loss" L in Equation (2). Taguchi refers to this kind of loss as a loss to society. In general, a loss in this sense may be due to a product that pollutes, is noisy, calls for excessive downtime, or results in a similar kind of burden. In some cases the loss value can be calculated easily (e.g., the cost of heat loss from a poor-quality window); in other cases, making an estimate of loss is more subtle.

We find the relationships between the loss and tolerance by using the quadratic loss function as follows. (see Figure 59). Let

ΔT_c = customer tolerance limit
L_c = loss to the customer when the force F deviates by ΔT_c from the target value T

From Equation (2), substituting $F = T + \Delta T_c$, we get:

$$L_c = k(\Delta T_c)^2 \tag{3}$$

The cost coefficient k is therefore:

$$k = \frac{L_c}{(\Delta T_c)^2} \tag{4}$$

Let us say that if the DCF is at the customer tolerance limit ΔT_c = 5 lb, the loss to the customer is L_c = \$50. The cost coefficient k is found from Equation (4) to be

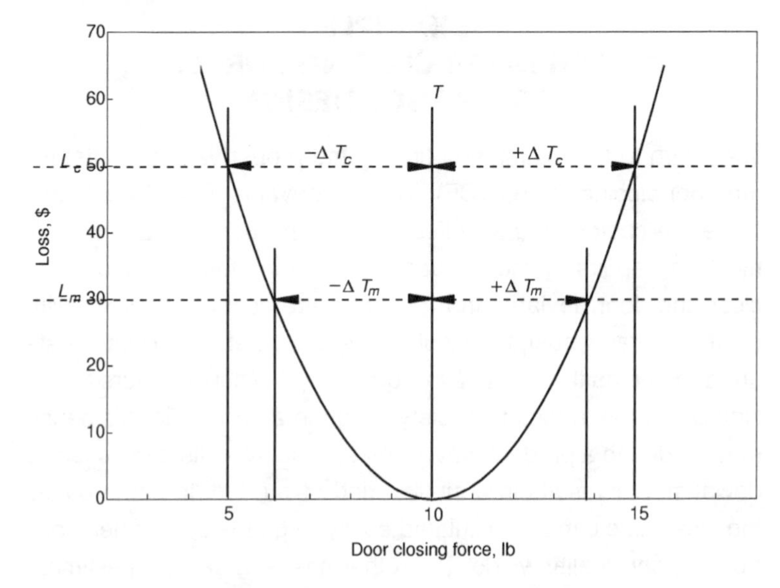

FIGURE 59. The quadratic loss function.

$$k = \frac{50}{5^2} = 2 \tag{5}$$

Suppose the loss (cost) to the manufacturer (or dealer) to repair the door mechanism is $L_n = \$30$. By spending this amount, the manufacturer reduces the probability that the customer will have to bring the car back for repairs. We find the manufacturer tolerance ΔT_m from the loss function curve, given by

$$\Delta T_m = \Delta T_c \left(\frac{L_m}{L_c}\right)^{1/2} = 5\left(\frac{30}{50}\right)^{1/2} = 3.9 \text{ lb} \tag{6}$$

Thus the upper and lower limits for the DCF should be set by the manufacturer to $T \pm \Delta T_m$; that is,

$$\text{UL} = T + \Delta T_m = 13.9 \text{ lb}$$

$$\text{LL} = T - \Delta T_m = 6.1 \text{ lb}$$

The loss function, loss values, and limits are shown in Figure 59. These tolerance limits are more satisfactory for the customer and more economical for the manufacturer than those originally proposed, i.e., 15 lb and 5 lb, respectively.

● ● ● ● ●

SUMMARY

A number of techniques, such as QFD and FMEA, can be applied for improving a product. The QFD technique fosters teamwork and breaks down barriers between functional groups. Its successful implementation may require fundamental changes in the organization and in people's attitudes. This may be the most difficult obstacle to its use. The perspective on QFD of Steiner is perhaps the most illuminating. At GE, QFD has been applied in the lighting, turbines, and appliance divisions. The number of questions raised by the procedure seemed to overwhelm the users: Who are the customers? What do they want? How good are the competitors' products? QFD forces the design people, and even the manufacturing people, to become aware of users' wishes. It is a discovery process that brings out an incredible amount of information contained in even the simplest products and processes. It is the management of this information that is QFD's greatest challenge.

Taguchi methods are used to develop robust designs. Traditionally, statistical methods are applied for quality control near the end of the manufacturing process; if a certain percentage of samples passes inspection, the whole batch is deemed satisfactory. The Taguchi approach uses statistics at the design stage. This aim is achieved by adjusting system parameters, leading to higher quality without added cost. A robust system is less sensitive to environmental and other factors.

There are several types of factors which affect the performance of a product or process:

- Those not under the designer's control
- Those the designer could control but would rather not, since it would be too expensive
- Those controlled without expensive modifications

The engineer's aim is to find the best combination of controllable factors that will make the system insensitive to factors which are not controllable or are too expensive to control. The engineer is thus able to design quality into the product, rather than leave it to inspection to check for below-grade products.

The Taguchi approach to manufacturing is "aim for the target" and not "stay within the specifications." A product that is just within the specifications and is shipped is not significantly better than one that falls just outside the specifications and is rejected.

BIBLIOGRAPHY

Clausing, D. *Total Quality Development.* New York: ASME Press, 1994. A definitive text on product development. Covers concurrent engineering, QFD, decision making, design and production planning, and management.

Day, R. G. "Using Quality Function Deployment in the Product Development Process." *J. Applied Manufacturing Systems* 6(1):28–39 (Fall 1993).

Fowler, T. C. *Value Analysis in Design.* New York: Van Nostrand Reinhold, 1990. Provides the results of several case studies of VA/VE.

Hauser, J. R., and D. Clausing. "The House of Quality." *Harvard Business Review* 66(3):63–73 (1988).

Miles, L. D. *Techniques of Value Analysis and Engineering.* New York: McGraw-Hill, 1961; also 2nd ed., 1972. The first book on VA/VE.

Pugh, S. *Total Design.* Reading, Mass.: Addison-Wesley, 1990.

Ross, P. J. *Taguchi Techniques for Quality Engineering.* New York: McGraw-Hill, 1988.

Roy, R. *A Primer on the Taguchi Method.* New York: Van Nostrand Reinhold, 1990.

Steiner, M. W. "In Search of QFD." *Proceedings, ASME DTM-91,* 191–194. New York: ASME, 1991.

Sullivan, L. P. "Quality Function Deployment." *Quality Progress,* June 1986, p 39.

Taguchi Methods: Selected Papers on Methodology and Applications. Dearborn, Mich.: ASI Press, 1988.

Taguchi, G. "Production of Vehicles and Quality Control." In *Taguchi Methods: Selected Papers on Methodology and Applications.* Dearborn, Mich.: ASI Press, 1988.

Taguchi, G., and D. Clausing. "Robust Quality." *Harvard Business Review* 68: 65–75 (January–February 1990).

BIBLIOGRAPHY

[illegible]

Cost Structures and Models

- **Need for Knowledge of Costs**
- **Components of Manufacturing Costs**
- **Types of Cost Breakdown—Cost Structures**
- **Types of Costing**
    - *Example: ABC versus Traditional Costing*
- **Relative Costs**
- **Effects on Costs During Concept and Embodiment Stages**
    - *Example: Effect of Concept on Cost of Gear Trains*
- **Effect of the Number of Parts**
    - *Example: Part Development Cost for Five Products*
- **Life Cycle Costs**
    - *Example: Optimum Life Cycle of Equipment*
    - *Example: Life Cycle Cost of a Journal Bearing*

NEED FOR KNOWLEDGE OF COSTS

Product costs are calculated by many different departments in a company: cost engineering, industrial engineering, design, and manufacturing, among others. The aims of these groups, the methods they use, and the uses to which their results are put are different. Cost engineering provides cost estimates based on finished designs. Industrial and manufacturing engineering groups calculate costs with a high emphasis on processing costs. Management uses these figures for making decisions on quantities and the product mix to be produced. Designers use (or should use) models to predict costs at the various design stages. These results help in making decisions during designing—when they are most effective.

While the emphasis in this book is on designers' knowledge of product costs, designers should realize that the accounting department prepares cost information in different forms, depending on the purpose for which it will be used. There are basically two types of accounting systems in use, which differ in the type of information they provide, as described below.

FINANCIAL ACCOUNTING

Financial accounting is used to prepare reports for external parties—e.g., stockholders, government agencies (contracting, rate setting)—and for tax and financial reporting.

MANAGEMENT ACCOUNTING

Management accounting is used for internal reporting to management for decision making, planning, and control.

Decision Making

Management uses the cost information to make decisions such as make-or-buy, for pricing the product, to decide on the product mix, and for similar activities relevant to the organization.

Planning and Control

Management also uses the cost information for planning and control. Cost information is used for preparing budgets, which are the instruments for formalizing plans. Performance of various units is judged on the basis of costs; thus, cost information becomes a means of exercising control.

Recognizing Costs During Design

Product costs need to be recognized early, i.e., during designing, when they can be controlled the most, as we saw in Figure 2. As a rule, however, cost information for a new product is not available at this point. This forms a paradox, as shown in Figure 60, particularly for *new designs*. Real knowledge of life cycle costs becomes apparent only after a product has been in use. For variant designs, on the other hand, where similar work has been done before, there is greater knowledge of costs, which can be extrapolated to a new product with a higher degree of confidence. In obtaining estimates of costs concurrently with the design process, there is a trade-off between the time involved in costing and the accuracy of the costs. Although in the earlier stages of design cost estimation can only be approximate, decisions made in its absence can be costly. Designing

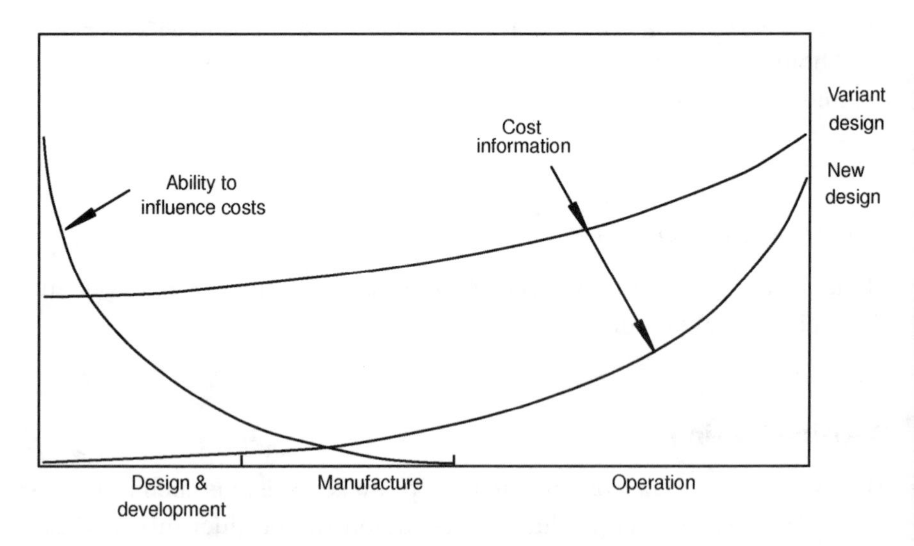

Figure 60. The designer's paradox—determining costs in the early stages.

for a cost goal is not possible unless such calculations are carried out simultaneously.

Figure 61 shows the data on costs incurred in individual activities and the potential for lowering product costs in each (Ehrlenspiel, 1985). For example, design and development account for only 7% of the product cost, but this phase can be responsible for 65% of the potential decrease in costs, through (for example), value analysis. See, for example, Fowler (1990). Compare the data in Figure 61 with those shown in Figure 2, the costs set at each stage.

The cost of a product is only one of the requirements on it. In a competitive marketplace it can be the decisive factor in the product's success or failure. The market price depends on factors both internal and external to the company. These include:

- Factory or total cost to the company
- Marketing and distribution costs
- Price of comparable products on the market
- Desired market share

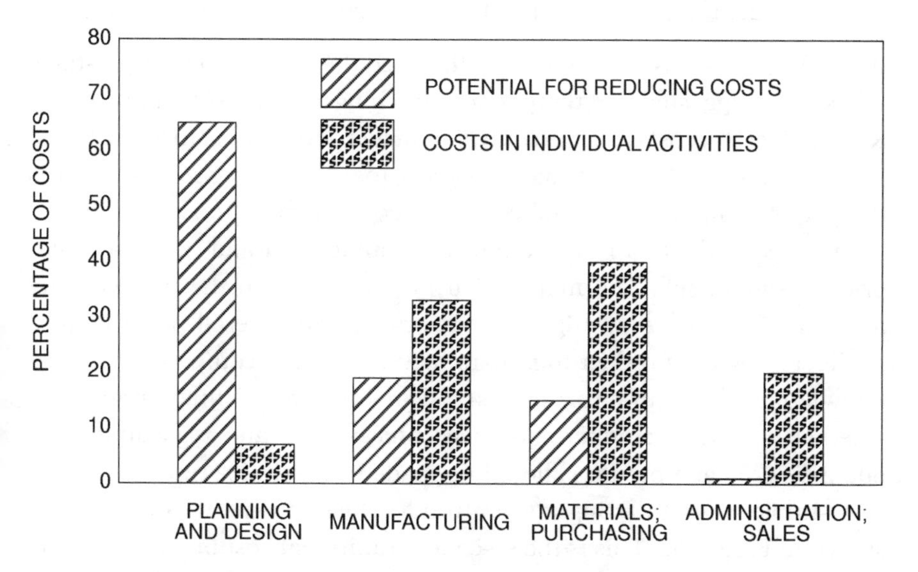

FIGURE 61. Importance of design in controlling product costs.

COMPONENTS OF MANUFACTURING COSTS

Total manufacturing costs can be classified in several ways. For example, they may be divided into

1. Material costs
2. Production costs

They may also be classified as

1. Direct costs, such as materials and labor, which can be assigned to the product
2. Indirect or overhead costs, such as heating and lighting of shops, which cannot be directly assigned

Further, the costs may be divided into:

1. *Variable costs*, which consist of direct costs and variable overhead costs. These include, for each unit produced, the material costs per unit and "processing" costs—direct labor and machine time.
2. *Fixed costs*, which remain constant over a period of time. These include, for each batch produced, the setup and tooling costs.

It is the variable costs that can be influenced the most at the design stage. It has been suggested that there is no such thing as fixed costs (Cooper and Kaplan, 1991b); costs are fixed only if managers do nothing to lower them.

It is important for a company to know manufacturing cost, since it is the major component of total cost. The company can thus know better whether a product can be profitably manufactured and, if so, at what price it should sell. The manufacturing cost helps in the make-or-buy decisions for parts that might be outsourced. The overall profit picture is affected by the product mix manufactured by the company. With the knowledge of manufacturing costs, decisions can be made regarding whether a certain product should be phased out, and regarding what quantities of each product should be made to maximize resources.

A simple way of costing a given product used by many companies is shown in Figure 62. This is the so-called traditional cost breakdown. The only costs considered direct costs are direct labor and materials. All other costs are lumped together as overhead. The overhead assigned to a product is

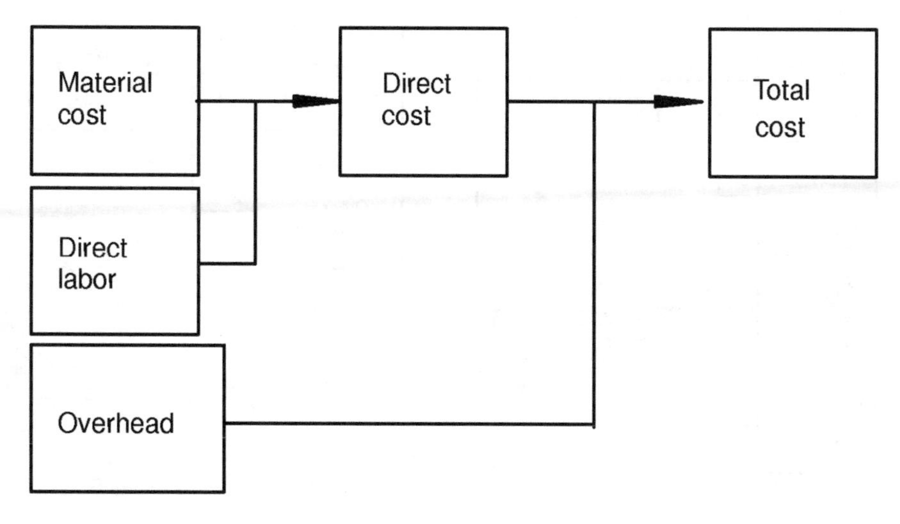

FIGURE 62. Simplified cost breakdown.

- Proportional to the direct labor, or
- Proportional to material cost, or
- Proportional to a combination of material and direct labor.

Such arbitrary assignment can lead to a significant misallocation of the overhead costs. This is discussed further in a later section.

Figure 63 shows a more detailed breakdown of various components of the total cost. The nomenclature and details vary among different authors and at different companies. The *customer*, on the other hand, is generally interested in not only the first or production cost but also total life cycle costs, which include the operation, maintenance, and disposal costs, as discussed later under "Life Cycle Costs" (and as illustrated in Figure 70.) The manufacturing cost, C_{mfg}, is given by

$$C_{\mathrm{mfg}} = C_{\mathrm{mat}} + C_{\mathrm{pr}} = C_{\mathrm{mat},\,d} + C_{\mathrm{mat},\,o} + C_{\mathrm{lab},\,d} + C_{\mathrm{fix}} + C_{\mathrm{pr},\,o} \tag{7}$$

where

$$C_{\mathrm{mat}} = \text{total materials cost}$$

consisting of

$$C_{\mathrm{mat},\,d} = \text{materials direct cost}$$

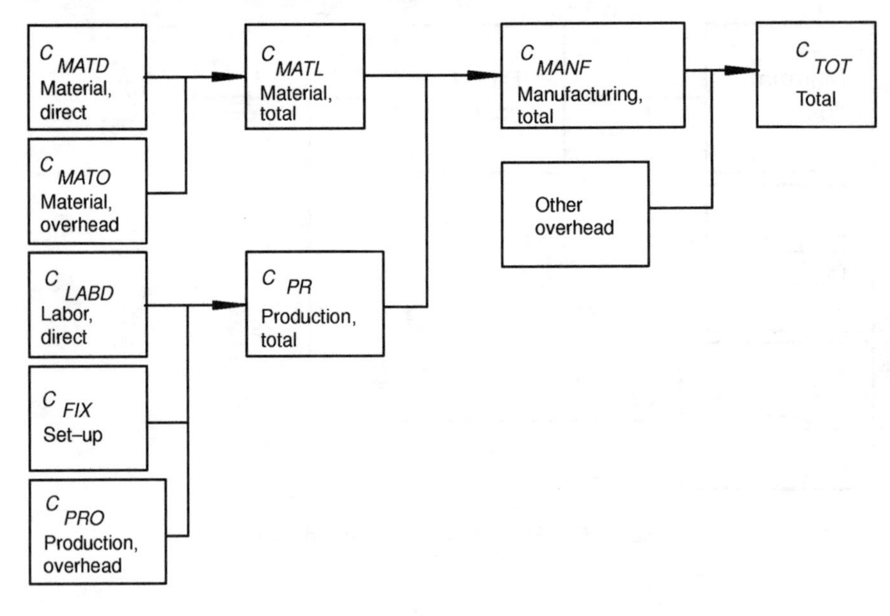

FIGURE 63. Detailed breakdown of product costs.

$$C_{\text{mat},\,o} = \text{materials overhead cost}$$

and where

$$C_{\text{pr}} = \text{total production cost including assembly}$$

consisting of

$$C_{\text{lab},\,d} = \text{direct labor cost}$$

$$C_{\text{fix}} = \text{machine, setup, or initial cost}$$

$$C_{\text{pr},\,o} = \text{production overhead cost}$$

The variable manufacturing costs $C_{\text{mfg},\,v}$, which can be influenced the most at the design stage, are given by

$$C_{\text{mfg},\,v} = C_{\text{mat},\,d} + C_{\text{lab},\,d} \tag{8}$$

Material costs include all purchased material—raw material, and semi-finished or finished parts or assemblies. As shown later, other costs, such as material overhead costs, can also be influenced to some extent at the design stage.

Figure 64 shows the results of a survey (Hendricks, 1989) regarding cost breakdown in seven different manufacturing industries. Direct labor cost is below 10% in two highly automated industries—computers and automobiles. The direct material cost typically ranges from 40 to 60%.

TYPES OF COST BREAKDOWN—COST STRUCTURES

The contributions to the cost of a product can be estimated in a variety of ways. These breakdowns are expressed in the form of cost structures. A cost breakdown is important for a number of reasons—e.g., to determine the areas of cost reduction, to allocate "value" to parts (Fowler, 1990), or for planning future products.

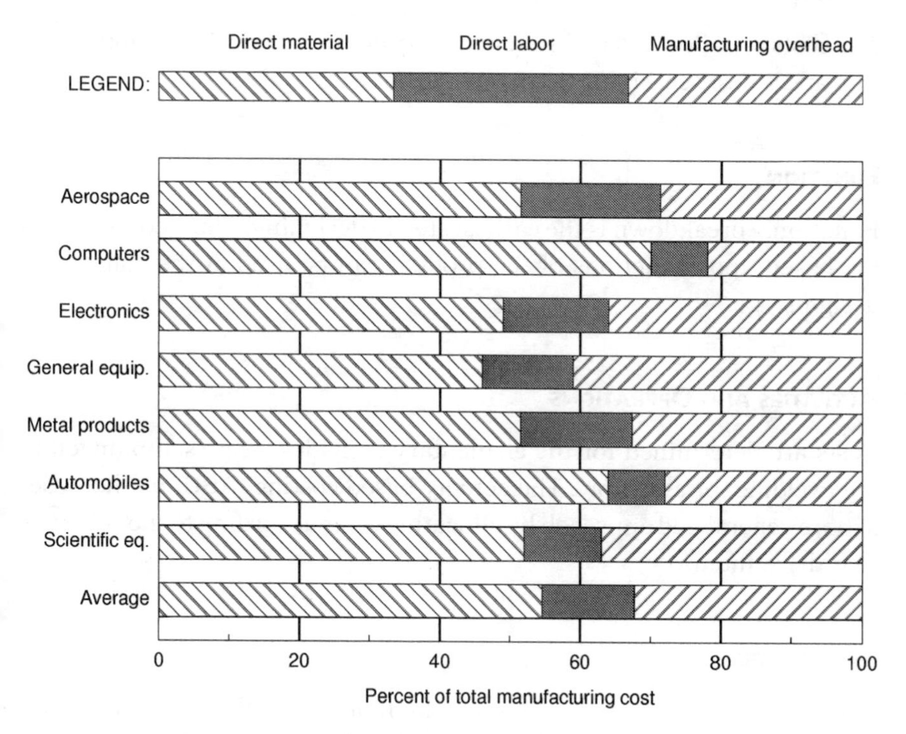

FIGURE 64. Cost breakdown in different industries.

A cost structure shows the breakdown of the product cost according to one of several criteria, such as parts, types of cost, functions, or production process. A cost structure brings into focus the proportion of cost contribution from the various viewpoints. Thus, a designer looking for ways of reducing costs can address the areas that offer the greatest possibilities and not worry about minor points. One advantage of cost structures is that they remain fairly constant over time, unless a major design change takes place.

The major forms of cost structures are the following:

Component

The cost of various product components is determined. Group technology can be used to advantage with this type of cost structure.

Department

The costs are allocated to different departments, e.g., design, manufacturing, and sales; see Figure 61.

Function

Functional breakdown is the earliest step in designing. Thus, cost estimating based upon functions of the product can be used at the conceptual design stage.

Activities and Operations

Costs are determined for the elementary activities—e.g., setup, machining, welding, or assembly. The activity-based costing (ABC) method uses a more generalized approach by allocating costs according to the activities that drive them.

Magnitude

Magnitude-based costing analysis (MBC) highlights the more important aspects of the design according to the chosen category. The designer can ascertain where costs can be reduced the most—by lower material cost, by combining functions, etc.

In actual use, cost structures can combine the attributes of two or more of the types described above.

TYPES OF COSTING

The form in which cost information is available in different departments (in cost engineering, design, or industrial engineering, or for use by management) varies greatly, as mentioned earlier. Quantitative cost information can be given in deterministic or statistical form.

Costing systems used by companies can be quite complex, using detailed cost tables. Or a company can use a very simple system in order to send a signal to the various departments, for example:

- To the design department to use fewer parts
- To manufacturing to reduce labor costs

As an example, a major electronic instruments manufacturer treats material cost as the only direct cost.

The major costing methods are described below in brief. This discussion is somewhat secondary to the main subject at hand, designing for low cost. Yet it is necessary for engineers and designers to be familiar with the terminology of costing. From the management perspective, the type of costing used is important, because it helps determine whether the overhead costs are allocated fairly among the various company products so as to reflect their true costs.

TRADITIONAL COSTING

In the traditional approach, cost per unit is found in terms of variable costs, e.g., labor, material, and the fixed overhead costs. The fixed costs are allocated on the basis of the presumption that products use overhead resources in proportion to the variable costs. Thus, the allocation may be proportional to the material costs or labor costs or machine time, for example. This method can lead to higher costs allocated to low-volume products than to products produced in mass quantities.

As an example, a company calculates the manufacturing cost by using the equation:

$$C_{mfg} = (A)C_{mat} + (B)C_{lab,\,d} \tag{9}$$

where $C_{lab,\,d}$ = direct labor cost (see Figure 62). The coefficients A and B can vary from 100 to 150% and 200 to 500%, respectively. Other companies use only the material cost, or only the direct labor cost, to determine the manufacturing cost.

Traditional costing methods date back to the days of the beginning of mass production when products were simple and few. The major part of the cost in those days was direct labor. Fixed costs accounted for only 10 to 20% of the total costs. In today's complex business and manufacturing environment, the overhead costs can no longer be so easily and simplistically allocated.

Activity-Based Costing (ABC)

ABC identifies the activities that drive costs by consuming resources (see Butler, 1994, and Cooper and Kaplan, 1991a). Cost drivers are items such as

- Number of units produced
- Cost of materials used
- Number of different materials used
- Labor hours
- Hours of equipment time
- Number of orders received

Resources consumed include

- Salaries
- Space
- Energy
- Time

ABC assigns activities cost to the units of production, called *cost objects*. The gist of the ABC method is illustrated in Figure 65, which shows that a number of different sources of indirect costs, caused by a number of different actions, are assigned to the different cost objects. The traditional costing method uses a common pool of overhead costs and assigns these on some simple basis, such as direct labor, to different products. The ABC method uses the more rational approach of determining how and why the overhead costs arise and divides the overhead into several pools.

Activities can be categorized into four classes (Cooper and Kaplan, 1991b). These are, in hierarchical order, from the lowest level up:

- Unit-level activities, related directly to the production rate. These include material, direct labor, machine costs, and energy.
- Batch-level activities, which take place for each batch that is processed, such as machine setup and teardown, purchase orders, inspection, and material handling.
- Product-sustaining activities that are associated with a given product as a whole, such as product and process engineering.
- Facility-sustaining activities, which cannot be allocated to any one product—building, utilities, general management.

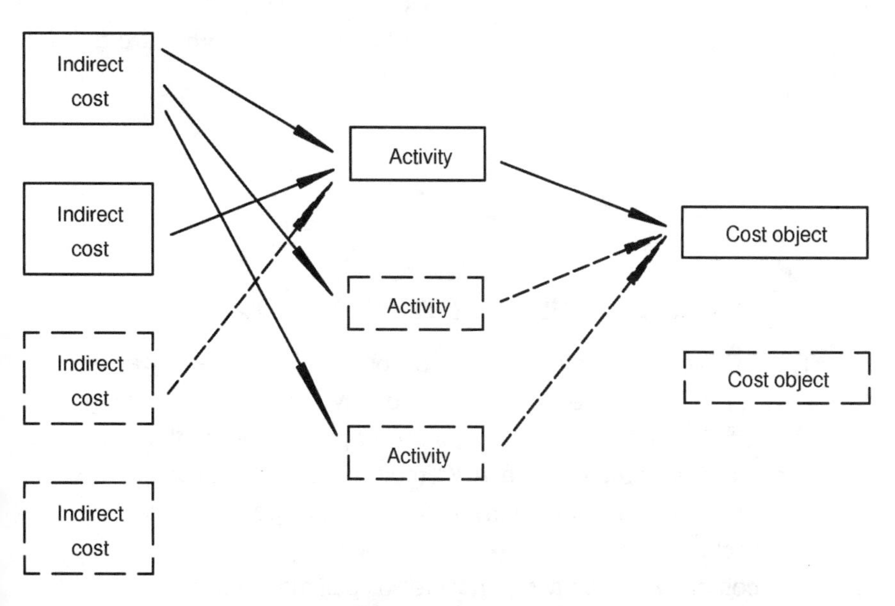

FIGURE 65. Allocation of costs in activity-based costing.

The main advantage of activity-based costing is that the indirect costs are more accurately reflected in the costs of the company's different products. Its main disadvantage is the difficulty of obtaining accurate information which would enable the proper allocations.

As an example (Ulrich and Eppinger, 1995), a company allocates costs on the basis of machine hours, setup time, number of production orders, number of material handling loads, and the number of new parts in a design, as follows:

$$C_{\mathrm{mfg}} = (A)C_{\mathrm{mat}} + (B)[(C)C_{\mathrm{lab},\,d} + aT_{MH} + bT_{\mathrm{set}} + cT_{\mathrm{pr}} + dN_{MH} + eN_{\mathrm{pt}}] \tag{10}$$

where

T_{MH} = machine hours
T_{set} = workpiece setup time
T_{pr} = production order count
N_{MH} = material handling load count
N_{pt} = new parts count

The company uses A and B in the range of 100 to 110% and the coefficient C in the range of 200 to 220%. The coefficients a through e are the unit costs of the respective items.

● ● ● ● ●

EXAMPLE
ABC VERSUS TRADITIONAL COSTING

The following example is based on one in Cooper and Kaplan (1991a) in which the unit-level, batch-level, product-sustaining, and facility-sustaining expenses are separately identified. The example will show that the ABC method allocates expenses to different products more fairly than does the traditional costing approach.

A company makes two parts; we will call these parts A and B. For the sake of simplicity, we will assume that the two parts are identical in all properties that affect costs. The only difference for our purpose is:

- Part A is made in lot sizes of 100.
- Part B is made in lot sizes of 1000.

The direct labor cost for each part is $2.50.

Allocation of Costs With Traditional Costing

The company finds that the total cost for 1100 parts is $27,379. Thus:

$$\text{Total cost per part} = \frac{\$27{,}379}{1100} = \$24.89$$

The company bases its overhead entirely on the direct labor:

$$\text{Overhead, based on direct labor} = 100\left(\frac{\$24.89}{\$2.50}\right) = 996\%$$

The company applies this overhead rate to both parts A and B. If the direct labor cost of either part changes, its new cost is calculated by using this overhead. Let us assume that the direct labor cost for part A drops to $2.45. The company finds that:

$$\text{New cost for part A} = (\$2.45)\left(\frac{996}{100}\right) = \$24.39$$

We will now compare the costs using ABC.

Allocation of Costs With Activity-Based Costing

Using the ABC method, the company determines, for each part, the unit- and batch-level expenses, as well as the product- and facility-sustaining expenses. According to this method, the manufacturing cost C_{mfg} is given by

$$C_{\text{mfg}} = C_{\text{mat},\,d} + C_{\text{lab},\,d} + \text{ULOH} + \text{BLOH} + \text{PSOH} \qquad (11)$$

where

ULOH = unit-level overhead
BLOH = batch-level overhead
PSOH = product-sustaining overhead

Equation (11) is an alternative form of Equation (7).

The details for part A are shown in Table 16. The material and direct labor costs per part are $8 and $2.50, respectively. Direct labor overhead at 120% of direct labor is charged. Machine-hour (MH) overhead is $20 per hour for 0.3 hour per part, i.e., $6.00. This yields a unit-level cost of $19.50 per part.

At the batch level, for 100 parts, two setups are required, each taking 4.5 hours, at $30 per hour (total = $270). There are two

TABLE 16 Costs for Part A Made in Lots of 100

	Per Part	For 100 Parts
Unit-Level Expenses		
Material	$ 8.00	$ 800
Direct labor	$ 2.50	$ 250
120% DL overhead	$ 3.00	
$20.00 MH overhead	$ 6.00	
Total unit-level per part	$ 19.50	
Number of parts = 100		$1950
Batch-Level Expenses		
2 Setups @ 4.5 h	$ 30.00	$ 270
2 Production runs	$ 95.00	$ 190
4 Material moves	$ 20.00	$ 80
Total per batch		$ 540
Product-sustaining expenses		
One product	$500	
		$ 500
Total product expenses		$2990
Facility-sustaining expenses		
10% value added		$ 219
Total, for 100 parts		$3209
Total cost/part	$ 32.09	

production runs, costing \$95 per run (total = \$190), and four material moves per batch at \$20 per move (total = \$80). Total expenses for the batch of 100 parts are \$540.

Product-sustaining expenses for this one part are \$500, which includes product and process engineering. The total expenses for the lot up to this point are \$2990.

The facility-sustaining expenses are calculated as 10% value added to the total cost less materials cost; i.e.,

$$\text{FSOH} = (\text{VA\%})(C_{\text{mfg}} - C_{\text{mat},\, d}) \tag{12}$$

where

FSOH = facility-sustaining overhead

VA% = value added, %

From Equation (12) we can write

$$\text{VA\%} = \frac{\text{FSOH}}{C_{\text{mfg}} - C_{\text{mat},\, d}} \tag{13}$$

In general, combining Equation (11) with Equation (13), we can write the value added percentage as

$$\text{VA\%} = \frac{\text{FSOH}}{C_{\text{lab},\, d} + \text{ULOH} + \text{BLOH} + \text{PSOH}} \tag{14}$$

For this example, we find the facility-sustaining expenses to be

$$\text{FSOH} = 10\% \times [\$2990 - (\$8 \times 100)] = \$219$$

Therefore, the total cost for 100 parts is \$2990 + \$219 = \$3209, or \$32.09 per part.

Part A is thus much more expensive than was calculated by the traditional approach.

The details for part B are shown in Table 17. For the sake of comparison, let us assume that part B has the same material and direct labor costs and the same machine usage. The difference is that in each batch, 1000 of these parts are produced, rather than 100 as for part A. Their unit-level expenses are the same.

At the batch level, for 1000 parts, 10 setups are required, each taking 4.5 hours, at $30 per hour (total = $1350). There are 10 production runs, costing $95 per run (total = $950), and 20 material moves per batch at $20 per move (total = $400). Total expenses for the batch of 1000 parts are $2700.

As shown in Table 17, product-sustaining expenses for this one part are also $500, which includes product and process engineering. The total expenses for the lot up to this point are $22,700.

TABLE 17 Costs for Part B Made in Lots of 1000

	Per Part	For 1000 Parts
Unit-Level Expenses		
Material	$ 8.00	$ 8000
Direct labor	$ 2.50	$ 2500
120% DL overhead	$ 3.00	
$20.00 MH overhead	$ 6.00	
Total unit-level per part	$ 19.50	
Number of parts = 1000		$19,500
Batch-Level Expenses		
10 Setups @ 4.5 h	$ 30.00	$ 1350
10 Production runs	$ 95.00	$ 950
20 Material moves	$ 20.00	$ 400
Total per batch		$ 2700
Product-sustaining expenses		
One product	$500	
		$ 500
Total product expenses		$22,700
Facility-sustaining expenses		
10% value added		$ 1470
Total, for 1000 parts		$24,170
Total cost/part	$ 24.17	

The facility-sustaining expenses are also calculated at 10% value added as follows:

$$\text{FSOH} = 10\% \times [\$22{,}700 - (\$8 \times 1000)] = \$1470$$

Thus, the total cost for 1000 parts is $22,700 + $1470 = $24,170, or $24.17 per part.

We note that ABC provides a very different and a more realistic picture of part costs. Part A, made in lots of 100, is 33% more expensive than part B.

Now, for the sake of comparison with the earlier example, let the direct labor for part A reduce to $2.45. In Table 16, if we make this change we find that the total cost of part A decreases from $32.09 to $31.97, that is, a reduction of $0.12, or 0.37%. This is significantly less than the 2% reduction provided by conventional costing. This is a more rational figure for the cost of part A. Figure 66 illustrates the percentage and absolute changes in cost by the conventional and ABC methods.

Thus, the ABC method helps us recognize which activities are driving up costs and what the real costs of the different products are. The above example has proven the well-known fact that larger lot size leads to lower cost. However, there are disadvantages to producing in large lot sizes. When a greater variety of products in smaller lot sizes are necessary, the ABC method points to where management must concentrate its efforts to keep costs down. Cooper and Kaplan suggest the following courses:

- Change product mix.
- Reduce resource consumption.
- Use fewer and more common parts.
- Customize near the end of the manufacturing cycle.

The last point is discussed further in Chapter X, under "Postponement."

• • • • •

LEGEND:

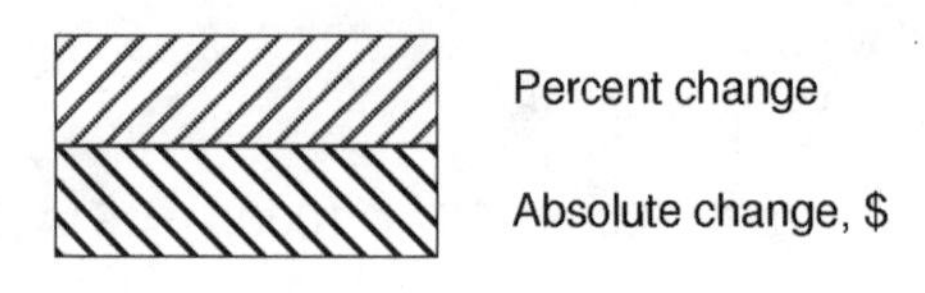

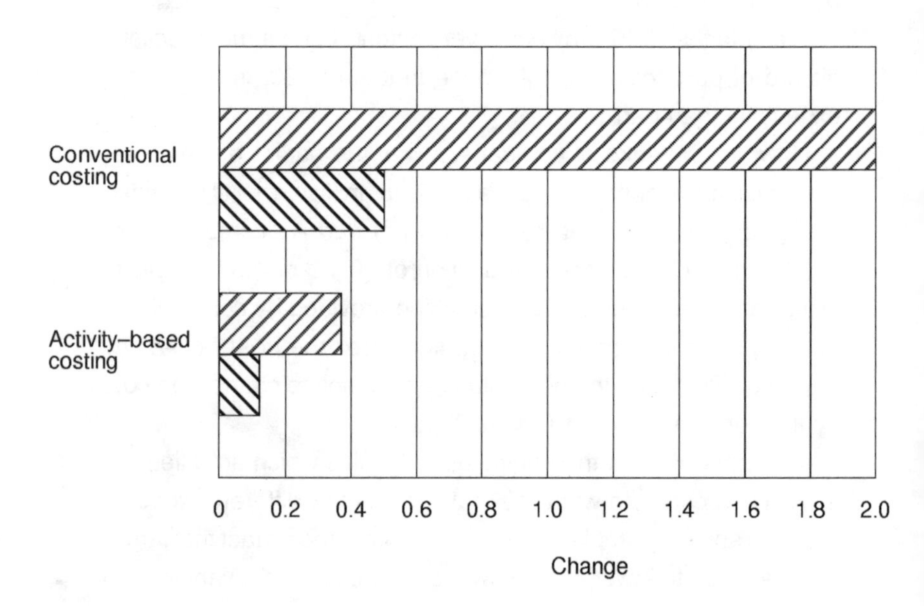

FIGURE 66. Change in part cost due to a 2% change in labor cost.

MAGNITUDE-BASED COSTING (MBC)

In the magnitude-based costing analysis (MBC) described by Ehrlenspiel (1985; he calls it ABC analysis), parts of a product can be divided into three categories; A, B, or C, according to some property, such as weight or cost. Parts in class A have the highest attributes; those in class C, the lowest; and parts in class B lie in the middle. In this analysis, the cost of A-class parts needs to be most accurately estimated, and of C-class, the least.

Table 18 shows the MBC classification of parts of a centrifugal pump, based on the manufacturing cost. The total manufacturing cost for the pump is $12,225. The manufacturing cost for each part is shown, both in dollars and as a percentage of the total cost. The manufacturing costs are further broken down into material, production, and assembly costs for parts made in-house. For purchased parts, the "manufacturing" cost includes the purchase price and assembly cost. Class A parts—the impeller and the housing—make up 82% of the total cost. Next, the class B parts—shaft, bearings, seals, and wear rings—make up 17%. Finally, class C parts—bolts, oiler, key, and gasket—total less than 1% of the total cost. Thus, when we aim for cost reduction, the greatest potential for cost savings lies in reducing the cost of class A parts.

TABLE 18 Cost Structure for Centrifugal Pump Showing MBC Classification

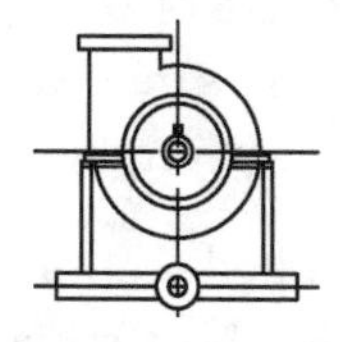

MBC Class	Part	Manufacturing Cost		Type of Cost, %		
		$	%	Material	Production	Assembly
A	Housing	5500	45.0	65	25	10
A	Impeller	4500	36.8	55	35	10
B	Shaft	850	7.0	45	45	10
B	Bearings	600	4.9	Purchased		
B	Seals	500	4.1	Purchased		
B	Wear rings	180	1.5	35	45	20
C	Bolts	50	< 1	Purchased		
C	Oiler	20	< 1	Purchased		
C	Key	15	< 1	30	50	20
C	Gasket	10	< 1	Purchased		

RELATIVE COSTS

Relative costs are used for comparing the costs of different materials, configurations, designs, etc., that fulfill the same needs. Relative costs allow quick and rough calculations. They are expressed in terms of the cost of the most commonly used or lowest-cost entity, which serves as a base cost; the costs of all other entities are then normalized to the base entity. The designer can use relative costs for making quick decisions between competing entities. Relative costs change less with time than do absolute costs and are most suitable when the competing entities are similar—for example, when the choice is among different ferrous metals, castings of similar shape, or similar gears.

The relative costs of different types of materials change at different rates. Generally, the relative costs of mass-produced materials—e.g., structural steels—change more slowly than those of, say, plastics and specialty materials, where the rate of technology change is rapid.

Relative cost comparisons can be made on the basis of material, manufacturing process, function, etc. (Eversheim and Shuppar, 1977):

- *On the basis of material:* Rolled sections of various metals and alloys; casings of different metals; sheet metals; wires; plastics; etc.
- *On the basis of process:* Casting, welding, machining, different surface treatments, heat treatments, etc.
- *On the basis of function:* Different types of fastenings, springs, couplings, bearings, shaft/hub fastenings, etc.

EFFECTS ON COSTS DURING CONCEPT AND EMBODIMENT STAGES

INFLUENCE AT THE CONCEPT STAGE

During planning and conceptual design, the methods for task clarification and recognizing the focal points for lowering product costs are the most important. These are:

- Market studies
- Economic analyses
- Cost structures

At the design order stage it is advantageous, in the requirements list, to:

- Keep the number of demands low.
- Exclude any unnecessary demands.

After this, it is the concept that has the single largest influence on costs. The concept determines the arrangement of elements through the function structure and thus the flows of energy, material, and signals in the system. The physical effects determine how subfunctions are realized. The solution principles determine how complex the product will be.

Methods for lowering product costs during concept development are:

- Set a cost goal.
- Analyze similar products according to types of costs.
- Determine areas for controlling costs with cost structures.
- Decide what can or cannot be changed.
- Use fewer subfunctions, and mainly those easily realized.
- Use simple function structures.
- Strive for robust physical effects.
- Aim for function integration.
- Use the same type of energy.
- Use fewer active surfaces.
- Use simple motions.

• • • • •

EXAMPLE
EFFECT OF CONCEPT ON COST OF GEAR TRAINS

Figure 67, from Ehrlenspiel (1995), shows an example of how the concept affects costs. The upper graph shows that the weight per unit torque for a gear reduction decreases as its complexity, defined by the number of stages and parallel transmission paths, increases. The manufacturing costs per unit torque transmitted

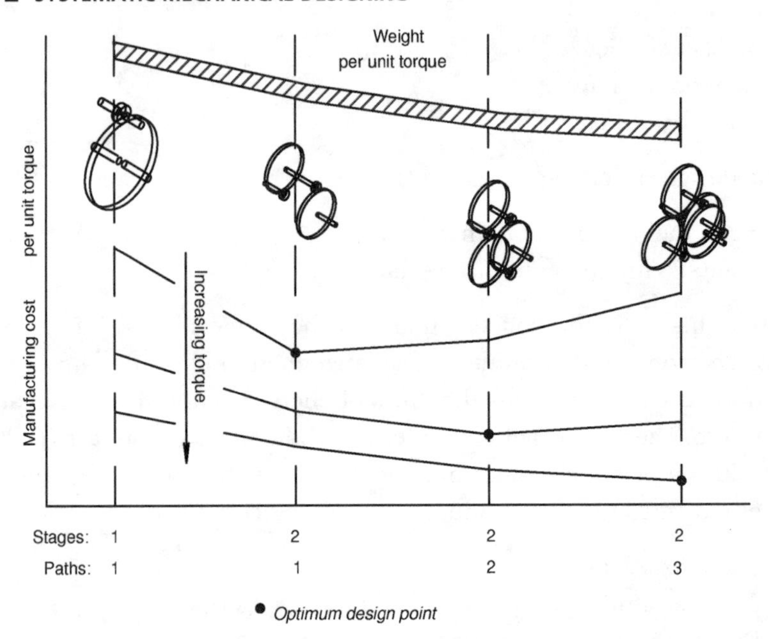

Figure 67. Gear train cost as a function of complexity. (Ehrlenspiel, 1985)

attain a minimum at different torque values for the various reductions shown. As the transmitted torque increases, it becomes more economical to use more stages and a greater number of parallel paths.

• • • • •

Influence at the Embodiment Stage

Next to the concept, embodiment has the greatest influence on product costs. The thrust of embodiment (and detail design) is toward the selection of materials, shapes, manufacturing processes, and systems. A number of individual topics are involved here: production quantities, sizes, layout, production processes, and assembly. Standardization affects product costs; it takes the form of use of standard parts, design of modular products, and manufacture of a product in a series of size ranges.

Steps that can be taken for lowering product costs during embodiment are:

- Analyze similar products according to manufacturing costs, and also life cycle costs.
- Ascertain areas where costs may be reduced; determine what can or cannot be changed.
- Develop knowledge of costs (for example, through the company database).
- Consider alternative concepts for secondary functions.
- Consider alternative embodiment (shapes, materials, motions).
- Use fewer parts, similar parts, same parts, and standard parts.
- Consider modular design or design in size ranges.
- Use standard material, or more cost-effective material.
- Use the most cost-effective material/manufacturing process combination.
- Minimize machined surfaces and surface treatments.
- Consider production in the company's own shop versus outsourcing.

Design for manufacture and assembly methods reduces manufacturing costs. These steps and methods will be discussed further in detail and illustrated by examples.

Even at the detail design stage, there is some cost reduction possible:

- Avoid overtolerancing.
- Adhere to standards.

EFFECT OF THE NUMBER OF PARTS

A recurring theme in design is the use of fewer parts, similar parts, same parts, and standard parts. Products are typically made of hundreds or thousands of parts. A good design should be simple and contain few parts. Problems caused by using a multiplicity of parts are many. These are ultimately reflected as higher costs in design, production, storage, assembly, bookkeeping, etc. Figure 68 depicts the evolution of a transmission for an earth-moving machine (Ehrlenspiel, 1985). In each redesign, the number of parts and thus the total cost have been successively reduced.

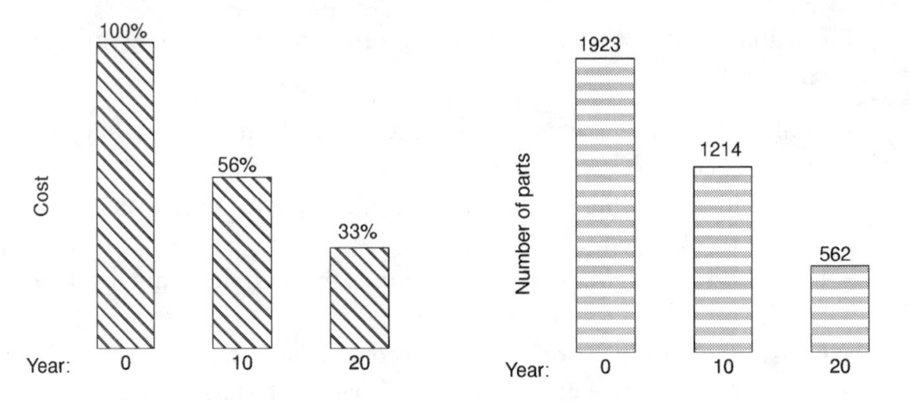

Figure 68. Reduction in cost and number of parts in a gear train. (Ehrlenspiel, 1985)

Cost of Part Development

Introducing a new part is expensive, in money and time. It entails one or more of the following:

- Design and development costs
- Looking for a vendor
- New tools and processes
- Decisions on materials
- Life cycle testing
- Associated overhead expenses such as record keeping

• • • • •

EXAMPLE
PART DEVELOPMENT COST FOR FIVE PRODUCTS

Data from the development of five different products were shown earlier, in Figure 9 (Ulrich and Eppinger, 1995). The products are:

- Jobmaster screwdriver, from Stanley Tools
- Rollerblade, Bravoblade in-line skates
- Hewlett-Packard DeskJet 500 printer

- Chrysler Concord automobile
- Boeing B-777 commercial aircraft

Figure 69 shows the number of unique parts in each of the products (first bar). If we divide each product's total development cost by its number of parts we get an idea of the development cost per part. This is shown as the second bar in the charts in Figure 69. The true development cost per part is, of course, lower than the values shown, since there are other activities involved in product development. Nevertheless, these numbers provide an estimate

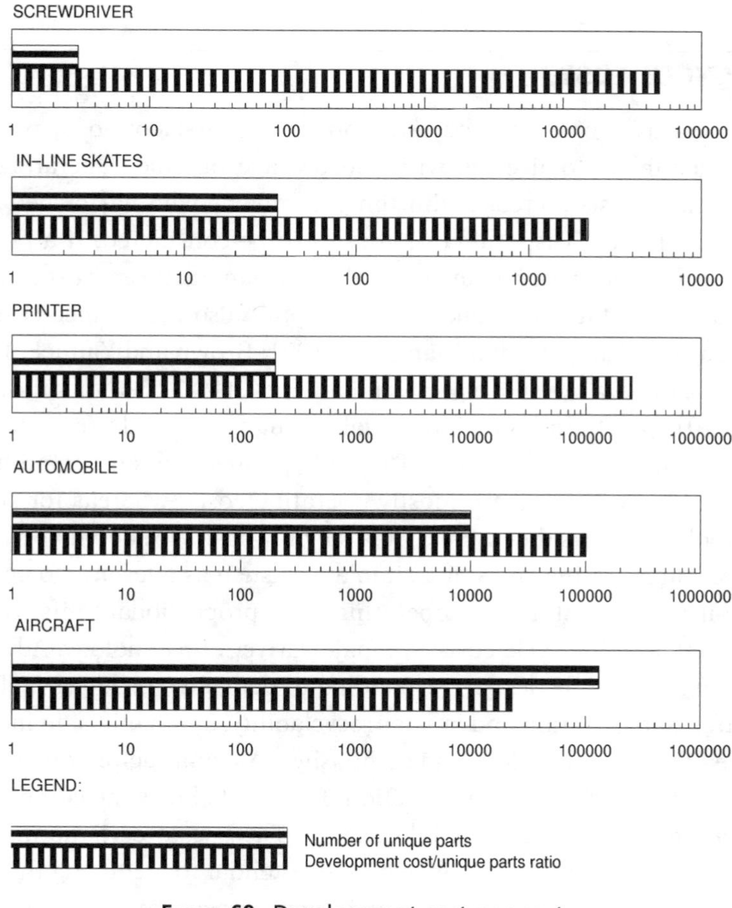

FIGURE 69. Development cost per part. (Ulrich and Eppinger, 1995)

of the order of magnitude of the part development cost—that it typically runs into thousands of dollars.

We shall look at the advantages of using fewer parts, similar parts, same parts, and standard parts when standardization is discussed in Chapter X.

• • • • •

LIFE CYCLE COSTS

Customers as a rule are interested not in the first cost of a product, but rather in its total costs with the passage of time. Operating and maintenance costs increase with time, as anyone who has owned a car or a home knows. Disposal costs must also be considered. If a product breaks down, the owner incurs cost. Thus, a manufacturer must estimate the true ownership costs, taking into account also the finance cost and salvage value (Fabrycky and Blanchard, 1991; Brown and Yanuck, 1985).

It is the life cycle cost of a product that interests the customer the most. Figure 70 shows the division into initial, maintenance, and operating costs for different types of products. Thus, for pumps and similar products, the total life cycle costs are mostly operating costs, whereas for simple hand tools and parts the initial cost is the only cost. We should be careful, however, in assessing costs: if a simple part such as a lever is to be used in an airplane, it will run up operating costs proportional to its weight.

Reduction in life cycle costs is a major driver of technology. Advancements in automobile design, responding to demands for lower fuel consumption and pollution and greater recyclability, are manifest in increasing use of nonferrous alloys and composites. A major factor contributing to losses in the drive system is friction. Figure 71 shows the components of friction loss in an automobile engine. The use of ceramic material coatings on the cylinder bores holds the potential for reducing this loss, thus reducing the operating and maintenance costs and extending the life of the engine ("Engine Performance Improvements," 1995).

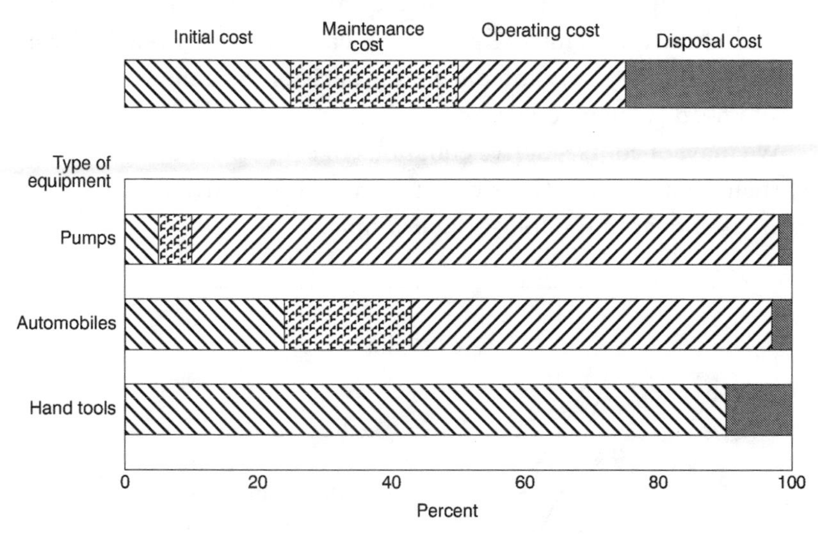

Figure 70. Cost structures for different types of products.

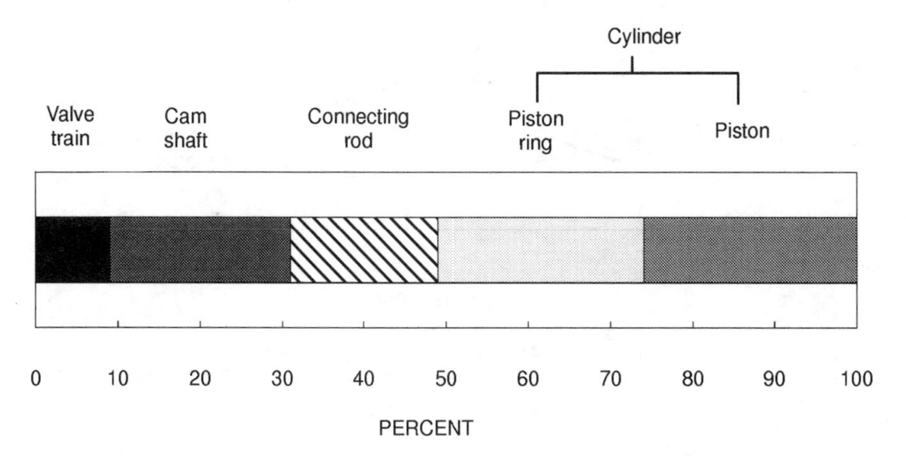

Figure 71. Components of friction loss in an engine.

Total Annual Cost

The initial cost of a product divided by the number of years it is owned gives the average annual capital cost. This cost thus decreases with the passage of time. If we add to this the operating and maintenance (O&M) costs, we obtain the average total annual cost.

The operating and maintenance costs, as a rule, increase with time. This is shown in Figure 72. Thus, typically, a time is reached when the total annual cost bottoms out and a decision on renewing the product would need to be made.

We may, of course, derive equations to express life cycle costs in terms of their components. However, it is more instructive to work in tabular form as shown in the following example.

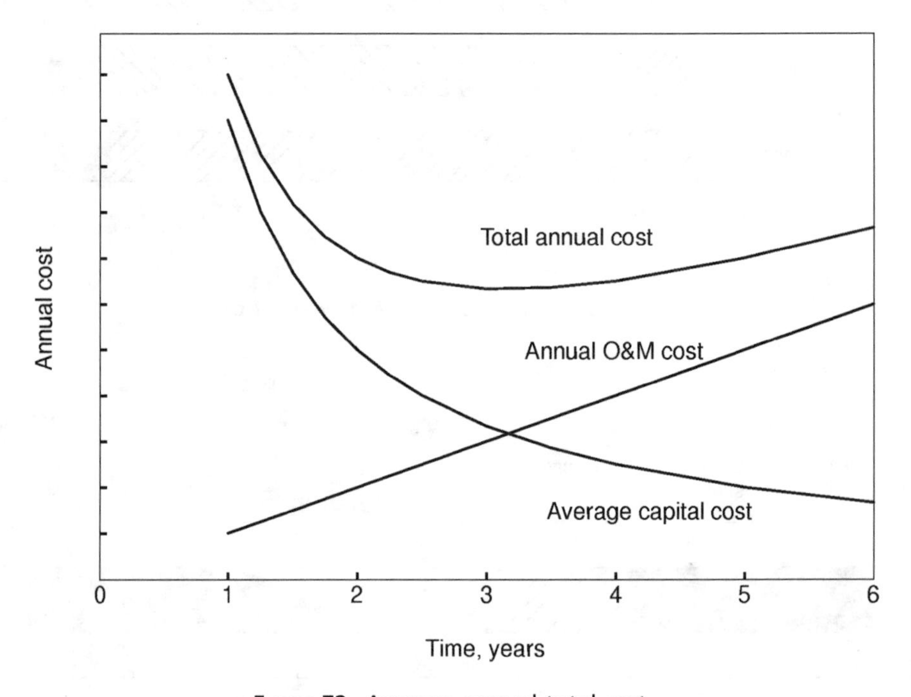

FIGURE 72. Average annual total cost.

• • • • •

EXAMPLE
OPTIMUM LIFE CYCLE OF EQUIPMENT

We wish to determine the average annual life cycle cost of a piece of equipment. In the first go-around, the time value of money and the salvage value are ignored. These are included in later computations. The following data are used:

Initial cost of the equipment C_{init}: $15,000

O&M costs for the first year: $2000

Annual increase in the O&M costs: $400

Table 19 shows the computations. Column 2 shows the annual O&M costs starting at $2000 in the first year and increasing at $400 per year thereafter, with cumulative value shown in column 3. The average annual O&M costs are calculated in column 4. Column 5 shows the initial cost divided by the number of years of ownership, i.e., the average capital cost. The average annual life cycle cost is in column 6, which is the sum of the previous two numbers. We see that the minimum occurs in year 9.

These results are, of course, not realistic. For example, the entire cost of the equipment ($15,000) has been charged off at the end of the first year, implying that it has no salvage value. Also, no financing costs have been taken into account.

TABLE 19 Average Annual Life Cycle Cost of Equipment

(1) Year	(2) O&M Cost per Year	(3) Cumulative O&M Cost Σ (2)	(4) Average O&M Cost (3)/(1)	(5) Average Capital Cost C_{init}/(1)	(6) Total Annual Cost (4) + (5)
1	2000	2,000	2000	15,000	17,000
2	2400	4,400	2200	7,500	9,700
3	2800	7,200	2400	5,000	7,400
4	3200	10,400	2600	3,750	6,350
5	3600	14,000	2800	3,000	5,800
6	4000	18,000	3000	2,500	5,500
7	4400	22,400	3200	2,143	5,343
8	4800	27,200	3400	1,875	5,275
9	5200	32,400	3600	1,667	5,267
10	5600	38,000	3800	1,500	5,300

We next look at the case where the salvage value of the equipment is considered. We assume that the salvage value decreases each year—in this example, by 17% per year. The calculations are shown in Table 20.

The first four columns are identical to those in the previous table. Column 5 shows the capital value at the end of the given year, reduced by 17% from the previous year. For example, $\$15,000 \times 0.83 = \$12,450$, etc. The capital cost in column 6 is found by subtracting the capital value from the initial cost. The average capital cost and the annual cost are found as before. In this case the lowest annual life cycle cost occurs in year 2.

These figures, although an improvement over the previous calculations, still do not account for the time value of money. This will be shown next in Table 21. Here we assume a finance charge of 12% of the annual balance and a three-year period over which the total borrowed sum of $15,000 is paid off.

The first four columns in Table 21 are again identical to those in Table 20. Column 5 values are found as follows:

The payment on the principal is $\$15,000/3 = \$5000$ each year.
The interest for the first year is $\$15,000 \times 0.12 = \$1800$.
The total payment for the first year is $\$5000 + \$1800 = \$6800$.

The interest for the second year is on the balance of $10,000; that is, $\$10,000 \times 0.12 = \$1200$.
The total payment for the second year is $\$5000 + \$1200 = \$6200$.
The cumulative payments by the end of the second year are $\$6800 + \$6200 = \$13,000$.

The interest for the third year is on the balance of $5000; that is, $\$5000 \times 0.12 = \$600$.
The total payment for the third year is $\$5000 + \$600 = \$5600$.
The cumulative payments by the end of the third year are $\$6800 + \$6200 + \$5600 = \$18,600$.

By the end of the third year, the principal has been paid off; thereafter, the cumulative payment remains at $18,600.

TABLE 20 Life Cycle Cost Including Salvage Value

(1) Year	(2) O&M Cost per Year	(3) Cumulative O&M Cost Σ (2)	(4) Average O&M Cost (3)/(1)	(5) Capital Value	(6) Capital Cost C_{init} − (5)	(7) Average Capital Cost (6)/(1)	(8) Total Annual Cost (4) + (7)
1	2000	2,000	2000	12,450	2,550	2550	4550
2	2400	4,400	2200	10,334	4,667	2333	4533
3	2800	7,200	2400	8,577	6,423	2141	4541
4	3200	10,400	2600	7,119	7,881	1970	4570
5	3600	14,000	2800	5,909	9,091	1818	4618
6	4000	18,000	3000	4,904	10,096	1683	4683
7	4400	22,400	3200	4,070	10,930	1561	4761
8	4800	27,200	3400	3,378	11,622	1453	4853
9	5200	32,400	3600	2,804	12,196	1355	4955
10	5600	38,000	3800	2,327	12,673	1267	5067

TABLE 21 Life Cycle Cost Including Salvage Value and Time Value of Money

(1) Year	(2) O&M Cost per Year	(3) Cumulative O&M Cost Σ (2)	(4) Average O&M Cost (3)/(1)	(5) Cumulative Payment	(6) Salvage Value	(7) Capital Cost (5) − (6)	(8) Average Capital Cost (7)/(1)	(9) Total Annual Cost (4) + (8)
1	2000	2,000	2000	6,800	0	6,800	6800	8800
2	2400	4,400	2200	13,000	0	13,000	6500	8700
3	2800	7,200	2400	18,600	8577	10,023	3341	5741
4	3200	10,400	2600	18,600	7119	11,481	2870	5470
5	3600	14,000	2800	18,600	5909	12,691	2538	5338
6	4000	18,000	3000	18,600	4904	13,696	2283	5283
7	4400	22,400	3200	18,600	4070	14,530	2076	5276
8	4800	27,200	3400	18,600	3378	15,222	1903	5303
9	5200	32,400	3600	18,600	2804	15,796	1755	5355
10	5600	38,000	3800	18,600	2327	16,273	1627	5427

Column 6 shows the salvage value at the end of the given year. However, since the equipment is not owned until it is paid off, its salvage value is zero for the first two years. At the end of the third year, the salvage value is found by taking the reduction of 17% for each of the three years:

$$\$15{,}000 \times (0.83)^3 = \$8577$$

Thereafter, the salvage value at the end of each year is found by reducing the previous year's value by 17%. For example, $8577 × 0.83 = $7119, and so on. The capital cost in the next column is found by subtracting the salvage value from the cumulative payment. The average capital cost and the annual cost are found as before. In this case the lowest annual life cycle cost occurs at the end of year 7.

Figure 73 shows the plots of the average annual life cycle costs for each of the three cases discussed above.

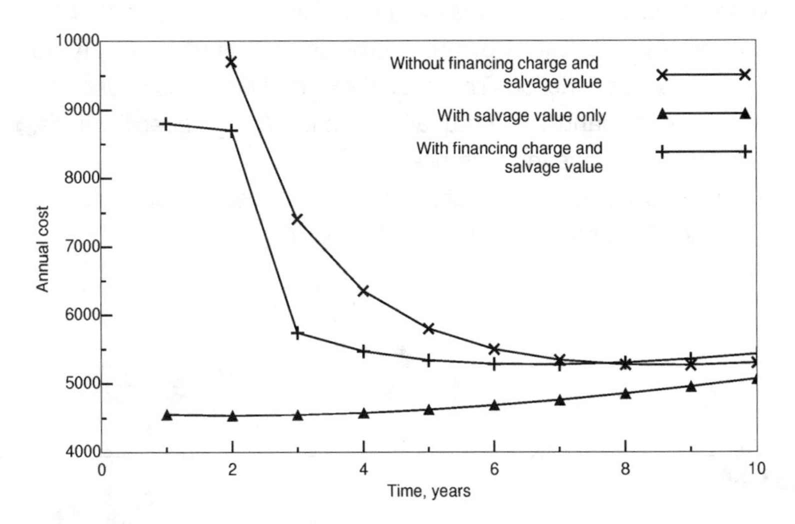

Figure 73. Life cycle costs with and without salvage value and finance charge.

• • • • •

EXAMPLE
LIFE CYCLE COST OF A JOURNAL BEARING

Tandon and Siereg (1992) show the analyses and calculations for determining the optimum life cycle cost for a journal bearing as a function of tolerance. It is a rather detailed modeling exercise. The authors express the total life cycle cost C_{LC} as

$$C_{LC} = C_M + C_O + C_{IQ} + C_{EQ} \tag{15}$$

where

C_M = manufacturing cost
C_O = operating cost
C_{IQ} = so-called internal quality cost
C_{EQ} = cost of "external quality"

The manufacturing cost C_M depends on the material properties and the required tolerance. The operating cost C_O depends on the energy loss due to friction and the lubricant cost. The "internal quality" cost is based on a scheme by Taguchi and consists of the costs of diagnosing the quality and making process adjustments combined with the internal failure costs found by applying Taguchi's loss function. The cost of external quality is caused by variations in manufactured quality from the design specifications, e.g., the variations in clearance.

Figure 74 shows the plots of the three cost components C_M, C_{IQ}, and C_{EQ}, and the total life cycle cost C_{LC}.

• • • • •

COST MODELS

Among the many methods for cost estimating during designing, the most noteworthy are those based on operations, weight, material, throughput

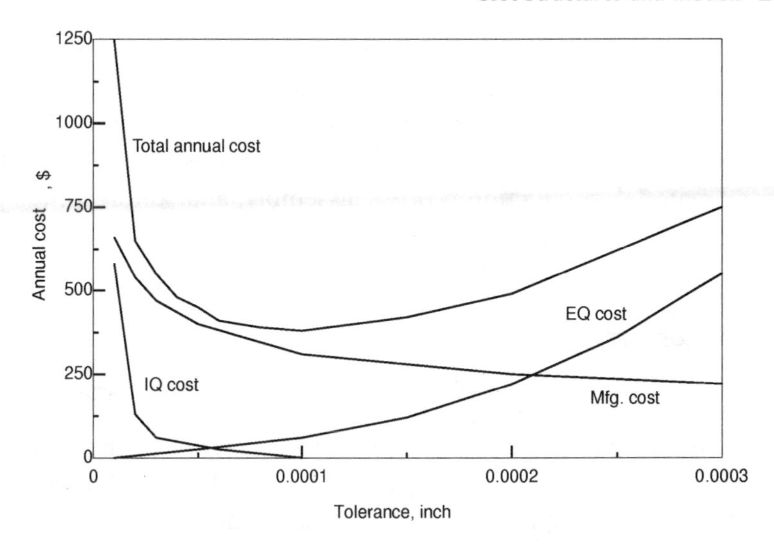

FIGURE 74. Bearing life cycle cost as a function of tolerance. (Tandon and Siereg, 1992)

parameters, physical relationships, regression analysis, and similarity laws.

The dependence of costs on product characteristics such as throughput, sizes, or weight can be found through statistical analyses. These analyses can be extensive and computationally intensive. Regression analysis shows this dependence in the form of coefficients and exponents in the regression equation. The data collected for this purpose are almost always presented first in graphical form and then converted into mathematical equations.

Similarity laws, on the other hand, are based upon physical relationships and expressed in the form of power laws. The exponents are derived from the equations for material costs, machining times, etc. The coefficients of the terms are found on the basis of some basic element, such as a part having nominal size or weight. These methods, which are often used in combination, are discussed below in brief. Quick cost estimation requires a simple method based on a defined cost entity which is at the designer's disposal *at the given design stage*.

A cost goal for the product must be set by the company and provided to the design group. The designer then sets up cost structures according to parts or other criteria, based upon similar products. Only if cost calculation is carried out in parallel with design steps is there a reasonable chance to hold to the goal. This will be illustrated shortly with an example. In the case of products for which no cost guidelines are available, a qualitative approach can be used in making decisions. By using relative costs for material, production process, assembly, and so forth, the optimum solution can be chosen.

Function Costing

Cost breakdown according to function is useful for several reasons:

1. It allows comparing design variants according to cost per subfunction. Although the method is more abstract than costing by parts, it has possible application at the concept stage.
2. It allows comparing the cost/benefit ratios of different functions.
3. Subsolutions can be combined to achieve lower function costs.

French and Widden (1993) show cost models for rolling bearings, induction motors, and linear actuators. The equation linking the cost and shape of a bearing is given as

$$C = G_m B(D + d) \tag{16}$$

where

B = breadth
D, d = outer and inner diameters, respectively, of the bearing
G_m = coefficient related to machining cost

The linear actuator model is

$$C = k_1 + k_2 F + k_3 F L \tag{17}$$

where F is the force and L the stroke length. The coefficients k_i (as well as G_m for the bearings) are given by the authors in tables.

Next we look at estimating the cost for each function in the example of centrifugal pump cost analysis given earlier (Table 18). These costs may be estimated as follows. The functions in the device are listed and each part is allocated the share of each function it fulfills. Table 22 lists the functions in the pump, and which parts are involved in fulfilling each of these.

We now reproduce and enhance Table 18 shown earlier. The addition is the allocation of functions for each part, as shown in Table 23. The last group of columns in Table 23 shows the estimates of of functions allocated to the various parts. For example, the impeller is assumed to serve 30% to transfer energy from the shaft to the liquid (function F2) and 70% to convert energy from mechanical to fluid dynamic form (function F3). By multiplying the cost of each part by the fraction it serves a given function, the total cost for each function can be calculated. The costs are shown in Table 24.

Let us look at function F1, "contain liquid." From Table 22 we see that it is fulfilled by the housing, seals, and gasket. Table 23 shows that the housing is allocated to the function F1 (50%) and to F6, "support parts" (50%). Thus, 50% of the housing cost of $5500 (= $2750) goes toward fulfilling the function F1. Seals and gasket are intended only to satisfy this function; hence, 100% of their costs ($500 and $10, respectively) are added to the cost of function F1. The last two columns of Table 24 show the cost of each function in dollars and as a percentage of the total manufacturing cost of the pump.

Cross (1989) gives an example of function cost analysis of an air valve. By eliminating parts that serve no function and by reducing the cost of the function "connect parts," a redesign resulted in reducing the cost by 60%.

TABLE 22 Functions Fulfilled by Parts of the Centrifugal Pump

Function	Function Description	Parts
F1	Contain liquid	Housing, seals, gasket
F2	Transfer energy	Impeller, shaft, key
F3	Convert energy	Impeller
F4	Connect parts	Bolts, key
F5	Increase life	Wear rings, oiler
F6	Support parts	Housing, shaft, bearings

TABLE 23 Cost Structure for Centrifugal Pump, with Function Cost Allocation

MBC Class	Part	Manufacturing Cost		Type of Cost, %			Function Allocation, %			
		$	%	Material	Production	Assembly				
A	Housing	5500	45.0	65	25	10	F1	50	F6	50
A	Impeller	4500	36.8	55	35	10	F2	30	F3	70
B	Shaft	850	7.0	45	45	10	F2	60	F6	40
B	Bearings	600	4.9	Purchased			F6	100		
B	Seals	500	4.1	Purchased			F1	100		
B	Wear rings	180	1.5	35	45	20	F5	100		
C	Bolts	50	< 1	Purchased			F4	100		
C	Oiler	20	< 1	Purchased			F5	100		
C	Key	15	< 1	30	50	20	F2	80	F4	20
C	Gasket	10	< 1	Purchased			F1	100		

TABLE 24 Calculation of Function Costs for Centrifugal Pump

Function	Part	% of Part Cost for Function	Part Cost, $	Function Cost of Individual Part, $	Total Function Cost	
					$	%
F1: Contain liquid	Housing	50	5500	2750		
	Seals	100	500	500		
	Gasket	100	10	10	3260	26.7
F2: Transfer energy	Impeller	30	4500	1350		
	Shaft	60	850	510		
	Key	80	15	12	1872	15.3
F3: Convert energy	Impeller	70	4500	3150	3150	25.8
F4: Connect parts	Key	20	15	3		
	Bolts	100	50	50	53	0.4
F5: Increase life	Wear rings	100	180	180		
	Oiler	100	20	20	200	1.6
F6: Support parts	Housing	50	5500	2750		
	Shaft	40	850	340		
	Bearings	100	600	600	3690	30.2

• • • • •

EXAMPLE
DESIGNING TO A COST GOAL

As an example of designing to a preset cost goal, we again consider a centrifugal pump. The total manufacturing cost is to be \$10,000, within ±5%. The first step is to set the cost goals for parts by referring to the cost structures of similar pumps. Using the magnitude-based costing (MBC) method, the parts are divided into categories A, B, and C according to manufacturing cost. The calculations are shown in Table 25. On the basis of the percentage of total cost of each part (column 3), the maximum allowable deviation is found from Figure 75 and shown in column 4. Column 5 shows the maximum cost of each part. By using cost models and methods described previously, the cost of each part in the

TABLE 25 Calculations for Designing to a Preset Cost Goal

(1) Part	(2) Part Cost Goal, $	(3) % Total Cost c_i	(4) % Allowable Deviation $e_{i, al}$	(5) Maximum Cost, $	(6) First Design Cost, $	(7) Decision	(8) Redesigned Cost, $
A1 (housing	4,500	45	7	4835	4,400	Acceptable	4,400
A2 (impeller)	3,500	35	8	3796	4,000	Redesign	3,600
B1 (shaft)	700	7	19	832	900	Redesign	750
B2 (bearings)	600	6	20	722	600	Acceptable	600
B3 (seals)	400	4	25	500	400	Acceptable	400
B4 (wear rings)	200	2	35	271	220	Acceptable	220
Σ C (key, bolts, gasket, etc.)	100	1	50	150	100	Acceptable	100
Total	10,000	100			10,630		10,070

Cost goal: \$10,000 (c_{tot}) ± 5% (e_{tot}); maximum cost: \$10,500.

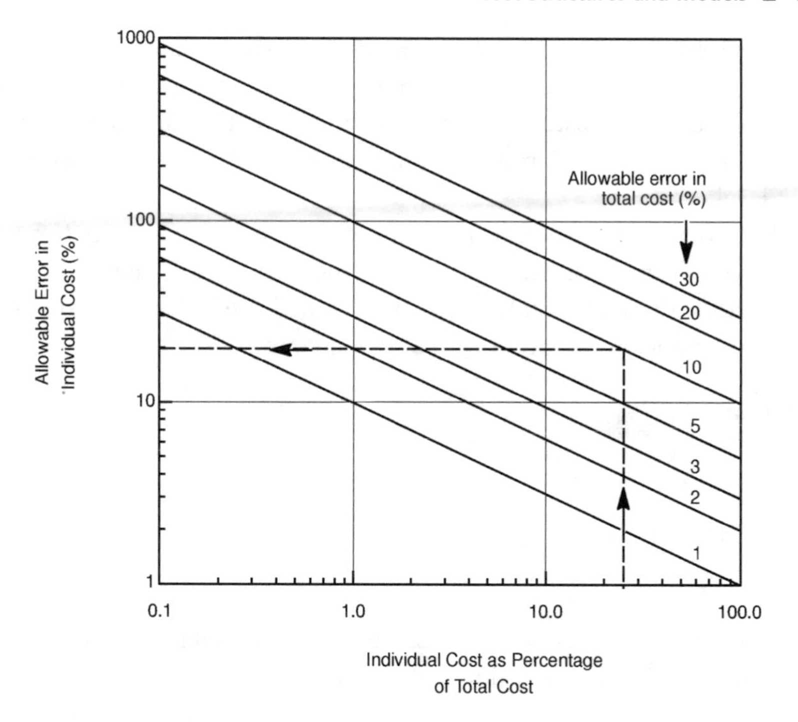

FIGURE 75. Allowable errors in cost estimation. (Ehrlenspiel, 1985)

initial design is calculated, shown in column 6. At this stage the cost of more expensive parts (category A) needs to be calculated more accurately, whereas category C parts require only a rough estimation. Comparing the figures in columns 5 and 6, we see that parts A2 and B1 exceed the limits and must therefore be redesigned. After redesign and calculation, the final cost estimates are shown in column 8. The total cost of the redesigned pump is $10,070, which is close to the preset goal.

• • • • •

PARAMETRIC COSTING

In parametric modeling, cost is expressed as a function of important design parameters. Parametric models are one of the few cost estimating techniques available at the conceptual design stage. The models use statistical regression techniques (see Appendix H) and cost databases that must be continually refined and kept up to date. Parametric cost equations are also known as cost estimating relationships (CERs). Although generally used at the project or product level, they have also been applied at the part level.

The different mathematical forms of parametric cost equations are given below. In these expressions C is the cost, the x's denote the cost parameters, the a's the statistically determined coefficients or exponents, and A the appropriate proportionality constant.

Linear model:

$$C = A\sum_i a_i x_i \tag{18}$$

This model, called the linear regression model, is used in situations where the data exhibit a straight-line distribution.

Generalized linear model:

$$C = A\sum_i a_i g_i(x) \tag{19}$$

This is a generalized form of the previous equation where the $g_i(x)$ are any of many functions that may be used, such as polynomial, exponential, or sinusoidal. A polynomial is a special case of this equation.

Product-exponential model:

$$C = A\prod_i x_i^{a_i} \tag{20}$$

This form of equation depicts a nonlinear regression model. By taking logarithms on both sides, the equation becomes a linear equation.

● ● ● ● ●

EXAMPLE PARAMETRIC MODELS IN THE AEROSPACE INDUSTRY

The following examples of parametric equations (Apgar and Daschbach, 1987) are used in the aerospace industry, which is typical of high-technology industries that require frequent design updates.

For a mechanical system, the development cost is given by

$$C_{\text{dev}} = K_D(W)^{wd}(\text{ND})(1 - \text{RD}) \tag{21}$$

where

K_D = development cost proportionality constant
W = weight
wd = weight exponent = 0.95
ND = fraction new design
RD = fraction redundant design

For a machined assembly, the first-piece manufacturing cost is

$$C_{\text{mfg}} = K_M(W)^{wf}(\text{TOL})^t(P)^p \tag{22}$$

where

K_M = manufacturing cost proportionality constant
wf = weight exponent = 0.92
TOL = tolerance
t = tolerance exponent = −1.4
P = number of parts
p = part count exponent = 1.1

For a mechanical subsystem, the annual maintenance cost is given by

$$C_{\text{maint}} = K_{MT}(\text{MTBF})(N)(\text{EF})\left(\frac{1}{OL}\right)^{ol} \tag{23}$$

where

K_{MT} = maintenance cost proportionality constant
MTBF = mean time between failures
N = quantity deployed
EF = environmental factor
OL = number of operating locations
ol = operating location count exponent = 0.8

● ● ● ● ●

EXAMPLE SIMPLIFICATION FROM THE PARAMETRIC MODELS

Parametric models derived by using regression analysis may be simplified to obtain related cost functions (Klasmeier, 1985). The variable manufacturing costs of a pressure vessel for high-voltage circuit breakers are given by:

$$C_{\text{mfg},\,v} = A + BD^{1.42}P^{0.94}L^{0.21}t^{0.17} \tag{24}$$

where A and B are constants, P is the pressure in the tank, and D, L, and t are the tank's internal diameter, length, and wall thickness, respectively. Since the diameter D is proportional to the conductor separation distance Δ, which in turn is proportional to the voltage V, we can write

$$V = c_1\Delta = c_2 D \tag{25}$$

Thus, if the thickness remains a minimum (constant), and since the length is independent of the voltage, Equation (24) simplifies to

$$C_{\text{mfg},\,v} = A_1 + B_1 V^{1.42} \tag{26}$$

● ● ● ● ●

Learning Curve

It is a well-known fact that we learn by repetition: if we perform the same operation over and over again, it takes us less and less time to do it (Gallagher, 1982). Of course, the amount of improvement decreases with the number of repetitions. This phenomenon was first recognized and used for cost estimating in the production of aircraft during the 1930s. It has been found that the reduction in time follows the power law $y = ax^s$. Commonly used numbers for estimating direct labor hours are that doubling the quantities reduces the time required by 20%. For example, if it takes one unit of time when the 5th unit is produced, it will take 0.8 unit of time when the 10th unit is produced. This is the so-called 80% learning curve. The equation for a learning curve is

$$T = T_1 N^s \tag{27}$$

where N is the unit number under consideration, T the units of time (or other resource) required to produce the Nth unit, T_1 the units of time for the first unit, and s the slope of the T versus N curve on log-log graph paper.

The value of s for a given P% learning curve can be found as follows. Let the time taken $T = T_a$ for unit $N = N_a$. For double the unit number, that is, $N = 2N_a$, the time is reduced to $T = PT_a/100$. Substituting these values in Equation (27) we get

$$T_a = T_1 N_a^s \tag{28}$$

and

$$\frac{PT_a}{100} = T_1 (2N_a)^s \tag{29}$$

Dividing the second of the above equations by the first, we get

$$\frac{P}{100} = 2^s \qquad \text{or} \qquad s = \frac{\log (P/100)}{\log 2} \tag{30}$$

The following table shows the values of the slope s for values of P from 100% (no learning) through $P = 50\%$.

P%	s
100	0
90	−0.152
80	−0.322
70	−0.515
60	−0.737
50	−1.000

The corresponding plots of T versus N are shown in Figure 76 for $T_1 = 100$ and P from 100% through 50%.

The data for a learning curve are found by observations on a given task. By using regression analysis (see Appendix H), we determine the slope of the line and T_1, the "theoretical first-unitpp" time. Data from one task may be used for another task only if the tasks are similar. The learning curve technique can be applied to a variety of jobs—from the purely manual to the purely mental. The slope of the learning curve depends not only upon the learning of the individuals, but also on factors such as improvements in production methods and other efficiencies in the overall process. Figure 77 (Stewart et al., 1995) shows the average learning values for some industries and operations.

Effect of Part Size and Quantities

Part size influences the setup, material, and production costs. Production in large quantities leads to reduction in unit cost. The fixed (initial) costs are divided up into two categories. First are startup costs for design, development, and prototype production. These relate to the total production quantity. Second are the one-time costs arising for the production of a lot size n.

The basic equation for calculating the manufacturing costs per unit for lot size n, C_{mfg}, if the initial costs consist only of setup costs, is

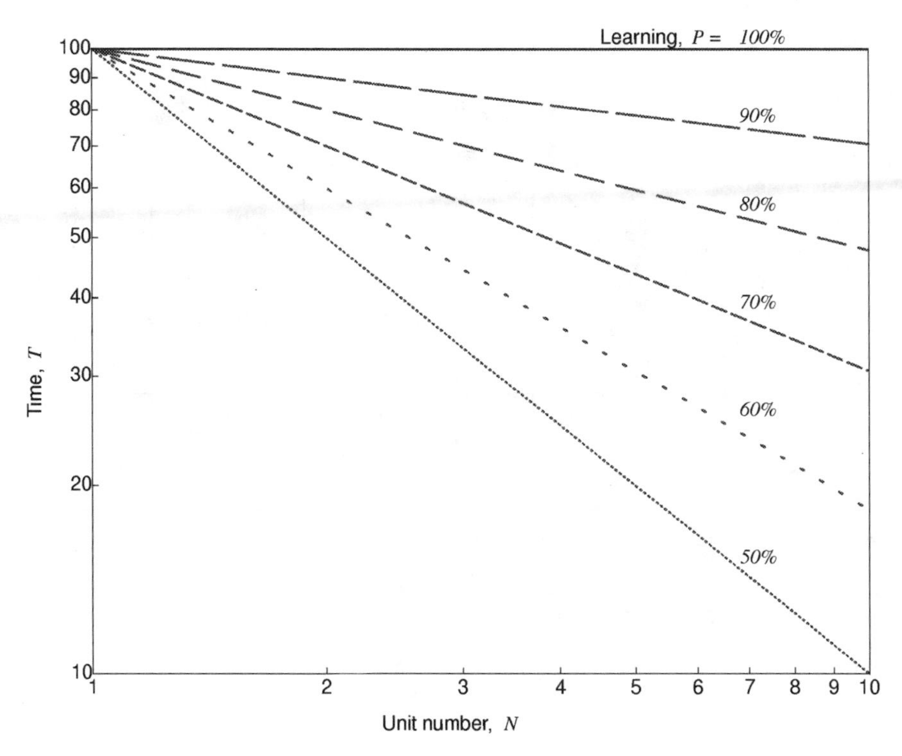

Figure 76. Learning curves.

$$C_{\text{mfg}} = \frac{C_{\text{fix}}}{n} + C_{\text{pr}} + C_{\text{mat}} \tag{31}$$

where

C_{fix} = fixed setup cost
C_{pr} = production cost
C_{mat} = total material cost

Part geometry also affects cost; the cost can increase with linear dimension, surface area, volume, or complexity.

As the quantities produced increase, different production processes, and perhaps different materials, can be used more effectively. Figure 78a shows the costs per unit of a part made of a certain material in low

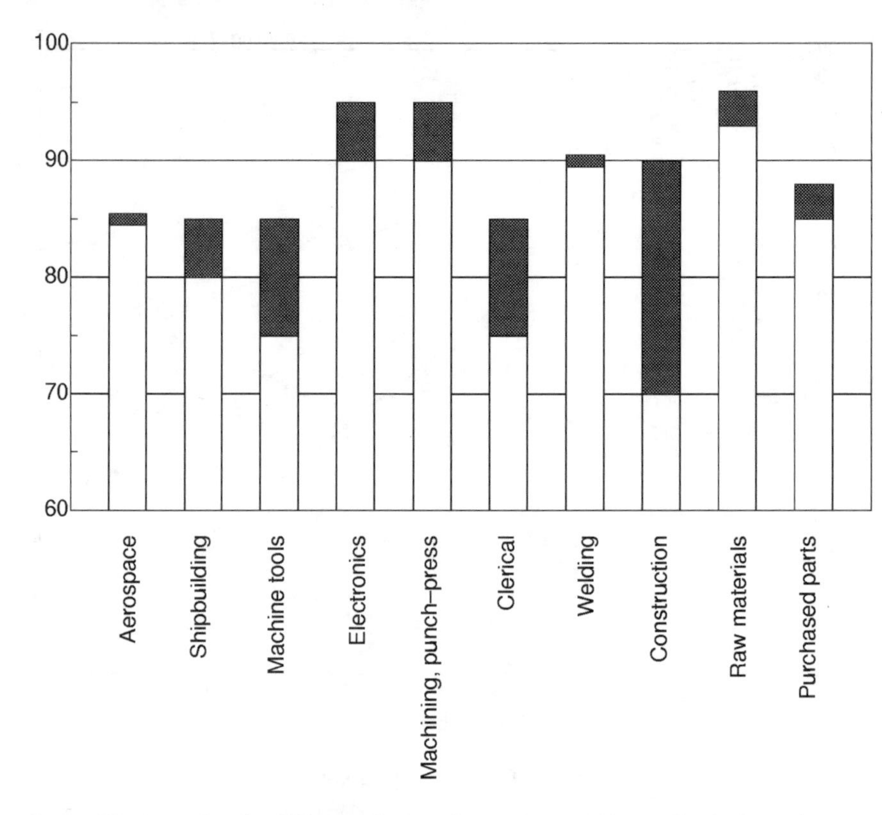

FIGURE 77. Learning in different industries and operations. Shaded portions show the ranges of variations for different industries and operations. (Stewart et al., 1995)

quantities. In Figure 78b, as the production quantities are increased, a new process is introduced which has correspondingly lower costs. Finally, in Figure 78c, a new, lower-cost material is introduced.

Figure 79 shows the trend in the relative costs of small turned parts produced in different quantities on different types of lathes (Ehrlenspiel, 1985). We will look again in Chapter IX at the effect of quantities made on the manufacturing process and cost.

COSTS BASED ON UNITS OF MEASURE

Weight-based calculation for estimating manufacturing costs is particularly suited for one-of-a-kind products and for all mass-produced items

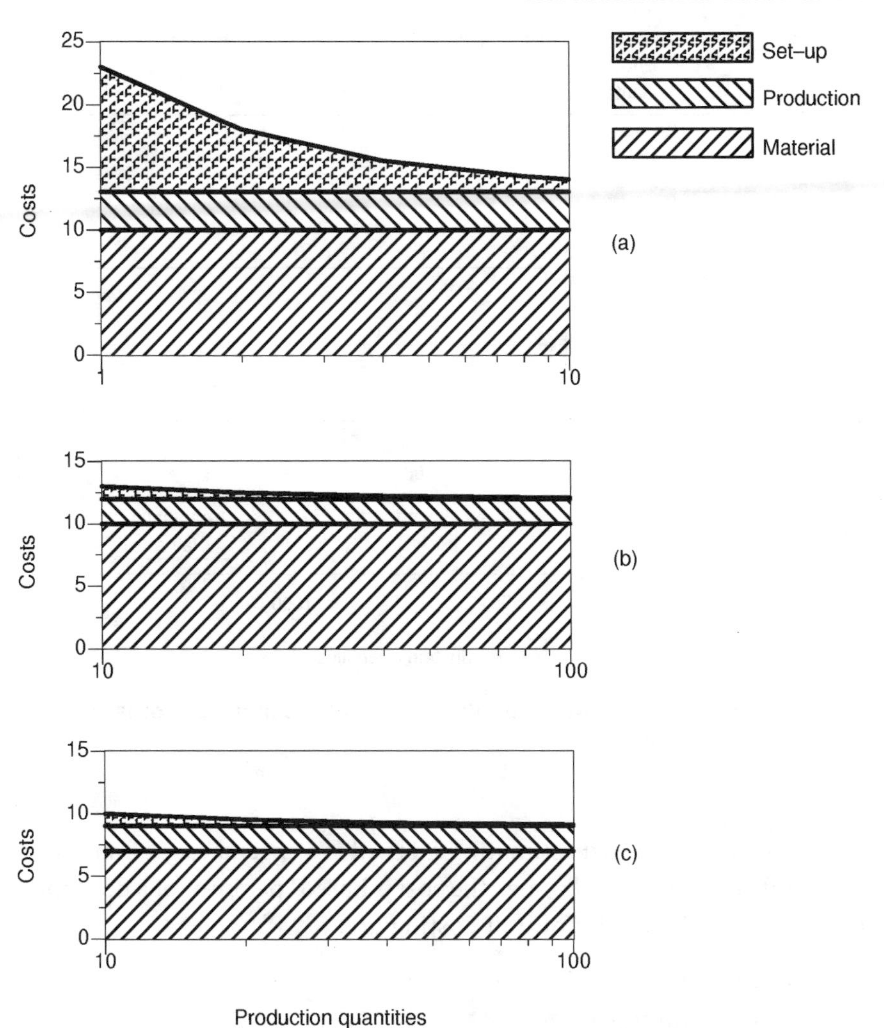

FIGURE 78. Effect on cost of production quantities, processes, and material.

if only size changes are made. The production costs per unit weight of similar products are proportional. At the low end are simple, heavy machines such as pumps and agricultural machines, at $5/lb. Complicated, high-technology products can range upward of $100/lb. Avionics has been estimated at $7000/lb (Rich and Janos, 1994). The weight-based method allows cost estimation for products for which material costs predominate and only minor changes in the product are involved.

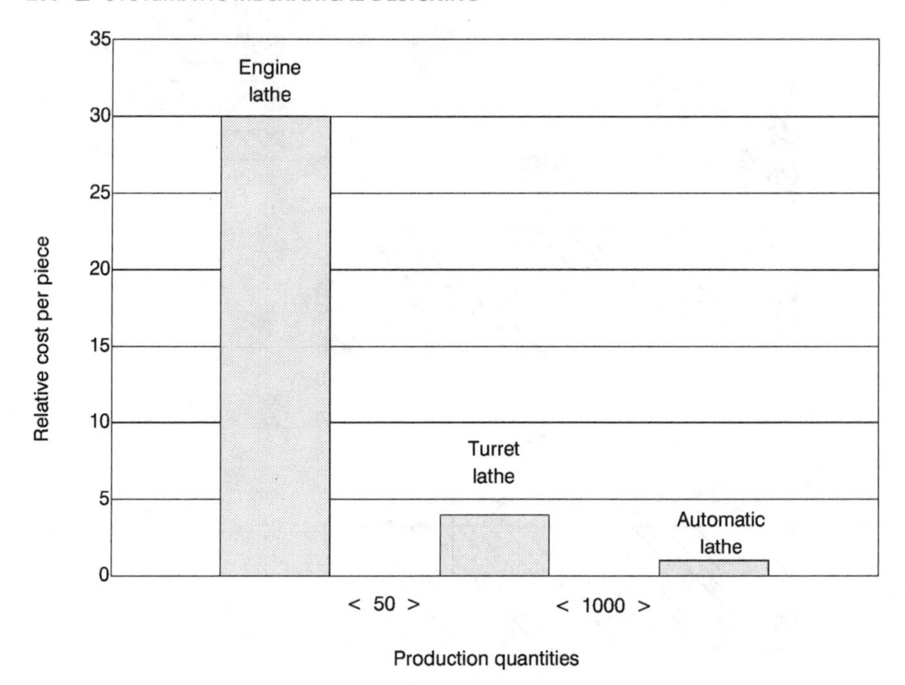

FIGURE 79. Effect on costs of production quantities and types of lathes. (Ehrlenspiel, 1985)

The weight-based cost estimating method is a special case of measurement unit costing methods—i.e., the cost is based "per unit" of a determining parameter. Examples of unit costing are:

- Cost of a motor per HP
- Cost of a pump per cfm

Unit costing models may be of any one of the mathematical types described earlier. The simplest type, linear, is given by the equation:

$$C = mx \tag{32}$$

where C is the cost and x the parameter on which the cost is based. The slope m is calculated by applying linear regression analysis on a number of observations. The cost of a "good" machine tool (1994 data) is estimated as \$4000/HP. The machining cost of various materials depends on, among other things, the energy requirement (as shown in Chapter IX and Figure 142).

The equation of the regression analysis line might be a line with an offset:

$$C = a + mx \tag{33}$$

Generally such equations are valid only in a limited range of the parameter x, with the coefficients a and m taking on different values for different ranges of x.

The unit cost model may also be a power law relationship, e.g.,

$$C = bx^{m} \tag{34}$$

where b and m are again determined by regression analysis.

The unit cost model may be based on more than one parameter, e.g.,

$$C = bx_1^{m_1}x_2^{m_2} \tag{35}$$

For example, the cost of a motor may be expressed by such an equation where x_1 is the HP and x_2 the rated motor speed.

Costs Based on Operations

Operation-based cost models were one of the earliest attempts in estimating manufacturing costs. Due to the type of information required, these models can be used effectively only in the final design stage. Dewhurst and Boothroyd (1987) describe cost models for machining and injection molding operations. Kiewert (1979) provides a model for cost of injection-molded thermoplastic parts.

The unit cost C of a printed wiring board assembly (PWBA) is given by Keys, et al. (1987) as:

$$C = \frac{C_M + C_{MT} + C_A + C_T + C_{TSR} + C_{RM} + C_F}{Y + Y_R(1 - Y)} + C_{FM} \tag{36}$$

where the costs are

C_M = materials
C_{MT} = materials testing
C_A = assembly
C_T = PWBA testing
C_{TSR} = troubleshooting and repair
C_{RM} = repair materials
C_F = fixed
C_{FM} = field maintenance

Y and Y_R are the initial test yield and the repair yield, respectively. They give further breakdown of these individual costs in terms of operations.

ACCURACY AND ERRORS DURING COST ESTIMATION

The total estimated costs of a unit (part, assembly, etc.) arise by summation of many separate items: costs of individual parts, production steps, and materials. If the individual estimated costs c_i have errors E_i which are uniformly distributed about the true cost C, the total estimated cost $c_{tot} = \Sigma_i^n c_i$ will exhibit a smaller total error E_{tot} than do the individual costs c_i. This comes about due to the partial cancellation of positive and negative errors.

Kiewert (1982) has shown that the total relative error for a unit is given by

$$e_{tot} = \left[\frac{\sum_{i=1}^{n} (c_i e_i^2)}{c_{tot}^2} \right]^{1/2} \tag{37}$$

where $e_i = E_i/c_i$ is the relative error in c_i. If the individual costs are equal ($c_1 = c_2 = \cdots = c_n = c$) and have the same relative error ($e_1 = e_2 = \cdots = e_n = e$), then Equation (37) simplifies to

$$e_{tot} = \frac{e}{n^{1/2}} \tag{38}$$

That is, the total error is $1/n^{1/2}$ of the individual error.

In order to determine the accuracy $e_{i,\,al}$ to which the individual cost c_i must be estimated so that the total cost c_{tot} has the error e_{tot}, Kiewert gives the equation (all quantities as percentages)

$$e_{i,\,al} = \frac{10e_{tot}}{c_i^{1/2}} \tag{39}$$

as shown in Figure 75 (Ehrlenspiel, 1985). Thus, if the allowable error in the total cost is to be $e_{tot} = 10\%$ on a cost $c_i = 25\%$ of the total cost, it must be estimated to within $e_{i,\,al} = \pm 20\%$. For a cost that is 1% of the total, the allowable error c_i is $\pm 100\%$.

SUMMARY

In this chapter we have discussed the various methods of cost breakdown and costing methods in use today. The advantages of using activity-based costing over traditional costing were shown by an example. Effects on costs at concept and embodiment stages were discussed. The effects of the number of parts on product costs and reliability were elaborated upon. This was followed by a discussion of life cycle costing. The next topic was function costing, with an example of designing to a cost goal. Parametric costing, the basis of many cost models used in both conceptual and embodiment stages, was described. Costs based on units of measure and on operations were described. Finally, a method of estimating errors in costing was given.

BIBLIOGRAPHY

Apgar, H. E., and J. M. Daschbach. "Analysis of Design through Parametric Cost Estimation Techniques." *Proc. ICED 87*, 2: 759–766. New York: ASME, 1987.

Empirical equations are developed relating cost parameters, such as costs and times, and physical or performance parameters, such as size, weight, or power.

Brown, R. J., and R. R. Yanuck. *Introduction to Life Cycle Costing*. Englewood Cliffs, N.J.: Prentice Hall, 1985.

Butler, A. C. "A Discussion of Accounting Theory from an Engineering Design and Manufacturing Viewpoint." *Design for Manufacturability 1994*, DE 67: 77–78. New York: ASME, 1994.

Cooper, R., and R. S. Kaplan. "Profit Priorities from Activity-Based Costing." *Harvard Business Review*, 69: 130–135 (May–June 1991a).

Cooper, R., and R. S. Kaplan. *The Design of Cost Management Systems*. Englewood Cliffs, N.J.: Prentice Hall, 1991b.

Cross, N. *Engineering Design Methods*. New York: Wiley, 1989.

Dewhurst, P., and G. Boothroyd. "Early Cost Estimating in Design." *J. Manufacturing Systems* 7(3): 183–191 (1987).

Ehrlenspiel, K. *Kostengünstig Konstruieren (Cost-Effective Designing)*. Berlin and New York: Springer Verlag, 1985.

"Engine Performance Improvements." *Automotive Engineering*, January 1995, pp. 15–20.

Ertas, A., and J. C. Jones. *The Engineering Design Process*. New York: Wiley, 1993.

Eversheim, W., and M. Shuppar. "Rechnergestütztes Erstellen und Aktualisierung von Relativkostenkatalogen" (Computer-Aided Preparation of Relative Cost Catalogs). *Ind. Anz.* 99: 958–960 (1977).

Fabrycky, W. J., and B. S. Blanchard. *Life-Cycle Cost and Economic Analysis*. Englewood Cliffs, N.J.: Prentice Hall, 1991.

Fowler, T. C. *Value Analysis in Design*. New York: Van Nostrand Reinhold, 1990.

French, M. J., and M. B. Widden. "Function Costing: A Promising Aid to Early Cost Estimation." *Design for Manufacturability 1993*, DE-Vol. 52: 85–90. New York: ASME, 1993.

Gallagher, P. F. *Parametric Estimating for Executives and Estimators*. New York: Van Nostrand Reinhold, 1982.

Hendricks, J. A. "Accounting for Automation." *Mechanical Engineering* 111: 64–69 (February 1989).

Keys, L. K., J. L. Balmar, and R. A. Creswell. "Electronic Manufacturing Process Systems Cost Modeling and Simulation Tools." *IEEE Trans. Component Hybrids and Manufacturing Technology* 10(3): 401–410 (1987). Electronic manufacturing process system cost modeling and simulation tools.

Kiewert, A. "Systematische Erarbeitung von Hilfsmitteln zum Kostengünstigen Konstruieren" (Systematic Working on Help Media for Cost-Effective Designing). Dissertation, Munich Technical University, 1979.

Kiewert, A. "Kurzkalkulationen und die Beurteilung Ihrer Genauigkeit" (Quick Calculations and the Evaluation of Their Accuracy). *VDI-Z* 124: 443–446 (1982).

Klasmeier, U. "Kurzkalkulationsverfahren zur Kostenermittlung beim Methodischen Konstruieren" (Quick Calculations for Costing during Methodical Designing). Dissertation, Berlin Technical University, 1985.

Li, M., C. A. Stokes, M. J. French, and M. B. Widden. "Function Costing: Recent Developments." *Proc. ICED* 93, 2: 1123–1129. Zurich: Heurista, 1993.

Pacyna, H., A. Hillebrand, and A. Rutz. "Kostenfrüherkennung für Gussteile" (Early Cost Estimation for Cast Parts). VDI Berichte *Konstrukteure Senken Herstellkosten—Methoden und Hilfen*, No. 457. Düsseldorf: VDI Verlag, 1982. Polynomial models for casting costs.

Rich, B. R., and L. Janos. *Skunk Works*. Boston: Little, Brown, 1994.

Sheldon, D. F., G. Q. Huang, and R. Perks. "Designing for Cost—Past Experience and Recent Developments." *J. Engineering Design* 2(2): 127–139 (1991).

Stewart, R. D., R. M. Wyskida, and J. D. Johannes, eds. *Cost Estimator's Reference Manual*. 2nd ed. New York: Wiley, 1995. This book has 19 chapters by various authors on cost topics—e.g., ABC, learning curves, parametric costing, design to cost, computer-aided cost estimating.

Tandon, M. K., and A. A. Siereg. "Manufacturing Tolerance Design for Optimum Life Cycle Cost." *Proc. ASME Manuf. Int. Conf.*, Dallas, March 29–April 1, 1992.

Thompson, W. "Computer Assisted Machining Economics." Technical Paper MM85-667. Dearborn, Mich.: Society of Manufacturing Engineers, 1985. Describes the application of parametric modeling to machining costs.

Ulrich, K. T., and S. D. Eppinger. *Product Design and Development* New York: McGraw-Hill, 1995.

Principles of Embodiment Design

The embodiment design phase involves the determination of shapes, motions or flows (if any), materials, and the manufacturing processes. Embodiment is one of the most complicated and difficult parts of design. This is the phase where the optimization (of shapes, sizes, materials, etc.) is carried out and considerable iteration and numerical analyses are performed. A great amount of concrete information regarding materials, processes, and costs is necessary.

Starting from the concept, the embodiment phase develops a definitive layout for the project in the form of general arrangement drawings. The chief considerations during preliminary design are shapes, principal material properties, and manufacturing processes. Requirements for the parts are developed, keeping in mind the system requirements. Part design is completed. Some preliminary computations are carried out at this stage to help in making decisions regarding general shapes, sizes, motions, and spatial arrangements. Several designs are developed and evaluated on the basis of technical and economic requirements. Prototypes are built in this phase, prior to the development of the final design.

RULES AND CHECKLIST

Pahl and Beitz (1988) provide the following three rules for embodiment design:

- Clarity
- Simplicity
- Safety

These rules are used to judge each of the items in the requirements list and the embodiment checklist. The main items are given below. These are the various "DFX" items. The term originated in the design literature of the 1980s—"design for X," where X can be M (manufacturing), A (assembly), C (cost), Q (quality), etc. These are all in a sense attributes of good designing.

Among the questions which should be asked related to the requirements list and the embodiment checklist are the following:

Function

- Are the main function and secondary functions fulfilled?
- What other functions are needed?
- Is there an unambiguous input(s)-to-output(s) relationship?
- Is the number of functions a minimum number?

Shape (Form) Design

- Check for strength, stiffness, expansion, corrosion, wear, stability.
- Check for robustness. Can loads be defined and calculated under all conditions?
- Aim for simple shapes—ease of analysis and manufacture.

Ergonomics

- Check the physical features and layout for potential for stress, fatigue, and injury.
- Check for ease of use.

Safety

- Check the system for safety of operation, of the parts and function.
- Check for safety of all potential users.

Manufacture

- Aim for ease of production and assembly.
- Know what production facilities are required and available.

Transport

- How is the product to be packed, transported, stacked?
- Look at sizes, weights, regulations.

Operation

- Check for ease of operation, unambiguous signals, simple layout of controls.
- Check for vibration, noise, and chemical, and thermal pollution.
- Ensure adequate training of personnel.

Maintenance

- Check for ease of maintenance and availability of spare parts.
- Ensure adequate training of personnel.

Environment

- Consider pollution of the environment during manufacture, operation, and disposal.
- Does the design allow for ease of reuse, recycling, and remanufacturing?

Costs

- Can the anticipated cost target be achieved?

EMBODIMENT PRINCIPLES

We shall mention the chief principles of embodiment here. They will be elaborated upon and applied in the cited and given examples.

Function Integration and Separation

By integrating several functions into one part, we reduce the number of parts. We have discussed this in Chapter 6. Function separation (division of tasks between two or more parts) can also be used at times to improve designs.

Forces and Deformations

The proper geometry and the choice of material lead to designs which have:

- Uniform strength, for material saving
- Stiffness or flexibility, as desired
- Matched deformations—desirable in the case of joints
- Balanced forces—e.g., double-helical gears, cone clutches

Self-Help

The principle of self-help refers in design to those cases where the aim is to make solutions self-reinforcing, self-balancing, or self-protecting.

EMBODIMENT PROCEDURES

There are certain requirements that have a dominant impact at the embodiment stage. These are the requirements that determine the size of the product, the arrangement of its components, and the materials used.

Size of the product is influenced by requirements such as capacity, throughput, interface with adjacent devices or supports, or ergonomics. Material properties (and, therefore, choice) are dictated by factors such as loading, environment, and aesthetics. These are discussed in Chapter VIII in more detail. A device's interfaces with humans and with other devices and supports determine its spatial envelope. These can now be shown on a drawing to scale.

EMBODIMENT VARIANTS

It was shown earlier that at the conceptual stage, concept variants could be generated by rearranging function structures, by varying physical effects, and by choosing different solution principles. Likewise, at the embodiment stage variant designs can be obtained by considering variations in the three basic embodiment quantities: shapes (surfaces), motions, and materials. Table 26 shows some of the possibilities that can aid in generating new ideas. Examples that follow will show the application of this scheme.

• • • • •

EXAMPLE
DESIGN VARIANTS OF A COUPLING

An application of this classification scheme was shown in the variable speed transmission example in Chapter 3. A highly creative example of this scheme is given by Ehrlenspiel and John

TABLE 26 Classification Scheme for Embodiment

Criterion	Examples
Surfaces	
Type	Point, line, surface, body
Shape	Line—straight, curve, etc.
	Surface—circle, triangle, rectangle, etc.
	Body—cylinder, cube, rectangular, etc.; hollow, solid, grooved, etc.
Position	Axial, radial, vertical, horizontal, etc.; parallel, sequential
Size	Small, large, narrow, wide, etc.
Number	Simple, double, multiple
Motions	
Type	None, translation, rotation, etc.
Nature	Steady, nonuniform, oscillation, etc.; planar, spatial
Direction	Along an axis, about an axis
Magnitude	Displacement, velocity, etc.
Number	Simple motion, composite motion
Constraint	Free, guided
Material	
State	Solid, liquid, gas
Type	Rigid, elastic, viscous
Loading	Axial, bending, shear
Shape	Solid, granular, powder

(1987), in which new coupling designs are developed by creating variants of an Oldham coupling. The basic Oldham coupling, which is able to compensate for axial offsets, consists of two halves with slots and a sliding block. Ehrlenspiel and John show the following sequences of variations:

- Size of slots; position of slots; number of slots
- Method of connection between the halves and slider to parallel; nature of coupling from rigid to flexible (springs); from flexible solid to fluid (hydrostatic)
- Nature of coupling motion from sliding to rolling; position of rollers; nature of coupling from rigid to flexible (ring)

These steps are shown in Figure 80, where some of the variants are depicted. By using a classification scheme as shown earlier in this book for solutions search, further variants are generated. See, for example, Figures 16 and 17. Ehrlenspiel and John show the final design as a patented coupling which can compensate for radial, angular, and axial offset.

• • • • •

EXAMPLE
DESIGN VARIANTS OF A YOKE

We will now illustrate the many variants that are possible at the embodiment stage with the example of a simple mechanical element—a yoke. The function of this yoke is to constrain two pins as follows:

- Pins' axes are mutually perpendicular.
- Pins' axes are offset a certain distance.
- Pins are allowed motion along and about their axes.

The pins to be constrained are shown schematically on the left side in Figure 81. Their shape and the constraint requirements determine the *active surfaces* of the yoke, shown on the right side in Figure 81. These are the inner surfaces which constrain the pins, i.e., the surfaces on which the pins bear, which carry the loads, etc.

We now look at the possible shapes for the yoke—i.e., the other surfaces, which provide shape to the element—the manufacturing process(es), and the material.

Figure 82 shows some of the embodiment possibilities. These are classified here primarily according to the manufacturing method. The other classification schemes used are:

- Raw material used (shapes, chemical composition)
- Made in one part versus made in several parts, with permanent versus separable joints

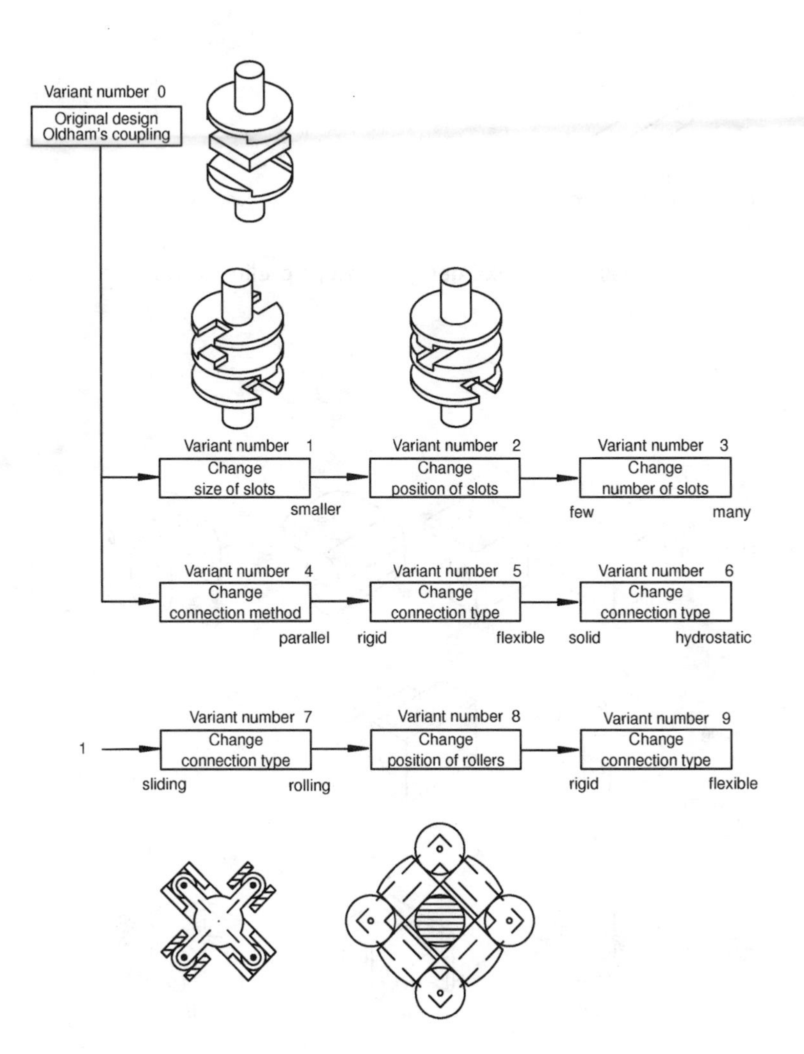

FIGURE 80. Variants of Oldham coupling. (Ehrlenspiel and John, 1987)

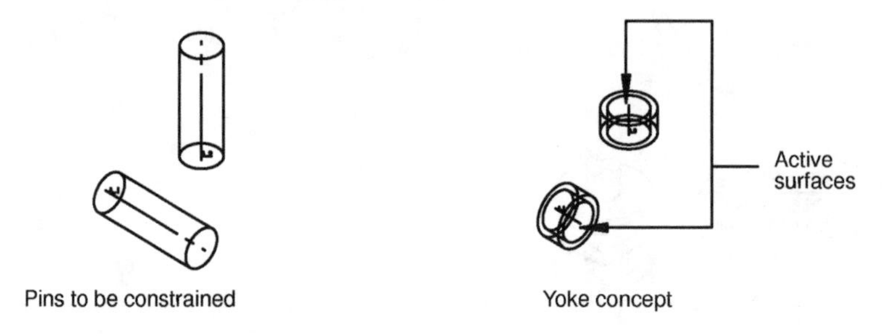

Figure 81. Yoke design concept preliminaries.

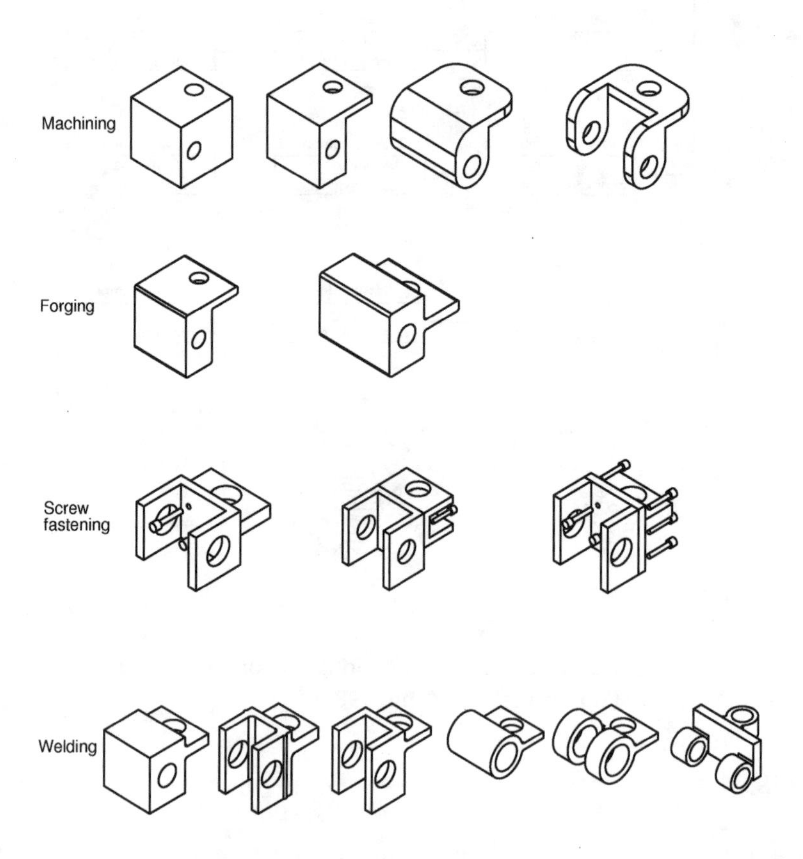

Figure 82. Embodiment variants for yoke.

The process possibilities shown in Figure 82 are:

- Machining—from a block
- Forging
- Stamping
- Welding of pieces of various shapes

Screw fastening of pieces of various shapes

There are, of course, other possible manufacturing methods, including:

- Castings of various types, using metals
- Plastic injection molding

Which is the best alternative? The answer to this question depends on other requirements, such as loads applied, environment, or quantities required. These and other topics will be addressed in the following discussion. If only one of a kind is needed, machining from a block is likely the best choice. If a process other than machining is used, then whether the holes will have to be machined depends on the tolerances required and the feasible tolerances of the material and the manufacturing method.

• • • • •

PART DESIGN

Traditional design methodology considers the processing requirements after the design is complete. The advantages of concurrent product and process design include higher quality, shorter time to market, lower costs, and optimized production. The design of parts brings about the interaction of three important factors: the part geometry (shapes and sizes), materials selection, and choice of appropriate manufacturing process—the embodiment "triad"—as depicted in Figure 83. Only the major requirements are shown in this figure. A methodical design procedure is required which takes into account all of these factors and determines the optimum combination, leading to the final part design (Durham and Hundal, 1991). Part design begins with defining the requirements for the

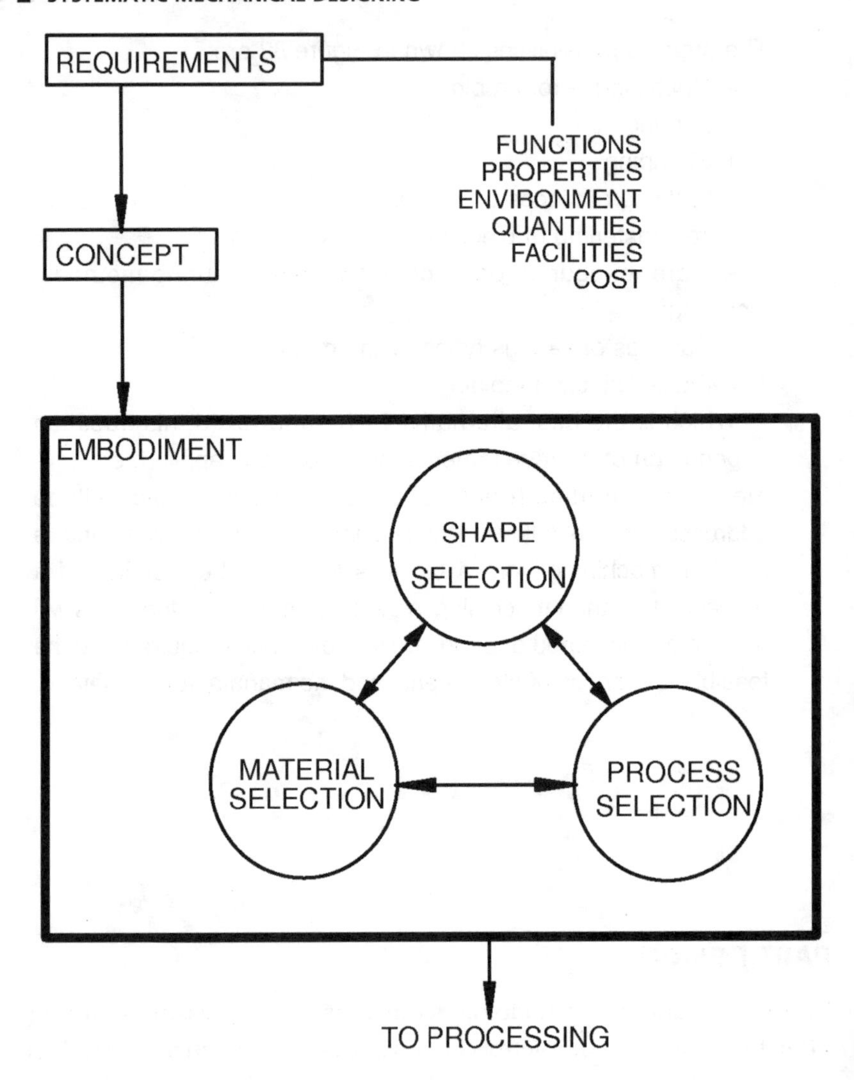

FIGURE 83. Concurrent selection of shape, material, and process.

part and developing a concept for it. Only then is a rational embodiment of the part possible.

The first stage of methodical design includes the problem definition and the generation of a list of requirements. The most important requirements are those which pertain to the product's use, i.e., the function(s). Additional requirements that must be considered include those that are relevant to the selection of the manufacturing process, the material, and the final geometry. These include any specific properties—e.g., hardness or conductivity—the quantities in which the part is to be made, the facilities available, and, of course, cost. The function of the part can be the transmission of energy (mechanical, thermal, fluid) the purely aesthetic, or some combination of these. In order to maintain as free a selection base as possible, function should determine only the rough shape of the component. Geometry should be specified only for those essential shapes or dimensions required to satisfy the function.

General Considerations

Figure 84 shows the steps in part design in more detail. The function determines the rough shape of the part. Shapes of the part interfaces are determined by the constraints on the interfaces, e.g., contacting surfaces,

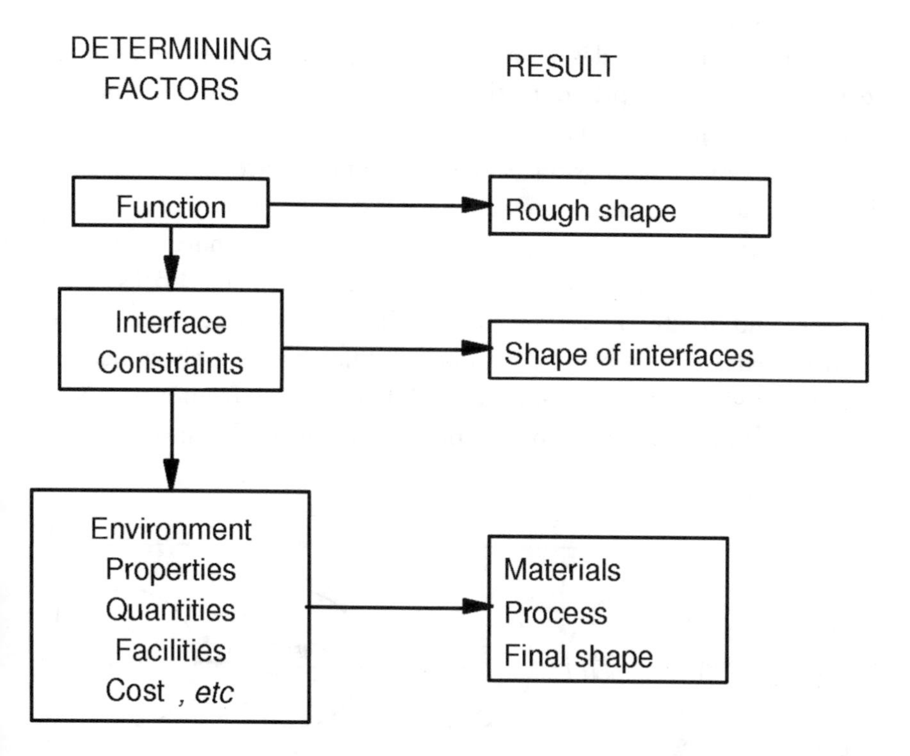

Figure 84. Steps in part design.

energy input/output surfaces—in short, all the active surfaces at the interfaces. The final shape, the manufacturing process, and the material are then determined by considering (1) the environment in which the part is to be used, including all types of loading, (2) the properties it must have (strength, hardness, corrosion resistance, etc.), (3) the quantities in which it is to be manufactured, (4) the facilities available for its manufacture, and (5) all of the above balanced with the desired cost of the part.

The systematic procedure for part design (Matousek, 1963) is the sequence given in Figure 85. Initially, a material is chosen, followed by a manufacturing process. Then shape design principles are applied for the chosen material and process combination to obtain the optimum shape. The final determinant of the design is the part cost—manufactured cost, or life cycle cost (Hundal, 1993). The feedback loops in Figure 85 show the iterative nature of design. In this instance a number of material/manufacturing process combinations are considered before a final decision is made.

The choice of material is dictated by the requirements. A process is then selected that is appropriate for the material. The compatibility between processes and materials is discussed in Chapter IX; see Figure 147 and the accompanying discussion. Shape design procedures are available for different manufacturing processes (and materials). The final decision is based on the cost. Therefore, for a given concept we have a choice of materials; for each material, one or more manufacturing processes may be used; and finally, the corresponding shape design rules are followed. Figure 86 shows the multiplicity of possibilities for the embodiment of a part. For a given material M_i it is possible to choose from a number of process combinations P_i. Thus, P_i could denote a combination such as:

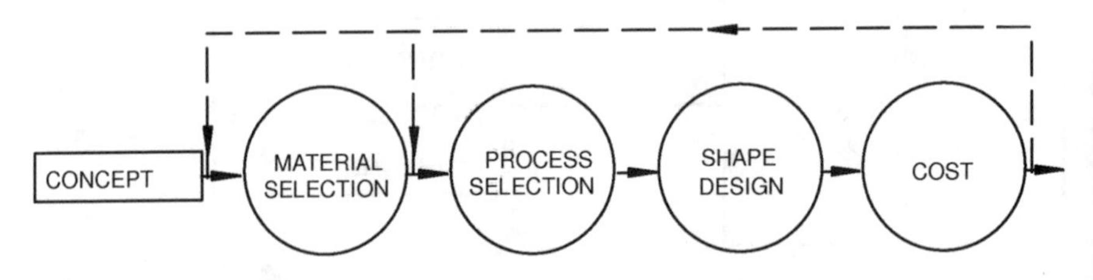

FIGURE 85. Systematic procedure for part design.

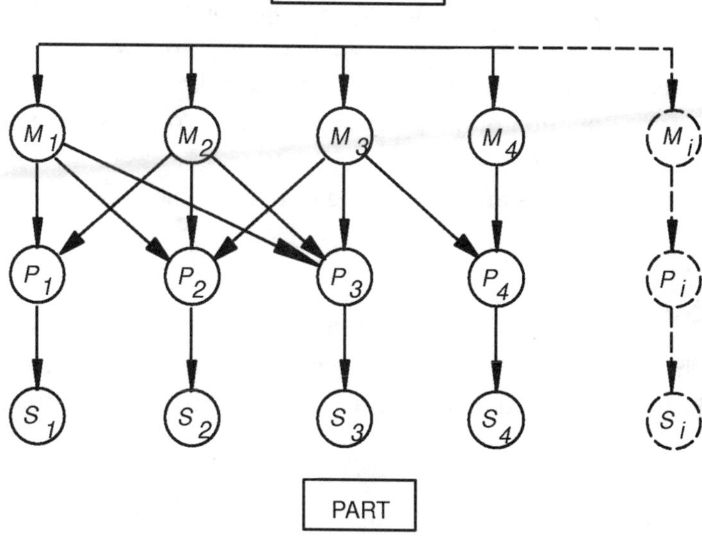

M_i MATERIAL
P_i PROCESS COMBINATION
S_i SHAPE

Figure 86. Choice of materials, processes, and shapes in embodiment design.

$$P_i = \text{forging} + \text{machining} + \text{surface treatment}$$

For each material and process combination there is an optimum shape S_i dictated by the requirements and shape design principles. It is recognized here that assembly may be required to produce a part.

We now take a brief look at shape design. The other two activities involved in embodiment—material and process selection—will be discussed in Chapters VIII and IX. We will find that each depends on the other two, as well as on other design factors. This will be followed by a design example in order to elaborate on their interaction.

Shape Design

Since shapes are (to the eye) the most obvious results of designing, it is often believed that the major design activity is the design of shapes, or form design. However, as anyone except a beginning student knows (and from the foregoing discussion), shape design is only a part of the overall design effort. There is an old saying—form follows function—that originated from the study of natural objects and phenomena. In human-made objects, function plays the decisive role not only in the concept but also in the embodiment.

Major factors affecting shape design will now be discussed.

Concept

The chosen concept defines the part in a sketch form, showing only the essential functions. Thus, it has a significant effect on the final shape.

Loading

Shape design principles call for an efficient flow of energy through material. This implies a uniform stress in the case of mechanical loading and a uniform heat flow in case of thermal loading.

Material

The two types of material properties that have the greatest influence on shape design are properties of strength and flexibility and processibility. The designer aims to achieve shapes that utilize the most favorable properties and minimize the effects of the least favorable properties. Let us consider cast iron as an example. From the physical property point of view, it is stronger in compression than in tension; thus, we aim for minimizing tensile stresses. From the processibility viewpoint; casting can yield intricate shapes; we use castings when such shapes are needed.

Figure 87 shows two examples of appropriate use of material properties in embodiment. Parts are designed differently for materials that are stronger in tension than in compression, and vice versa. The low modulus of elasticity of plastics is used to advantage in reducing the number of parts by integrating functions into fewer parts, e.g., the "living hinge" and the snap fit.

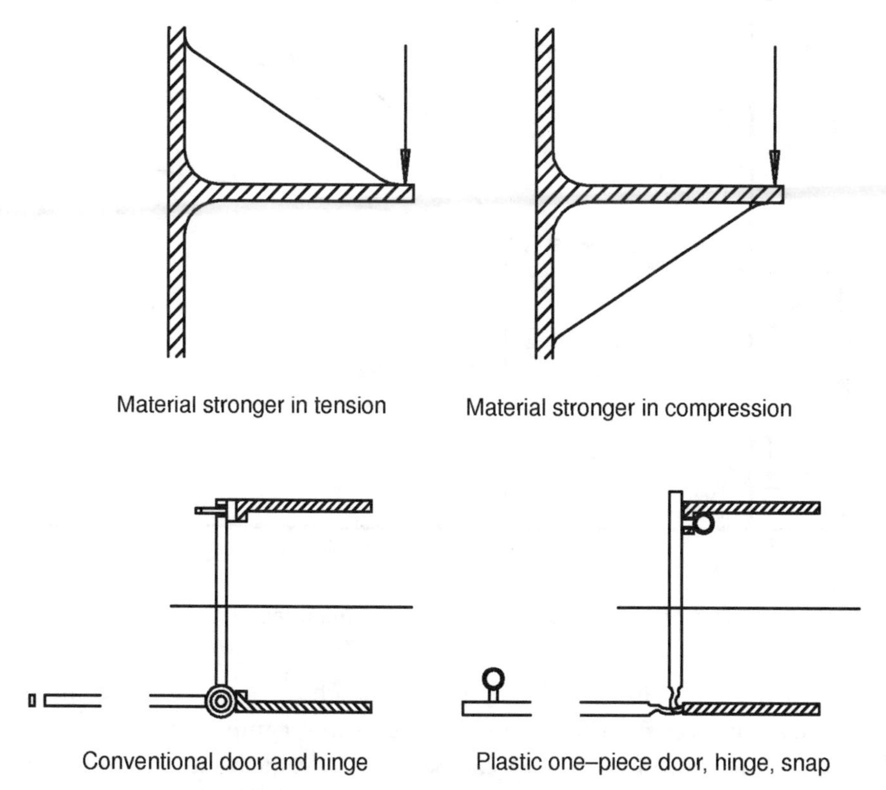

FIGURE 87. Effect of material properties on embodiment.

Manufacturing Process

As mentioned earlier, the choice of material narrows down the processes that may be used, as well as the shapes that may be economically produced. The final choice is dictated by (1) whether the required properties can be obtained and (2) the cost of the part. Various textbooks (Kalpakjian, 1992; Matousek, 1963; Pahl and Beitz, 1988) give design rules, hints, and examples of "right/wrong," "good/poor" procedures.

Figure 88 shows a simple example of how the quantities produced influence the method used to manufacture a part. Suppose a given part can be made by molding or machining a solid piece. Machining costs are characterized by low fixed costs of jigs and fixtures and high variable cost of labor for producing each part. On the other hand, molding involves high initial costs of making the patterns and mold, followed by low per-

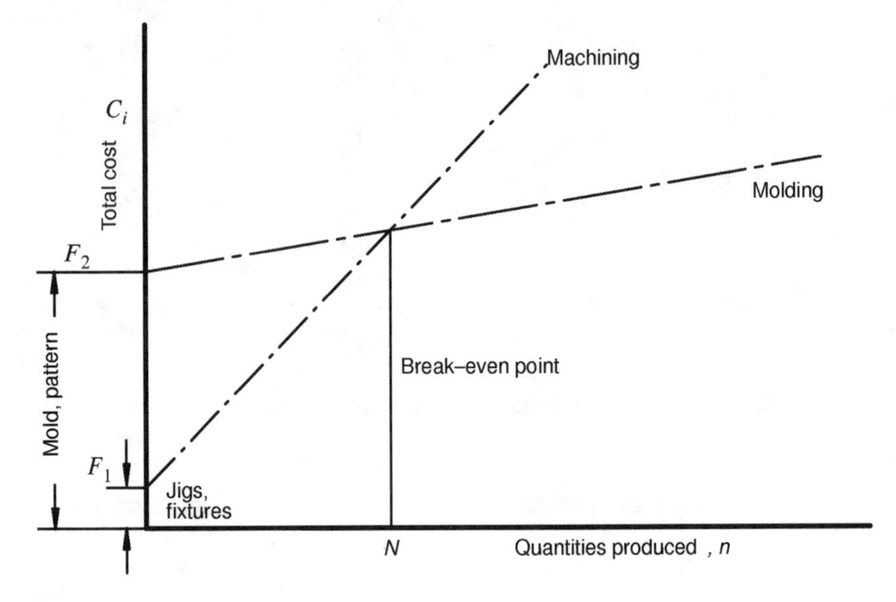

FIGURE 88. Effect of quantities produced on the manufacturing method.

part costs. Thus, for small quantities it is cheaper to use machining, whereas for large quantities molding yields lower unit cost.

The data shown in Figure 88 can be used to decide between two manufacturing processes. Let F_i, V_i, and C_i be the fixed costs, the variable costs per unit produced, and the total costs of producing n units, respectively, for processes $i = 1$ and $i = 2$; then

$$C_1 = F_1 + V_1 n \qquad \text{and} \qquad C_2 = F_2 + V_2 n \tag{40}$$

The break-even quantity N can be found by setting $C_1 = C_2$ in Equation (40) and solving:

$$N = \frac{F_2 - F_1}{V_1 - V_2} \tag{41}$$

Space and Size

Space requirements and the size of a part are generally dictated by the surrounding parts. The smaller the part—the less material used—the cheaper it is. (We cannot directly scale up from a small part when designing a larger part. Larger parts tend to be of a lattice type, with hollow spaces.)

Standardization

Standardization reduces product costs by minimizing the design costs for parts and assemblies. The number of different parts required is smaller, and therefore so is the size of inventories. It also reduces the time to market and improves reliability. Designers may resist standardization in the belief that it restricts their freedom. However, lack of standardization would lead to chaos.

Surface Properties

If surface treatments such as hardening or plating are foreseen, the shape of the part must be such as to make it easier for such treatments to be applied and minimize the effects of possible shrinkage and distortion.

• • • • •

EXAMPLE
COST OF PRODUCING A GEAR BY DIFFERENT PROCESSES

Figure 89 shows the manufacturing cost of a gear as a function of quantities produced, made by three different processes: machined out of a blank, cast and machined, and forged and machined. The high cost of forging dies drives up the cost per unit for low lot sizes, but forging becomes advantageous for high-volume production. The cost structures for the three methods are shown in Table 27.

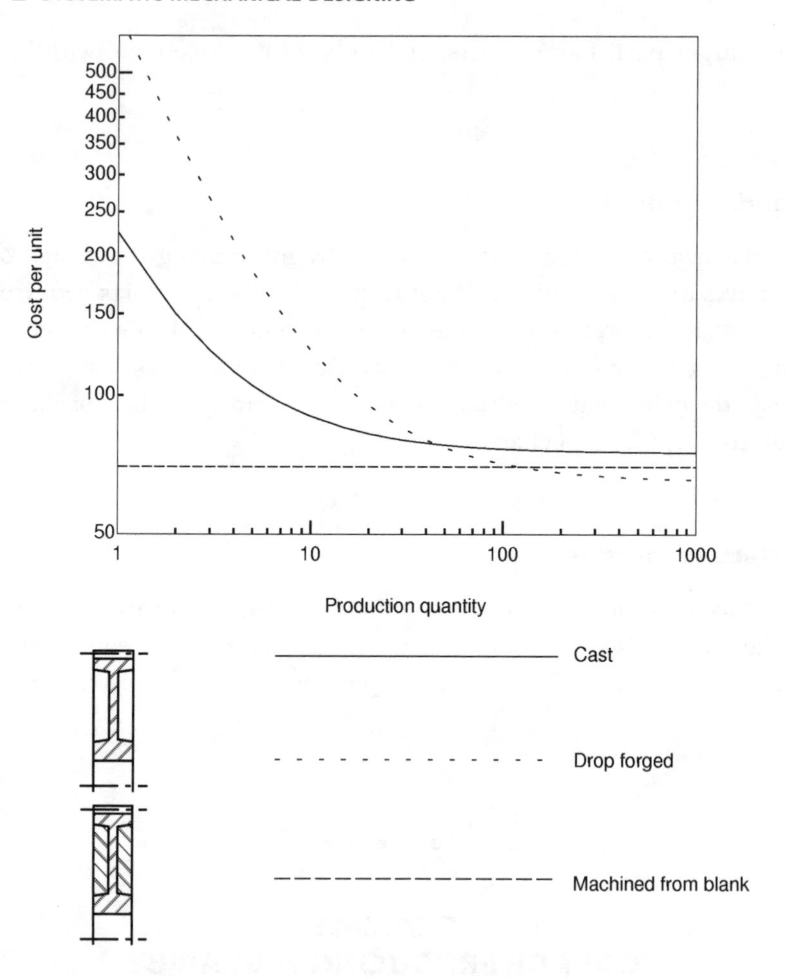

FIGURE 89. Gear costs for different production methods.

TABLE 27 Cost Structures for Three Production Methods for Gear

	Cast	Drop-Forged	Machined From Solid
Pattern, die, or jig	150	600	0
Material	15	15	20
Labor	15	5	0
Machine	45	45	50
Material + labor + machining	75	65	70

The number of units of a given design that becomes the most economical is called the *economic order quantity* (EOQ). Specific examples are given later of computing the cost by different manufacturing methods.

• • • • •

MANUFACTURING CONSIDERATIONS IN DESIGN

By manufacturing we understand the following operations:

- Production of individual parts
- Assembly of parts

Thus, production cost may also be expressed as the sum of two costs:

$$C_{\mathrm{pr}} = C_{\mathrm{pt\ pr}} + C_{\mathrm{pt\ as}} \quad (42)$$

where

$C_{\mathrm{pt\ pr}}$ = cost of part production
$C_{\mathrm{pt\ as}}$ = cost of part assembly

Of all the DFX practices, design for manufacturing has the greatest impact on a product's characteristics. Design for manufacturing directly includes the design of parts for production and assembly. Indirectly, it also implies designing under the constraints of time, cost, and quality. As was shown in Figure 61 in Chapter VI, manufacturing activities account for 30 to 40% of the total manufactured cost of a typical mechanical product.

While manufacturing considerations are already present to a certain extent during the conceptual phase, it is during the embodiment phase that concrete steps must be taken so that the product can be manufactured in short time, at low cost, and with high quality. As was stated earlier, the design must proceed with active participation by manufacturing, cost engineering, and purchasing (see Figure 44 in Chapter IV). Improved cost estimates, which are possible at this stage, allow alternative designs to be evaluated and the most cost-effective design to be achieved.

Reducing the manufacturing costs implies, according to Equation (42),

- Reducing the part production costs, and
- Reducing the assembly costs

Reducing Part Costs

Part design was addressed in a previous section. The most important considerations regarding costs of parts are:

- Materials (and their costs)

which are closely related to:

- Manufacturing processes
- Number of different fabrication steps
- Quantities produced
- Use of standard parts (see the section "Effect of the Number of Parts" in Chapter VI)

Fewer fabrication steps are possible, for example, by using near-net-shape forging and casting, sheet metal stamping, and plastics molding and using materials that do not require protective coatings. Materials and manufacturing processes are discussed in later chapters.

Reducing Assembly Costs

Assembly costs arise primarily out of:

- Time required for assembly
- Equipment and tools required
- Workers' dexterity, in the case of manual assembly

Boothroyd and Dewhurst (1987) have described the various proven measures for design for assembly. Assembly requires paying attention to:

- Retrieval of parts from a supply, for example, from a bin or conveyor
- Handling the parts
- Insertion or mating of the parts

Thus, the number of parts and fasteners and the shapes of parts play the decisive role in reducing assembly costs.

Boothroyd and Dewhurst provide a method to determine if the existence of a part is justified as a separate entity, or if it should be eliminated. The three criteria applied to make this determination are:

- *Relative movement.* Does the part need to move relative to adjacent parts? Small movements that can be accommodated by elastic deformation are not counted.
- *Different material.* Does the part have to be made of a different material for a specific reason, e.g., corrosion or wear?
- *Separability.* Is it ever necessary to separate the part from the assembled product for any reason, such as maintenance?

Figure 90 shows the logic for deciding whether a part is essential or can be eliminated. By applying this test the number of parts can be reduced to the theoretical minimum. Integrating functions into fewer parts has several advantages:

- Reduced assembly time and tooling
- Better control over dimensions and tolerances
- Reduced number of dies or molds

However, an estimate of the total costs must be made prior to such a decision.

Boothroyd and Dewhurst give a measure, which they call the "DFA index," of the assembly efficiency of a product:

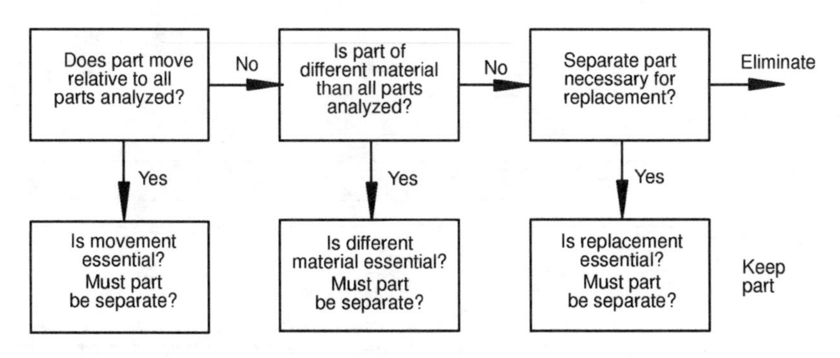

FIGURE 90. Criteria for essentiality of a part.

$$\text{DFA index} = \frac{N_{\text{min, th}} T_h}{T_{\text{as}}} \tag{43}$$

where

$N_{\text{min, th}}$ = theoretical minimum number of parts in the product (found by applying the above test)

T_{as} = estimated total assembly time

T_h = theoretical minimum time required for handling and insertion of a part "perfectly suited for assembly" (an average value of 3 s may be used)

The lower this efficiency value, the higher the potential for improving the design, as far as assembly time is concerned. Boothroyd and Dewhurst provide assembly times for typical components. The ranges of assembly times for screws, snap fits, pins, and springs are shown in Figure 91.

On the basis of the number of components, we can obtain a measure of the expected improvement by reducing their number. This improvement potential is IP given by

$$\text{IP} = \frac{N_{\text{act}} N_{\text{min, th}}}{N_{\text{act}}} \tag{44}$$

where N_{act} = the actual number of parts

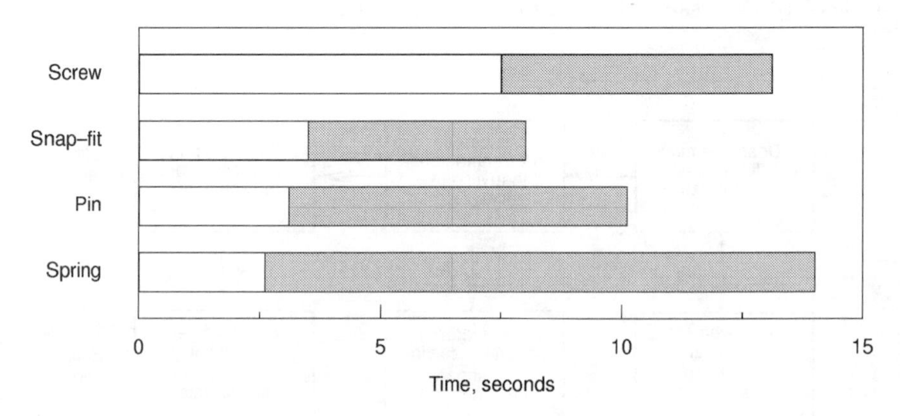

Figure 91. Range of assembly times. Shaded portions show ranges of variation for different parts and operations.

Depending on the improvement potential, a design may be classified on a scale from outstanding to poor, as shown in Figure 92.

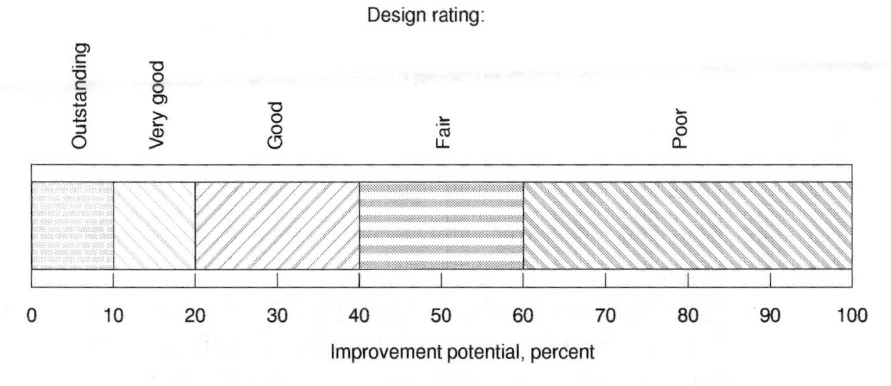

FIGURE 92. Design rating based on number of parts.

• • • • •

EXAMPLE
DESIGN OF A PRINTER FOR EASE OF ASSEMBLY

The IBM Proprinter is an example of a product designed for quick and easy assembly. Table 28 shows the data on the number of parts, number of assembly operations, and assembly times for the Proprinter compared with the older Epson MX80 printer. As we can see, the differences in these items between the two printers

TABLE 28 Comparison of Assembly Times of Two Printers

Item	Epson MX80	IBM Proprinter
Number of components and subassemblies	150	32
Number of assembly operations	185	32
Total assembly time	30 min	3 min

are dramatic. The theoretical minimum number of components for the Epson printer was found to be 40. Thus, the improvement potential is

$$IP = \frac{150 - 40}{150} = 73\%$$

which implies a poor design from a parts viewpoint.

The designers of the Proprinter used the following procedures for its design:

- Design is layered for parts to be placed on top of one another.
- Parts are designed so they cannot be installed if the previous part is missing or improperly inserted but will only fit into their right position.
- Each part is firmly attached by a robot. Thus, no loose ends are left that must be held down for later assembly operations.
- Extension springs are eliminated from the design and replaced by a cantilever spring molded into the side frame.
- Screws and other fasteners are eliminated and replaced by snap fits.
- The belt drive is eliminated because of its difficulty of assembly. It is replaced by a helix drive system, which is easier to assemble.
- The part count is minimized by combining multiple functions into single parts. One side frame replaces 14 parts.

● ● ● ● ●

GUIDELINES IN DESIGNING FOR ASSEMBLY

The general points to observe in designing for assembly are:

- Aim for the minimum number of parts. Find the theoretical minimum number of parts. Aim changes in design toward this number.

- Use the minimum number of fasteners. Use snap fits where possible.
- Design with a dominant base part or chassis. Such a part allows quicker assembly by minimizing the number of joining operations.
- Avoid having to reposition the assembly during the process.

To ease retrieval of parts:

- Set up the assembly procedure for the feed method that will be used: manual (for short runs), robotic (medium runs), or automatic machine (large lot sizes).
- Design parts to avoid tangling and nesting. Avoid using parts that are too flexible.

To ease handling of parts:

- Parts with the most symmetry require the least amount of orientation.
- End-to-end symmetry allows either end to be inserted.
- Axial symmetry avoids having to turn the part unnecessarily.
- Parts that intentionally have no symmetry should be made easy to orient by providing distinguishing features.
- Parts that are nearly alike should have some distinguishing feature, e.g., color coding.

To ease mating of parts:

- Allow assembly by requiring straight-line motions along one axis only. Ideally, this should be the vertical axis (z axis). Straight-down movements are the easiest to make.
- Enable one-hand assembly. This implies an easier-to-assemble design.
- Enable no-tools assembly. Having to use tools in order to handle the parts is extra work.
- Provide features to ease insertion and allow self-alignment, e.g., chamfers.

The above points are illustrated by the designs in Figure 93.

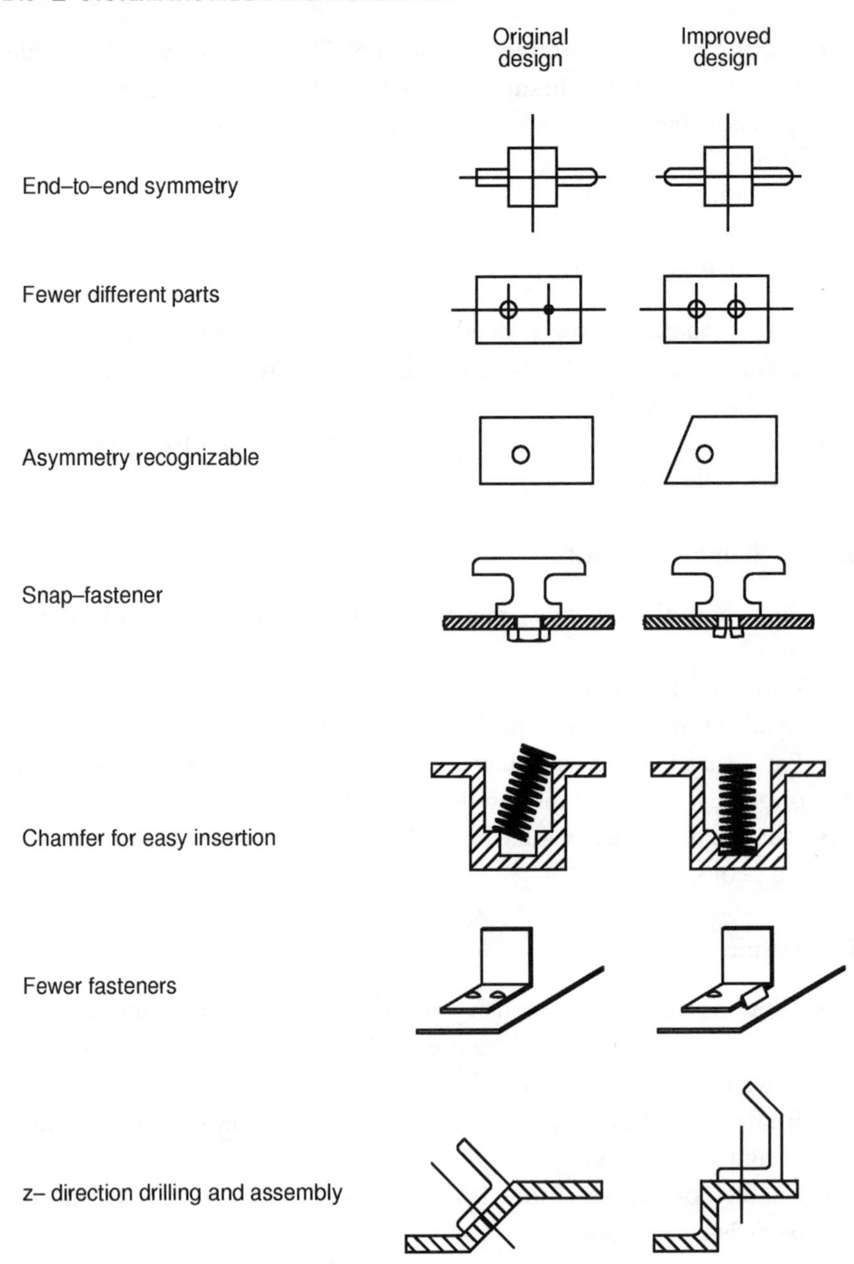

FIGURE 93. Examples of designing for easier assembly.

• • • • •

EXAMPLE
AIRCRAFT RAM AIR DOOR ASSEMBLY

The Douglas Aircraft Company redesigned the ram air door assembly, an inlet passage by which outside air enters to supplement the cabin air-conditioning system, for the MD-11 aircraft (Ashley, 1995a). By using design for assembly techniques, the company achieved the reductions shown in Table 29.

TABLE 29 Comparison of Ram Air Door Assembly Designs

Item	Before	After
Number of parts	2172	1383
Number of assembly operations	4038	2649
Weight reduction	—	107 lb

• • • • •

EXAMPLE
AIRCRAFT WASTE PIPE AND WIRE HARNESS BRACKET

As another example, the Douglas Aircraft Company redesigned the waste pipe and wire harness bracket for the MD-90 aircraft (Ashley, 1995a). By using design for assembly techniques, the company achieved the reductions shown in Table 30. The original design is shown in Figure 94a. Bolts and nuts are not shown in this figure. The bracket after redesign is shown in Figure 94b.

TABLE 30 Comparison of Waste Pipe and Wire Harness Bracket Designs

Item	Before	After
Number of parts	15	3
Number of assembly operations	210	8
Weight	2.1 oz	0.8 oz
Total assembly time	46 min	3 min
Total product cost	$64	$4.74

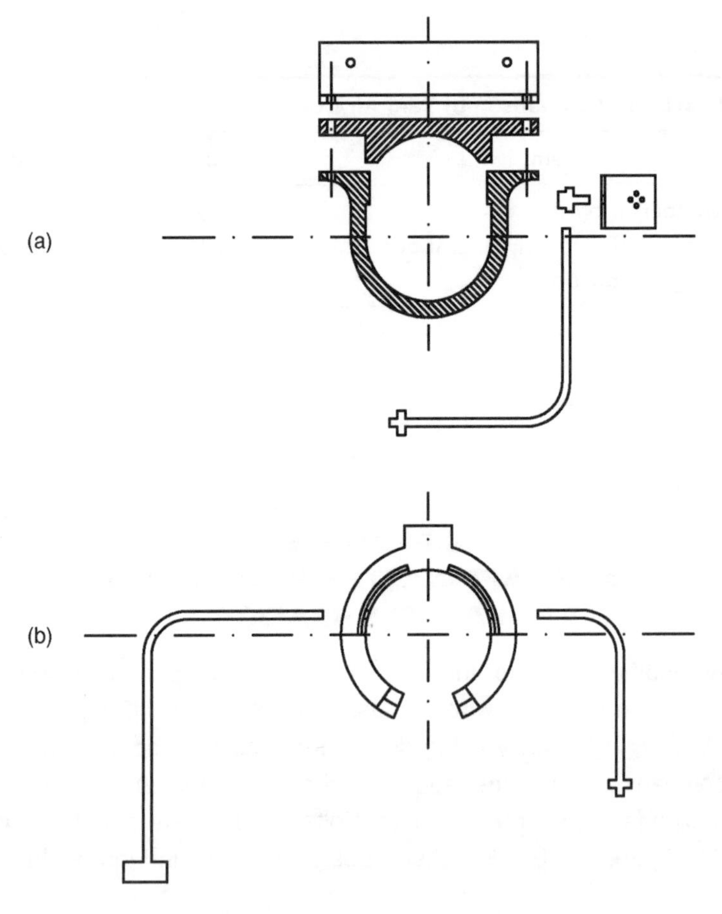

Figure 94. Waste pipe and wire harness bracket before and after redesign.

Other Examples

The design literature of the mid-1990s abounds with references to gains in productivity attained by using design for manufacturing and assembly (DFM&A) techniques. Carter (1995) presented a number of examples from the experience at General Motors. Table 31 is an excerpt from his paper.

Carter also shows significant savings in the production machines and tooling when they were redesigned using DFM&A principles. He cites examples of injection mold tooling, the brake/fuel bundle fixutre in the assembly plant, and the oil hole drill in the engine plant.

The Chrysler Corporation produced a new small car in 1994—the Neon—which replaces the older Shadow and Sundance models. Table 32

TABLE 31 Savings at GM by Using DFM&A

Application	Item	Savings
Condenser, radiator, fan	Piece cost	14%
	Assembly time	37%
Front bumper system	Number of parts	49%
	Assembly time	56%
Cross member cradle retainer	Number of parts	17%
	Assembly time	35%
Front door	Number of parts	50%
	Unit cost	13%
Rear independent suspension	Number of parts	37%
	Assembly time	30%

(Carter, 1995)

TABLE 32 Example of New Small-Car Design at Chrysler

Item	Old Model	New Model
Total number of parts	>3000	<1500
Sheet metal scrap produced	45%	33%
Average number of die operations on a stamped part	>3	1.8

summarizes the design advances of the new model that reflect lower cost and more efficient manufacturing.

PROTOTYPE DEVELOPMENT

Prior to finalization of the design and subsequent manufacturing, a prototype of the device is almost always prepared; see Figure 3 in Chapter II.

A prototype, or model, can encompass a wide range of possible embodiments—from a full-scale working device (100% embodiment) to an analytical model (0%; no embodiment). In the first instance, the physical model is said to be comprehensive; on the other hand, the designer may wish to focus on only a few features of the device at a time. A model that emulates the product only in size and appearance is called a mockup and is used for verifying fits and clearances. Models of parts can also be used as patterns for producing various types of castings.

During the product development process, prototypes serve different purposes at different stages. A physical model provides the "feel" and verifies the workability in a way that can seldom be achieved in an analytical or computer model. This type of model is particularly useful for demonstrating the product to customers as the project proceeds through its various phases. As the development proceeds, the experimental, alpha, beta, and preproduction prototypes are built, incorporating increasing levels of detail and consumer feedback (see Ulrich and Eppinger, 1995). Developments in CAD solid modeling (see Appendix G) have been used to advantage to produce computerized mockups of complicated devices such as airplanes and submarines. These models can be used effectively to verify the openings and clearances for personnel and equipment.

Clausing (1994) warns of the "hardware swamp" that is formed when too many prototypes are made and not enough resources are on hand to test and debug them. He recommends an early optimization of robust functionality to minimize the number of prototypes to be made. Robust functionality ensures that product and process performance will remain insensitive to the variation of a few important design parameters over a wide range.

Prototypes must be built in high-quality shops with highly trained personnel, rather than in job shops. In this way prototypes are developed in shorter time and are of higher quality, both of which lead to a reduction in product development time.

Rapid Prototyping

Models of parts can, of course, be made with traditional materials and manufacturing processes. In the last few years "rapid prototyping" processes have been developed, by which complex shapes can be produced in relatively short times (Burns, 1993). Rapid prototyping (RP), also known as desktop manufacturing, automated fabrication, tool-less manufacturing, and free-form fabrication, is playing an increasingly significant role in the time- and cost-driven development of products. Because process time is significantly reduced, more design iterations are possible, which reduces product development risk. Lower risk translates into greatly reduced program costs. In some cases cost savings of up to 90% have been documented (Ashley, 1995b).

Developments in rapid prototyping include not only the process itself, which generally involves sintering, layering, or deposition of material, but also improved software and materials technology. RP parts are no longer made just for appearance. They are already in use for form and fit studies, casting and tooling patterns, and as fully functional prototypes for mechanical testing, where they must withstand real-world physical loading to demonstrate casting and tooling patterns.

A summary of the dominant processes used in RP at the present time is given by Ashley (1995b). Commercial RP systems available to small consulting design firms, manufacturers, and service bureaus in the United States, as of this writing, are:

- *Stereolithography (STL).* This process uses liquid polymer which is cured by an ultraviolet laser beam. After a layer of solidified resin is produced, another coat of liquid is applied for the next layer. The platform moves vertically. Support structures are necessary to prevent overhangs from collapsing. Half of all installed systems are of this type.

- *Solid Ground Curing (Solider).* The Solider process uses photo-curable polymers and masking to produce complete models hardened by ultraviolet light. The vat moves both vertically and horizontally. Parts built are surrounded by wax; thus, no support structures are needed.
- *Selective Laser Sintering (SLS).* Selective laser sintering involves the use of meltable powders which are heated by a laser beam. The laser beam generates a slice of the CAD object. Additional powder is applied after each layer is fused. Because the surrounding powder supports the parts, there is no need for support structures.
- *Laminated Object Manufacturing (LOM).* Laminated object manufacturing uses heat-activated adhesive-coated paper. The object is formed by cutting one section of the paper at a time. Each succeeding layer is bonded to the previous layer.
- *Fused Deposition Modeling (FDM).* Fused deposition modeling involves a heated nozzle through which a thermoplastic is extruded. The material is deposited in layers on a table which is capable of x–y motion in its plane.
- *Three Dimensional Plotting.* This is a liquid-to-solid inkjet plotter. The inkjet system moves on an x–y drive carriage and deposits both thermoplastic and wax materials on the build substrate. The drive carriage also controls a milling system for maintaining z-axis dimensioning of the model by milling off the excess vertical height of the current layer.
- *Ballistic Particle Manufacturing.* A piezoelectric inkjet head sprays out droplets of a heated thermoplastic.
- *Desktop Rapid Prototyping.* This software slices the models and arranges the slices for cutting. These units use a sharp knife to cut the cross sections in either paper or plastic. The system uses a registration board for assembly of the slices into a completed prototype. The registration board contains two registration systems to ensure proper alignment of the slices.
- *Direct Shell Production Casting (DSPC).* The process produces the ceramic molds for metal castings directly from 3-D CAD designs. It uses 3-D printing technology to produce the ceramic casting molds using a layer-by-layer process.

New processes and companies are continually appearing; thus, the above list must be regarded as being a snapshot of the moment.

The selection of an RP process for a given application depends on the user's requirements as regards:

- Functionality desired
- Cost of the process
- Turnaround time

The restrictions in each of the processes of accuracy, materials, geometry, and size must also be taken into account. In the immediate future an increasing number of "office-friendly" systems, using nontoxic materials, are being introduced for the designer's desktop. An advantage of using an RP process is that it allows many more iterations of a part to be made and evaluated in a given amount of development time. Usually, the use of RP can be justified only if a quick turnaround is possible. In fact, it is becoming increasingly more common for a designer to make changes in a CAD model of a part, and the part to be resubmitted to the RP queue, before secondary finishing operations are completed on the previous version.

SUMMARY

This chapter started out with a discussion of the rules, checklist, and procedures for embodiment design. Two examples showed how variants may be generated at the embodiment stage. Part design was discussed, including shape design principles. The remainder of the chapter was devoted to designing for the ease of assembly. Guidelines and several examples of design for ease of assembly were given. Finally, prototype development, including rapid prototyping, was discussed.

BIBLIOGRAPHY

Ashley, S. "Cutting Costs and Time with DFMA." *Mechanical Engineering* 117(3): 74–77 (1995a).

Ashley, S. "Rapid Prototyping Is Coming of Age." *Mechanical Engineering* 117(7): 62–68 (1995b).

Boothroyd, G., and P. Dewhurst. *Product Design for Assembly.* Wakefield, R.I.: Boothroyd Dewhurst, Inc., 1987.

Burns, M. *Automated Fabrication: Improving Productivity in Manufacturing.* Englewood Cliffs, N.J.: Prentice Hall, 1993.

Carter, M. F. "Design for Manufacturability in General Motors." Paper presented at the 1995 National Design Engineering Conference, Norwalk, Conn., Reed Exhibition Companies.

Clausing, D., *Total Quality Development*. New York: ASME Press, 1994.

Durham, D. R., and M. S. Hundal. "A Way towards Methodical Design of Metal Castings." *Proc. ICED 91*, 1: 536–539. Zurich: Heurista, 1991.

Ehrlenspiel, K., and T. John. "Inventing by Design Methodology." *Proc. ICED 87*, 1: 29–37. New York: ASME, 1987.

Hundal, M. S. "Rules and Models for Low-Cost Design." *Design for Manufacturability 1993*, DE 52: 75–84. New York: ASME, 1993.

Kalpakjian, S. *Manufacturing Engineering and Technology.* Reading, Mass.: Addison-Wesley, 1992.

Matousek, R. *Engineering Design—A Systematic Approach.* London: Blackie, 1963.

Pahl, G., and W. Beitz, *Engineering Design—A Systematic Approach.* Berlin and New York: Springer Verlag, 1988.

Ulrich, K. T., and S. D. Eppinger. *Product Design and Development.* New York: McGraw-Hill, 1995.

Material Selection and Economics

The majority of engineering materials consist of solids from one or more of the three basic groups: metals, polymers (commonly called plastics), and ceramics. A combination of two or more of these forms a composite material. This concept is shown schematically in Figure 95. Materials from each of the basic groups have certain properties which are shared by most of the members of that group, although there can be exceptions. The three groups' chemical origins and manufacturing processes are different; thus, their chemical and physical properties also differ significantly.

Table 33 shows some of the important properties of materials in the different groups. We should note that what may be shown as a favorable property may become undesirable under a different set of conditions.

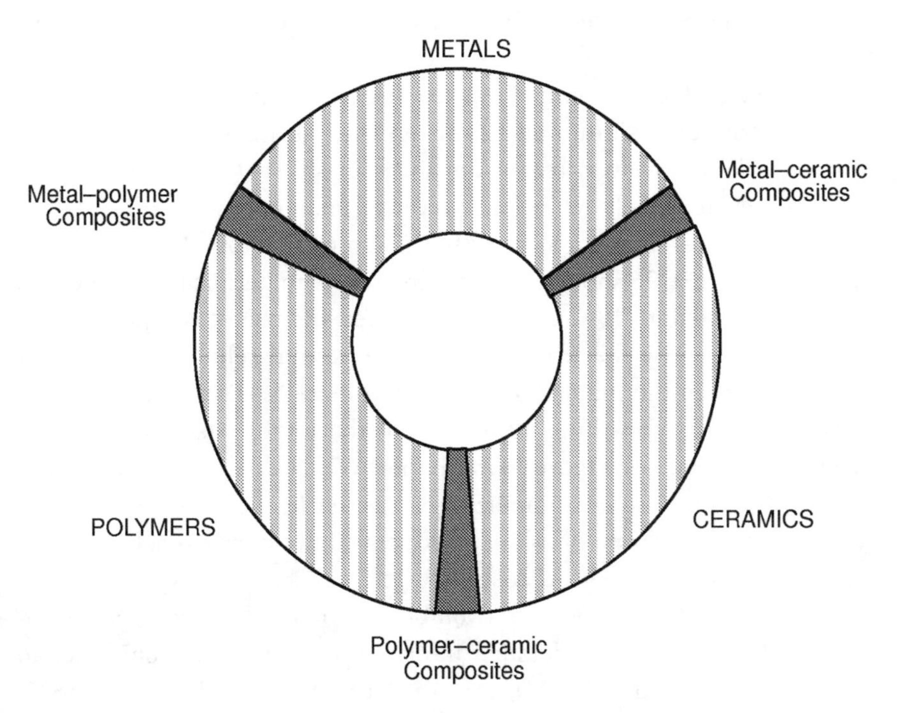

FIGURE 95. Three basic engineering material types.

TABLE 33 Properties of the Basic Material Groups

Material Group	Favorable Properties	Unfavorable Properties
Metals	Strength, toughness, electrical and thermal conductivity, use in wide range of temperatures	High density
Polymers	Low density, easy formability, high electrical resistance, resistance to chemicals	Low resistance to heat, low strength, poor dimensional stability
Ceramics	Withstand high temperatures, hardness	Brittleness

Composite materials are formulated to make the best use of the properties of their constituent materials.

The continuously changing picture due to improvements in materials and manufacturing processes requires that the designer be conversant with the latest developments. Consider the case of the beverage can. For a number of years aluminum has taken over an increasing percentage of the market. However, recent advances in sheet steel properties and manufacture are likely to increase its use. Such competitive forces are healthy. They can be seen at work in the aircraft, automobile, and other industries. Figure 96 shows the changing use of materials in commercial aircraft (Kalpakjian, 1992). We see an increasing use of composites and titanium, driven by the market forces of cost and reliability. A similar trend appears in the materials used in automobiles (Dieter, 1991).

Material costs constitute 40 to 50% of the total costs in the general machine industry (see Figure 64 in Chapter VI). They are higher for mass-produced items, where rationalization has led to the lowering of other costs, and lower for one-of-a-kind products. Material costs include direct material costs and overhead costs.

There is extensive literature available on materials science and engineering. For further and detailed information, the reader may refer to Askeland (1994), Khol (1994), the American Society of Metals (ASM) handbooks, and Van Vlack (1989), among other books.

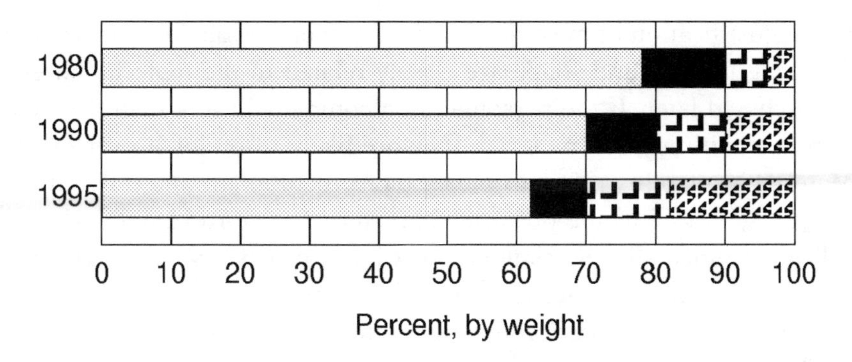

FIGURE 96. Changes in material usage in airframes of commercial aircraft. (Kalpakjian, 1992)

MATERIALS REQUIREMENTS

The selection of materials for a given application, out of the thousands available today, is one of the most crucial steps in embodiment design. The choice is dictated by the requirements. Requirements that govern the choice of materials are:

- Chemical environment
- Physical loading (forces, heat, etc.)
- Quantities required
- Shape
- Manufacture
- Expertise available
- Material availability
- Useful life
- Disposal, recycling
- Cost

The justification of most of these requirements is obvious. The manufacturing process and shape are closely related to the material chosen, as discussed later. If the personnel in a company have a high degree of expertise with a given material, they are likely to prefer it over others. Material availability refers to the ease of procurement and reliability of suppliers, as well as the physical characteristics—for example, whether rolled, extruded shapes, plate or sheet form, or cast as ingots. Disposal and recyclability of materials is becoming increasingly critical. Cost, is of course, the final criterion for the optimality of most designs.

The properties which determine how well some of the above requirements are satisfied can be classified as follows:

Mechanical properties:

- Density
- Porosity
- Strength (tension, compression, shear, bending, torsional, buckling)
- Fatigue characteristics
- Elastic properties (moduli, elastic limit, etc.)
- Impact strength
- Toughness
- Hardness
- Wear characteristics
- Surface friction
- Internal friction

Other physical properties:

- Specific heat
- Melting point
- Thermal conductivity
- Effect of temperature on other properties
- Electrical conductivity
- Magnetic properties
- Optical properties

Processibility: suitability of the material for

- Machining
- Casting
- Forging
- Deep drawing

- Welding
- Soldering
- Brazing

Systematic Selection of Materials

At the concept stage, the choice of material can be narrowed by rating the materials according to the principal properties of interest and other pertinent requirements, as we will see later in an example. Matousek (1963) shows a graphical scheme for making the initial selection of the material, or, rather, rejecting unsuitable materials. The principal properties and their values (on a scale of 1 to 10) are marked off on the axes of a graph to indicate the minimum requirements for the given application. On this graph are plotted the property values of candidate materials. Any material falling below the minimum value of any of the requirements is rejected at this stage. Figure 97 shows an example of the use of such a graph.

The initial go/no-go decision can also be made by using a process similar to the initial concept selection (see Figure 18 in Chapter III). Figure 98 shows such an application.

An alternative approach is to start with the one or two most important requirements of the material, e.g., strength per unit weight, conductivity, corrosion resistance, or formability. Once the number of materials is narrowed down, we then bring in the requirements next in importance.

MODERN ENGINEERING MATERIALS

Figure 99 shows the major engineering materials in use today, classified according to their chemical composition. Nonferrous materials are typically more costly than ferrous alloys. However, they offer advantages of light weight, resistance to chemicals, and conductivity. Figure 100 shows cost by volume of nine common materials (Kalpakjian, 1992).

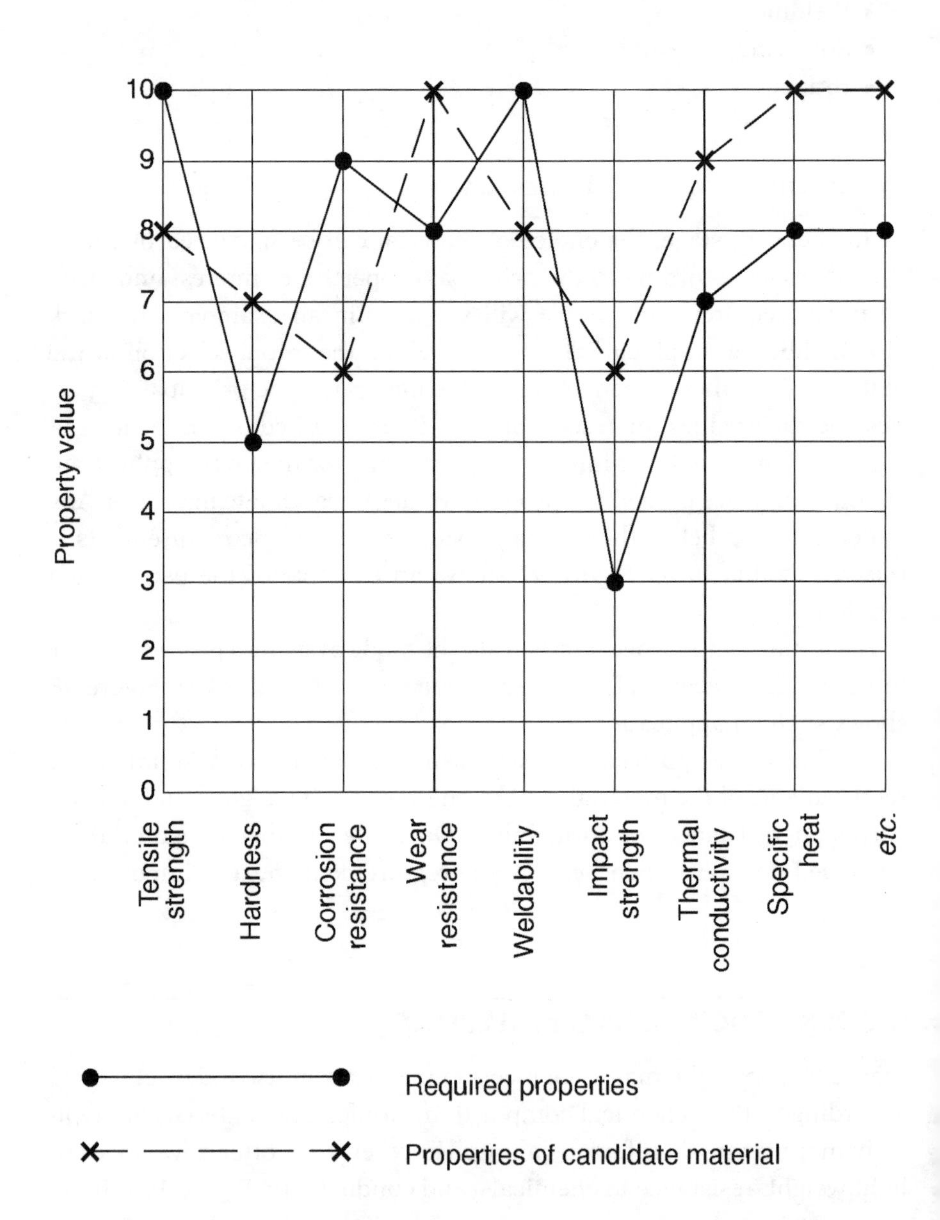

FIGURE 97. Judging a material by the required properties.

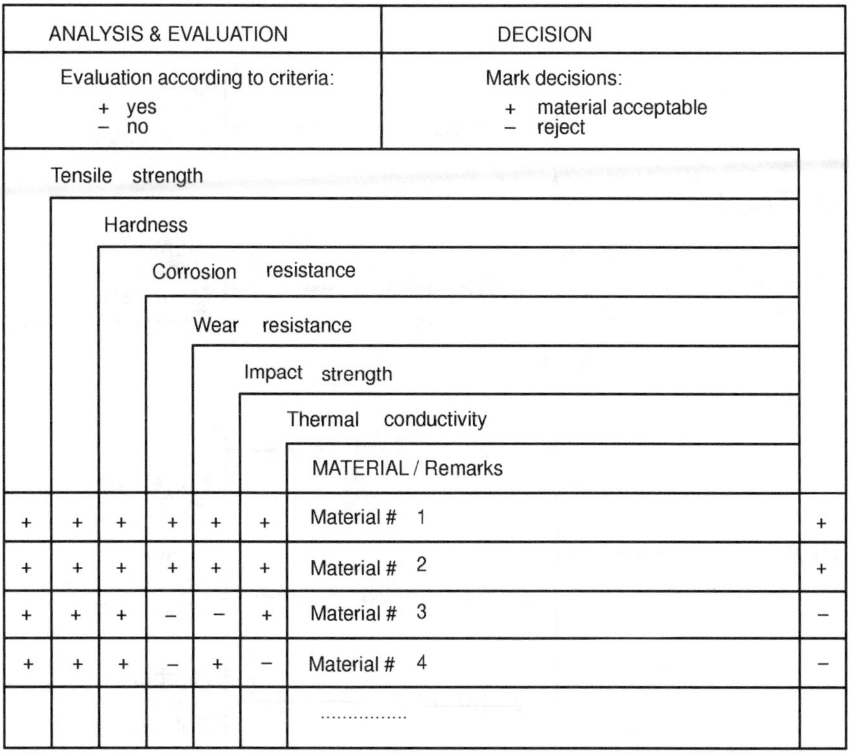

ANALYSIS & EVALUATION							DECISION
Evaluation according to criteria: + yes − no							Mark decisions: + material acceptable − reject
Tensile strength	Hardness	Corrosion resistance	Wear resistance	Impact strength	Thermal conductivity	MATERIAL / Remarks	
+	+	+	+	+	+	Material # 1	+
+	+	+	+	+	+	Material # 2	+
+	+	+	−	−	+	Material # 3	−
+	+	+	−	+	−	Material # 4	−
						....	

FIGURE 98. Initial material selection matrix.

METALS

FERROUS METALS AND ALLOYS

Ferrous metals and alloys constitute the largest portion of engineering materials in use today in mechanical products. These are divided up into cast irons and steels of various kinds.

Cast Irons

Cast irons are ferrous alloys of iron, carbon (2 to 4.5%), and usually silicon. Their designations and properties derive from the method of their

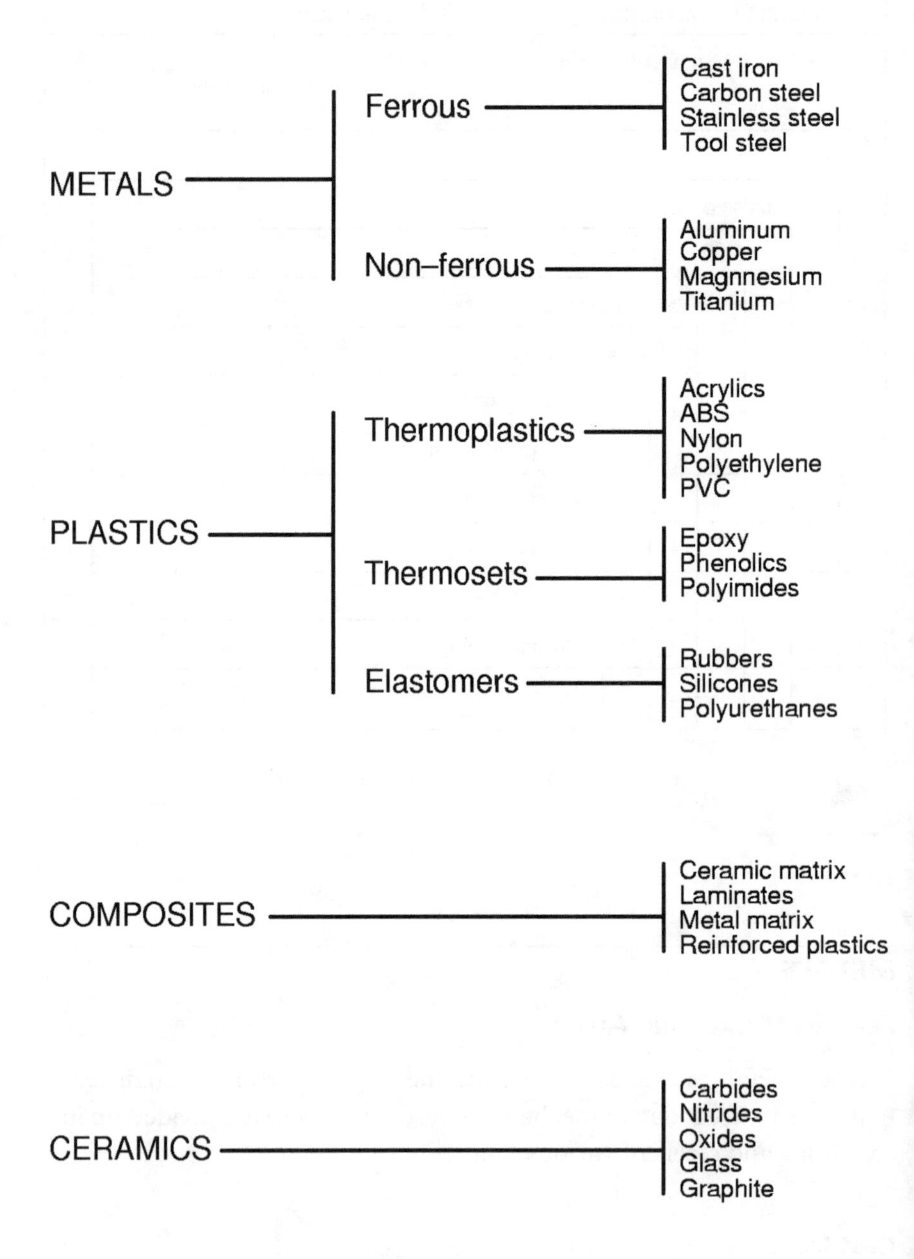

FIGURE 99. Major engineering materials in use today.

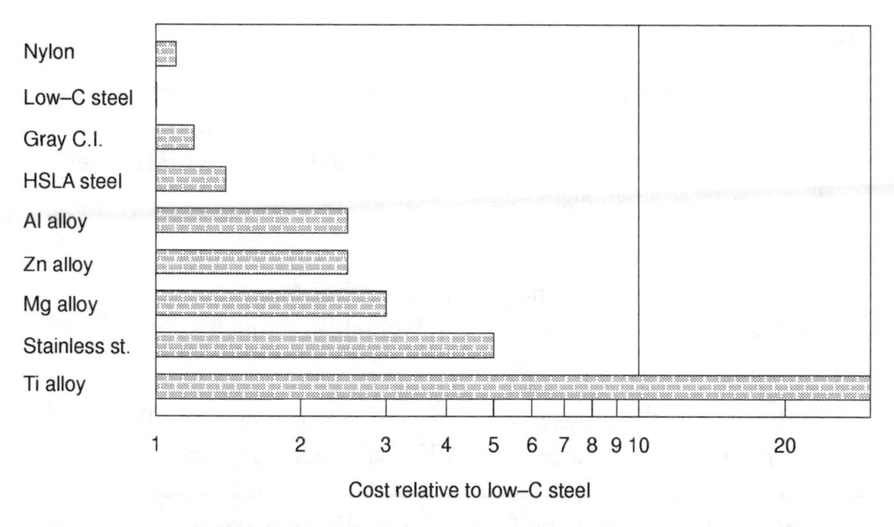

FIGURE 100. Relative costs of materials by volume. (Kalpakjian, 1992)

solidification. The major types are gray iron, white cast iron, malleable iron, ductile or nodular iron, and compacted graphite iron.

Gray iron is used for engine blocks, machine bases, gears, and brake disks and drums. It is easy to cast and machine, but difficult to weld. It has good internal damping.

White cast iron, due to its extreme hardness, is used in wear-resistant applications, such as liners, brake shoes, nozzles, and crushers. It is easy to cast, but difficult to machine and weld.

Malleable iron is obtained by annealing white cast iron. It is used to make railroad, farm, and construction equipment, miscellaneous hardware, and bearings for heavy machinery. It is easy to cast and machine, but difficult to weld.

Ductile iron is used for pipe, heavy duty gears, and engine crankshafts. It is easy to cast and machine, but difficult to weld.

Compacted graphite iron is used for engine blocks and heads. Its easy to cast and machine and weld.

Availability. While cast irons are typically supplied in ingot form for melting and casting into molds, they are also available as long-cast profiles from which flat parts may be cut.

Steels

Steels—(plain) carbon and alloy—are used for a wide variety of applications. The types of steels and their designations are shown in Table 34. The most common alloying elements are listed below, with their effects on the properties of steels:

Carbon

- Increases: strength, hardness, wear resistance
- Reduces: ductility, toughness, weldability, machinability

Chromium

- Increases: toughness, wear resistance, corrosion resistance, hardenability, depth of hardness, high-temperature strength

TABLE 34 AISI-SAE Numbering System for Steels, with Typical Applications

AISI-SAE Number	Primary Alloying Element	Secondary Alloying Element	Applications
10xx	Plain carbon		Car bodies, camshafts, lock washers, tubing, springs
11xx, 12xx	Free machining	(S,P)	Screws, nuts
13xx	Mn		
31xx, 33xx	Ni	Cr	Transmission chains, crankshafts, nuts
40xx	Mo		Springs, gears
41xx	Cr	Mo	Aircraft fittings
43xx	Ni	Cr, Mo	Landing gear parts
44xx, 45xx	Mo		
46xx, 48xx	Ni	Mo	Bolts
47xx	Ni	Cr, Mo	
50xx	Cr		
50xxx	Cr	High C	
61xx	Cr	V	Springs
81xx, 86xx	Ni	Cr, Mo	
87xx, 88xx	Ni	Cr, Mo	Aircraft fittings
92xx	Mn	Si, Cr	Springs
93xx, 94xx	Ni	Cr, Mo	
98xx	Ni	Cr, Mo	

Manganese

- Increases: strength, hardenability, machinability, abrasion resistance
- Reduces: weldability
- Other: deoxidizes steel

Molybdenum

- Increases: toughness, wear resistance, creep resistance, hardenability, hardness, high-temperature strength
- Reduces: temperature embrittlement

Nickel

- Increases: strength, hardenability, corrosion resistance, toughness

Vanadium

- Increases: strength, abrasion resistance, toughness, high-temperature hardness
- Other: controls grain growth during heat treatment

We look now at the different types of steels.

- *Carbon steel:* Low carbon (< 0.3%), medium carbon (0.3–0.6%), and high carbon (> 0.6%). It is the most common general-purpose steel.
- *Alloy steels:* As shown in the list above, various alloying elements impart particular properties to steel, leading to a variety of alloy steels. Alloy steels may be wrought or cast. Other than stainless and tool steels, discussed below, the most common alloy steels are the high-strength–low-alloy (HSLA) structural steels. These steels are used for plates, bars, and structural shapes. High-strength sheet steels have high ductility.
- *Stainless steels:* Stainless steels have high chromium content (minimum 10 to 12%), which gives them corrosion resistance and ductility. Higher carbon content increases their corrosion resistance. Five AISI grades are 303, 304, 316, 410, and 416. Ferritic steels are easy to machine.
- *Tool steels:* Tool and die steels are special alloy steels used where high strength, wear resistance, and toughness are needed, particularly at

high temperatures. The alloying elements used are molybdenum, tungsten, and chromium.

Carbon and low-alloy steel castings are used for railroad wheels, gears, and general machine parts. They are easy to weld, but only fair in machinability and castability. High-alloy steel castings are used to make components that must withstand harsh thermal, wear, and chemical environments. Like carbon and low-alloy steel castings, they are easy to weld, but only fair in machinability and castability.

Availability. Steels are available in the form of plates, bars, tubes, wire, structural shapes, and ingots for casting.

Nonferrous Metals and Alloys

Nonferrous metals and alloys offer advantages over ferrous alloys of corrosion resistance, low density, higher thermal and electrical conductivity, ease of fabrication, and aesthetic appeal.

Aluminum

Aluminum and its alloys offer the advantages of low density, high strength-to-weight ratio, high thermal and electrical conductivity, ease of processing, and good appearance. They are nontoxic. Aluminum alloys may be wrought or cast; they are also available as extrusions. Aluminum alloys are finding increasing use in automobile structures. Cast aluminum alloys are used for automobile pistons, housings, manifolds, and other parts and for aircraft parts. They are easy to cast and machine, but only fair in weldability. Their designations, alloying elements, properties, and typical applications are given in Table 35.

Availability. Aluminum alloys are available in the form of plates, foil, bars, tubes, wire, structural shapes, and ingots for casting.

Magnesium

Magnesium alloys are the lowest-density engineering materials in use today. Magnesium alloys may be wrought or cast; they are also available

TABLE 35 Designation and Applications of Aluminum Alloys

Number	Alloying Elements	Properties, Applications
1xxx	Pure Al	Low strength, high conductivity, easy formability Sheet metal products
2xxx	Copper	High strength, heat-treatable Aircraft structures, automobile wheels
3xxx	Manganese	Easy formability Cookware, building hardware, chemical vessels
5xxx	Magnesium	Corrosion resistance, weldability Automobile and ship parts, sheet metal products
6xxx	Mg and Si	Corrosion resistance, weldability, machinability Truck and ship structures, furniture, hydraulic tubing
7xxx	Zinc	High strength, heat-treatable Aircraft structures, hydraulic equipment

as extrusions. The alloying elements are Al, Mn, Zn, and Zr. Because of their high strength-to-weight ratio, they are used in aircraft, racing vehicles, bicycles, portable power tools, and other applications where low weight is important. A new application is for tubular seat frames for automobiles. Magnesium alloy castings are used for housings, toys, etc. They are easy to cast and machine, but less easy to weld.

Availability. Magnesium alloys are available in the form of plates, bars, tubes, structural shapes, and ingots for casting.

Nickel

Nickel is best known as the alloying element in stainless steels, as well as a family of alloys known as superalloys. Nickel imparts strength, toughness, and corrosion resistance at high temperatures. As the base metal, nickel is alloyed with various elements for applications as shown in Table 36. Nickel alloy castings are used in thermally and chemically hazardous environments, such as in gas turbine blades and parts for chemical plant equipment. They are difficult to cast, machine, and weld.

TABLE 36 Designation and Applications of Nickel Alloys

Name	Alloying Elements	Applications
Inconel	Cr, Fe	Gas turbine, electronics, nuclear reactor parts
Monel	Cu	Water instrument parts
Duranickel	Al, Ti	Springs, diaphragms, molds
Hastelloy	Cr, Mo	Sheet metal for jet engines

Titanium

Titanium and its alloys offer the advantages of high strength-to-weight ratio, high strength at high temperatures, and corrosion resistance at room and high temperatures. The major titanium alloys and their applications are given in Table 37. Titanium is difficult to machine due to its low thermal conductivity, which causes heat buildup at the tool.

POLYMERS

More commonly known as plastics, polymers offer the advantages of easy manufacturability, high strength-to-weight ratio, good surface appearance, and corrosion resistance. Polymers are produced when the chemical

TABLE 37 Designation and Applications of Titanium Alloys

UNS Number	Alloying Elements	Applications
50250	Pure (99.5%)	Marine, heat exchangers, airframes, chemical equipment
54520	Al, Sn	Gas turbine compressor and steam turbine blades
56400	Al, V	Aircraft turbine blades, disks, fasteners, forged parts
58010	V, Cr, Al	Aerospace equipment, parts, fasteners

elements (mers) are linked to form large chains of molecules. An example of this process is the polymerization of ethylene to form polyethylene. The bonds within the long-chain molecules are strong, as opposed to the weak bonds between the adjacent long chains, though it is the strength of the latter that determines the overall strength of the plastic. Figure 101 shows the strength of some plastic materials, with and without reinforcement.

Plastics are divided into three primary categories: thermoplastics, thermoset plastics (thermosets), and elastomers. Thermoplastics, when heated, go through a glass-transition temperature and a melting temperature, above which they are easy to mold. When cooled, they regain their original properties. Thermosets, on the other hand, form one large cross-linked molecule upon polymerization, attain their permanent shape, and cannot be remelted. The curing of thermosets has been compared to the cooking of an egg—once done, it cannot be undone. Some plastics, e.g., polyesters and polyimides, are produced both as thermoplastics and as thermosets.

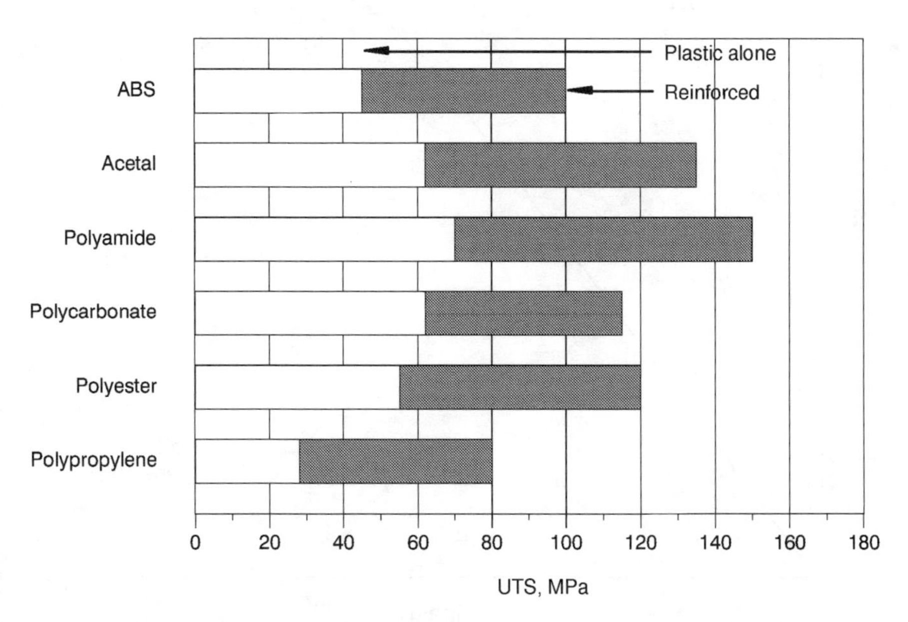

FIGURE 101. Strength of plastics. Shaded portions show the changes due to reinforcement.

Additives are used with plastics to improve their properties. These are fillers, plasticizers, coloring agents, lubricants, and flame retardants. Fillers, such as glass, cellulose, flour, or mica, help reduce the plastics' cost and improve their hardness, strength, dimensional stability, abrasion resistance, etc. Plasticizers, which are nonvolatile solvents, help improve the workability of plastics.

The behavior of plastics under loading varies from stiff and brittle for some thermosets, to soft and pliable for some thermoplastics (e.g., polytetrafluoroethylene, polyethylene), with other thermosets (e.g., nylon, ABS) lying in between, as shown in Figure 102. Reinforcing of plastics with glass or carbon fibers leads to significant changes in their physical properties. (See Figure 101, and the discussion under "Composites" later in this chapter.)

In working with both plastics and composites, it is of the utmost importance for designers to work with the manufacturers, who have special expertise in both the process and its economics.

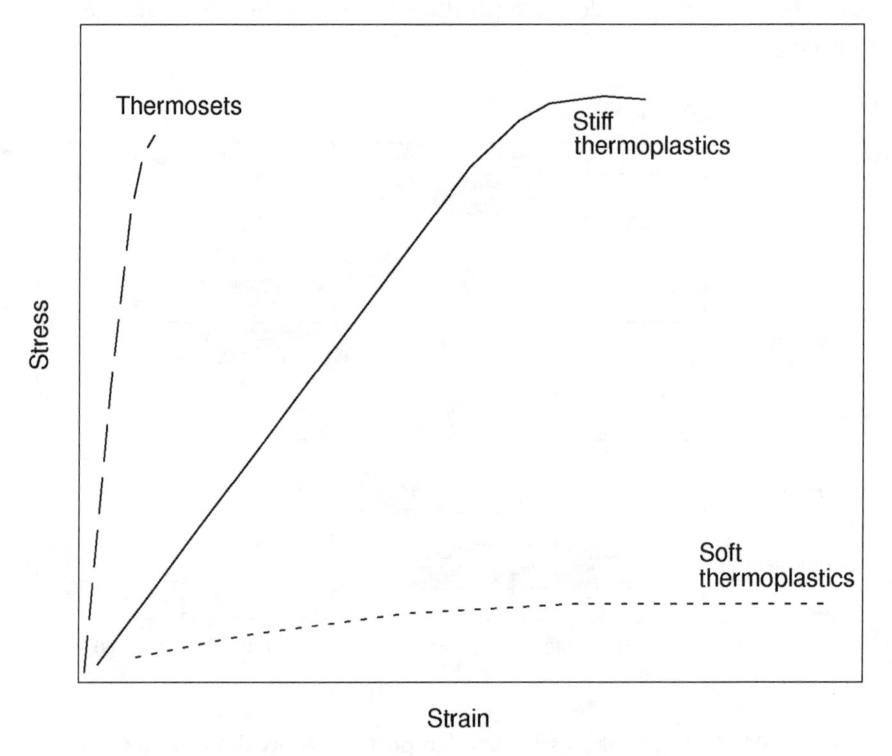

FIGURE 102. Stress–strain behavior of plastics.

Thermoplastics

Thermoplastics comprise over 90% of all polymers processed today. The long-chain molecules of thermoplastics allow them to be repeatedly melted and solidified. Compared with thermosets, thermoplastics offer higher toughness, easy processibility, and flexibility in design. We will briefly describe the characteristics and applications of the well-known plastics (trade names shown in parentheses).

ABS

ABS (acrylonitrile-butadiene-styrene) is amorphous; it can be processed by injection molding, extrusion, blow molding, foam molding, and thermoforming. Some grades can be electroplated.

- *Properties:* Tough, hard, rigid; low water absorption, high dimensional stability, high abrasion resistance
- *Applications:* Housings, handles, fittings, refrigerator door liners, panels

Acetal (Delrin, Celcon, Ultraform)

Acetal is crystalline; it can be processed by injection molding, extrusion, or blow molding. It is also available in rod and slab form, and is easily machined.

- *Properties:* Strong, stiff; high dimensional stability, low friction, abrasion-resistant, stable in water
- *Applications:* Cams, gears, bushings, valves, housings, wear surfaces

Acrylic (Plexiglas, Lucite)

Acrylic is available as sheet, rod, tube, and extruded sheet and film. It can be processed by injection molding and extrusion. Sheets are transparent, translucent, or opaque.

- *Properties:* Optically clear, hard, glossy; weather-resistant
- *Applications:* Window glazing, lenses, bubble tops, lighting fixtures

Fluoroplastics

The best known of the fluorocarbons is Teflon, used for its nonstick, nonadhesive, and low-friction qualities. Its generic name is polytetrafluoroethylene (PTFE). Other fluoroplastics include fluorinated ethylene propylene (FEP) and polyvinylfluoride (PVF). Inclusion of fibers improves wear and creep resistance.

- *Properties:* Nonadhesive, low friction; resistance to temperature, chemicals, weather, and electricity
- *Applications:* Nonstick coatings for cookware, process equipment, insulation, bearings, seals

Polyamides (Nylon, Kevlar)

Polyamides are semicrystalline. They can be molded or extruded. The most common varieties are nylon 6/6 and nylon 6.

- *Properties:* Tough, wear-resistant, low friction, high chemical resistance, hygroscopic
- *Applications:* Machine parts, zippers, tubing, medical equipment, toys

Aramid (Kevlar) fibers are used as reinforcement in other plastics.

- *Properties:* High tensile strength and stiffness
- *Applications:* Bulletproof vests, cables, reinforcement in tires

Polycarbonate (Lexan)

Polycarbonate is amorphous. It is available for molding and extrusion and as sheet and film.

- *Properties:* Optically clear; impact-resistant, resistant to chemicals, dimensionally stable
- *Applications:* Window glazing, windshields, machinery guards, electrical parts, medical equipment

Polyester (Dacron, Mylar, Kodel)

Polyester is a crystalline, high-molecular weight polymer. It is available for molding and as glass-reinforced. Alternative chemical constituents are polybutylene terephthalate (PBT) and polyethylene terephthalate (PET).

- *Properties:* Abrasion-resistant, low friction
- *Applications:* Machine parts, pumps, electrical parts

Polyethylene

Polyethylenes are used today in larger volume than any other thermoplastic. They have good electrical and chemical properties. The mechanical properties of polyethylene depend on density and molecular structure. According to density, polyethylenes are divided into low density (LDPE) and high density (HDPE). The ultrahigh–molecular weight variety (UHMWPE) has very different properties; it is of medium to high density. A cross-linked thermoset-like PE is also available.

	Type		
	LDPE	HDPE	UHMWPE
Properties	Tough, flexible, clear	Rigid, high strength, low-temperature impact resistant	Abrasion resistant, low friction, high impact strength, service to 200°F
Applications	Packaging, bags, food containers, toys, housewares	Food, chemical containers; fuel, water, waste tanks; highway barriers	Sheets; other applications requiring above properties

Polypropylene

Polypropylenes are translucent, milky-white, and colorable. They can be processed by injection molding, extrusion, and blow molding.

- *Properties:* Good mechanical, electrical, chemical properties; resistant to tearing

- *Applications:* Tanks, pipes, fittings, appliance parts, drinking cups, liquid containers

PVC (Saran, Tygon)

PVC (polyvinyl chloride) can be processed by injection molding, extrusion, compression molding, and blow molding. PVC can be mixed with other plastics to improve their properties. It can be rigid or flexible, clear or opaque.

	Type	
	Flexible	Rigid
Properties	Water-resistant, low strength	Tough, hard
Applications	Film, sheet, tubing, cable coating	Pipes, conduits, signs

Thermosets

Thermosets in general provide greater dimensional stability and resist higher temperatures than thermoplastics.

Epoxy

Epoxy plastics undergo very little shrinkage during curing, which is the process for forming thermoset resins.

- *Properties:* Excellent thermal, chemical, and electrical resistance; adhesive, hard, brittle, dimensionally stable; properties can be varied by using different combinations of resins
- *Applications:* Adhesives, electrical parts, tools, dies

Phenolics

Phenolics are low-cost, widely used thermosets. They can be processed by injection molding, extrusion, compression molding, and transfer molding. Their color is black or brown, and they are used in composites containing reinforcing fibers and fillers.

- *Properties:* Brittle, rigid, dimensionally stable; good heat, chemical, and electrical resistance
- *Applications:* Electrical components, knobs, appliance housings

Polyester

Polyester, with fiberglass reinforcement, is the commonly known "fiberglass" plastic (FRP). It can be hand-processed for large parts in medium quantities, or compression-molded from sheet molding compound. It can also be pultruded, cold press-molded, and resin injection-molded (resin transfer-molded).

- *Properties:* Weather-resistant; good thermal, electrical, and environmental properties
- *Applications:* Boat hulls, pool shells, auto bodies, furniture, fishing rods

Polyimides

Polyimides are available as laminates, shapes, and molded parts. They can be processed by injection molding, extrusion, compression molding, and transfer molding, as well as by powder metallurgy-like methods.

- *Properties:* Cold, heat, and fire resistant; low friction; creep resistant at high temperatures
- *Applications:* Bearings, seals, piston rings, aerospace parts, sports equipment

Elastomers

Elastomers are polymers that are soft and flexible, have a highly kinked structure, and, most important, can undergo large deformations and return to their original shape. The term *rubber* is used interchangeably with elastomer. Vulcanization of rubber, with sulfur, produces cross-linking, giving the object a permanent, though flexible, shape. Elastomers display hysteresis upon loading and unloading, and are therefore useful for absorbing noise and vibration energy.

Synthetic rubbers overcome the shortcomings of natural rubber. Synthetic rubbers more resistant to heat, gasoline, and chemicals are butyl, styrene butadiene, ethylene propylene, and polybutadiene. Synthetic rubbers more resistant to oil are neoprene and urethane, silicone, and nitrile rubbers.

Natural Rubber

Natural rubber, produced from latex, has high friction, abrasion, and fatigue resistance, but low resistance to oil, sunlight, heat, and ozone. It is used for seals, erasers, coupling inserts, and shoe soles.

Silicones

Silicones can used up to 600°F, in applications such as seals, electrical parts, and thermal insulation.

Polyurethane

Polyurethane can be processed by thermoplastic or thermoset methods. It is used for seals, diaphragms, cushions, gaskets, and parts for automobile bodies. It is strong, stiff, hard, and resistant to abrasion, cutting, and tearing.

CERAMICS

Ceramics fill the need for materials that must withstand higher temperatures than metals. Typical applications for ceramic materials are in ovens, gas turbine and rocket engines, and dies for hot extrusion. Ceramic materials are, in general, hard and brittle, and thus they are commonly used as abrasives and for cutting tools. New developments in ceramics are toward increasing their toughness and toward easier fabrication.

Ceramic matrix materials in use are silicon carbide and nitride, and alumina. Ceramic matrix composites not only withstand higher temperatures than metal matrix composites, but they also resist corrosion at high temperatures.

The most common ceramic materials are:

Oxides

- Aluminum oxide (alumina) is the most commonly used oxide ceramic. It is used for electrical and thermal insulation and as cutting tools.
- Zirconium oxide (zirconia) and a more recent development, partially stabilized zirconia (PSZ), are tougher and have low friction, high wear resistance, and low thermal conductivity. Their applications are for cylinder liners, extrusion dies, and valve bushings in automobile engines.

Carbides

- Silicon carbide has low friction and high resistance to wear, corrosion, and thermal shock. It is used in engines and as an abrasive.
- Tungsten and titanium carbides are used in dies and cutting tools.

Nitrides

- Silicon nitride is used in engines due to its resistance to thermal shock and creep, its low thermal expansion, and its high thermal conductivity.
- Titanium nitride is used as a coating on cutting tools.
- Cubic boron nitride is the hardest known material next only to diamond. It is used for cutting tools and as an abrasive.

COMPOSITES

Composite materials, a class of so-called engineered materials, are a physical combination of two or more materials. By combining different materials, we can take advantage of the best properties of each. A most obvious combination of properties is light weight and high strength. Steel-reinforced concrete is a long-existing example of a composite material. Some naturally occurring composite materials are wood and bone.

A composite material consists of a base or matrix material and the reinforcing material. The most common composite materials are the reinforced plastics, which have a plastic matrix with glass, graphite, or other fibers. The matrix forms the continuous phase of the composite and supports and protects the fibers, and provides the toughness and ductility. The most common matrix materials are epoxies. Their strength increases from 2 to 10 times upon reinforcement. Other plastics used for composites are polyesters, phenolics, fluorocarbons, and silicones.

The fibers form the dispersed or discontinuous phase of the composite and provide the strength and stiffness. The commonly used fiber materials (with their distinguishing properties) are fiberglass (inexpensive), graphite (light, strong, stiff), aramid (very tough and strong), and boron (strength, resistance to high temperatures). A distinguishing feature of reinforcing fiber materials is that they have a much higher tensile strength/density ratio and tensile modulus/density ratio than common materials such as steel.

Applications of composite materials are many and increasing day by day. They range from aircraft structures to sports equipment such as snowboards, sailboards, and skis. In aircraft and rockets, composites are used in helicopter blades, landing gear doors, floor beams and panels, and tail sections, for example. Other applications include helmets, pressure vessels, auto body panels, driveshafts, and pipes.

Other emerging technologies show the application of metal and ceramic matrix. Metal matrix composites withstand higher temperatures. Matrix metals used are aluminum, magnesium, titanium, lead, and copper, with fibers of graphite, boron, alumina, and silicon carbide. Their applications are in space structures and jet engines.

COSTS BASED ON MATERIALS

Material costs constitute 40 to 60% of the total costs in the general machine industry (see Figure 2 in Chapter I). They are higher for mass-produced items, where rationalization has led to the lowering of other costs, and lower for one-of-a-kind products. Material costs include direct material costs and overhead costs. Direct costs are the product of material volume and the cost per unit volume. Methods are available for reducing each of these during embodiment. See also Chapter VII, on part design.

Direct Costs of Material

We restrict ourselves here to mechanical and thermal loading. For mechanical loading, the basic principle of shape design is to aim for uniform stress (see Matousek, 1963). Two corollaries of this are:

- Avoid using material where it serves no purpose.
- Avoid bending and torsion.

In the latter case, specially shaped parts are needed to achieve uniform stress:

- Place the material so it follows the lines of force.
- Use shapes that provide uniform stress under the given conditions.

Bending and Torsion

The sections that provide optimum stress distribution in nonaxial loading may be conical, cylindrical, or spherical—not necessarily rectangular. The I-section and the closed hollow sections provide the highest stiffness in bending. Figure 103 shows the stiffness for various cross sections of the same cross-sectional area (normalized to the solid square). The numerical values of section properties are shown in Table 38. The hollow cylindrical shape provides the best use of material under a variety of loading conditions. Short, direct paths provide the most stiffness; use long, indirect paths if flexibility is desired, as attested to by the shapes of most springs. For maximum heat flow we need to ensure the highest heat transfer coefficient and provide the most surface area (see, e.g., Holman, 1981) for a given temperature difference.

Figure 104a shows a conventional bell-crank lever in which the arms are subjected to bending. In Figure 104b, the simple design achieves uniform stress by causing axial force and thus using less material and/or a simpler cross section.

Shape and Material Variation

In order to obtain uniform stress distribution in the material, the traditional procedure, earlier in this chapter, is to use a material with uniform

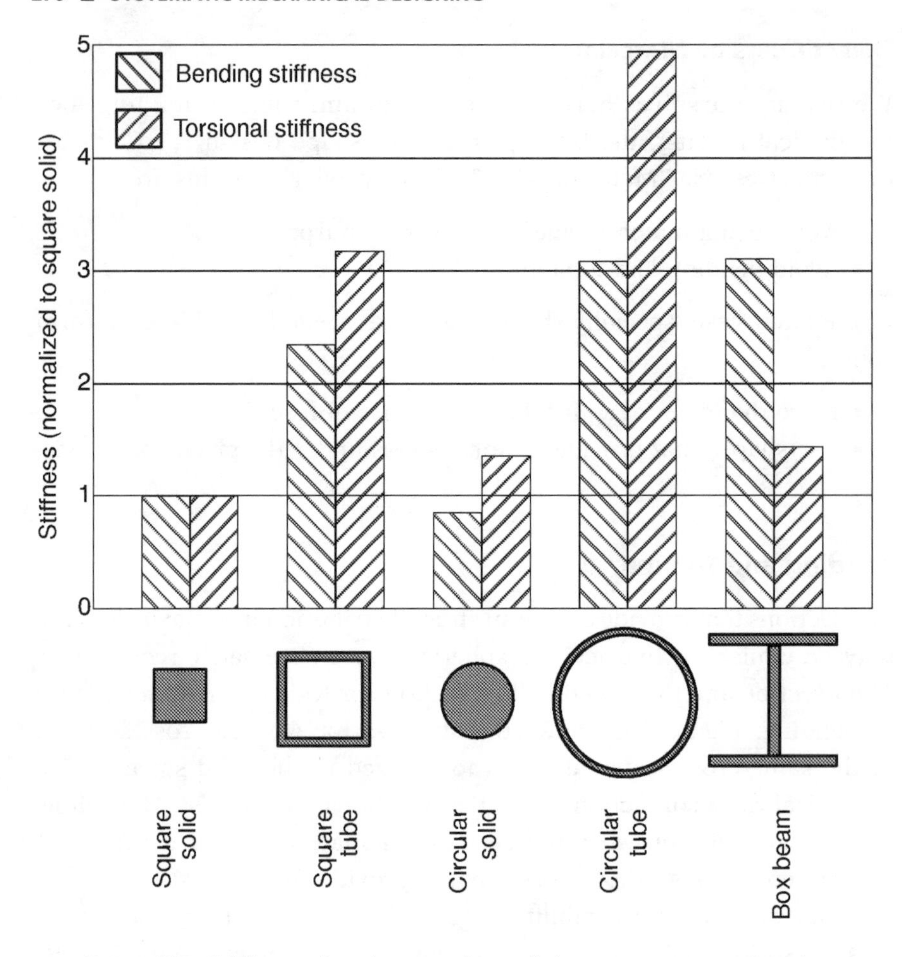

FIGURE 103. Stiffness of various sections of the same cross-sectional area.

properties with an appropriate shape. A somewhat similar result may be achieved by using materials with differing properties—e.g., metal-reinforced concrete or other conventional composite materials. Figure 105 shows the two possibilities for the case of bending.

Force Flow

Juvinall and Marshek (1991) and Marshek (1987) have discussed the topic of force flow at some length. It is helpful to visualize the lines of force

TABLE 38 Properties of Various Sections

	Section				
	Square Solid	Square Tube	Circular Solid	Circular Tube	Box Beam
Dimension	Side	Outer side	Radius	Outer radius	Width and depth
Size	1.00	1.51	0.56	1.17	1.71
I/c	0.17	0.39	0.14	0.51	0.52
I/c, normalized	1.00	2.35	0.85	3.09	3.11
T/J	0.21	0.66	0.28	1.03	0.30
T/J, normalized	1.00	3.18	1.36	4.95	1.44

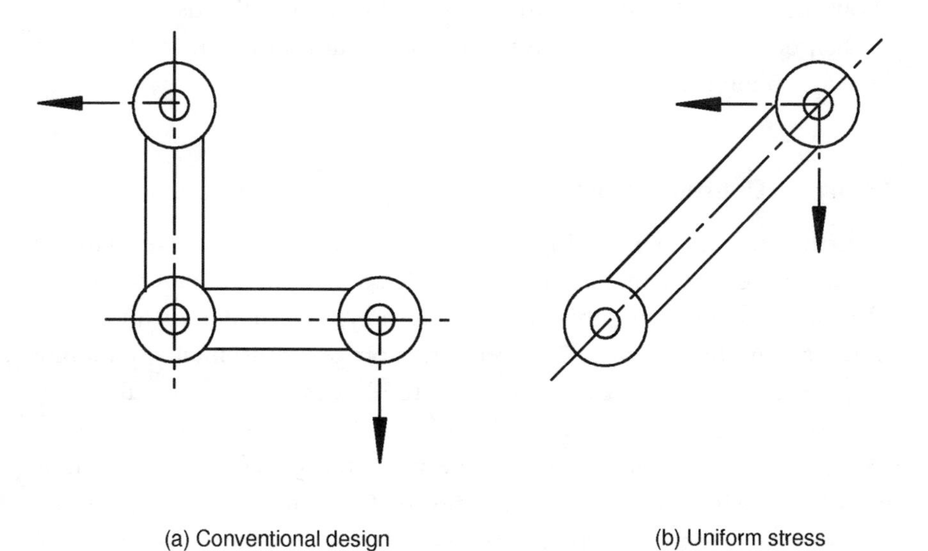

FIGURE 104. Minimum-weight design of lever.

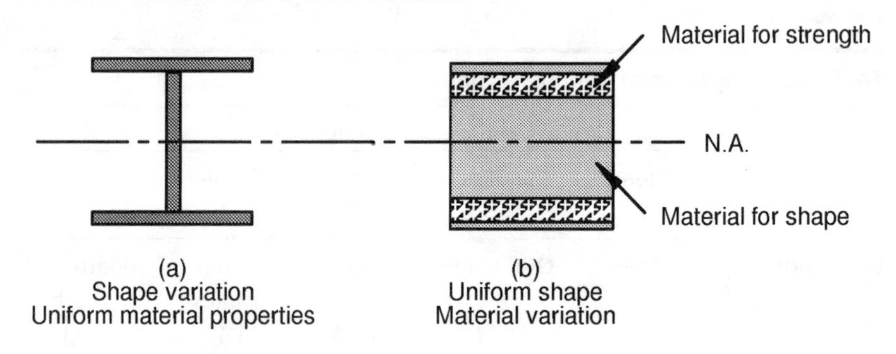

FIGURE 105. Shape versus material variation.

(stress) from the points of application of forces to the support points. By using this technique we can arrive at part shapes that are small and light. The method can be applied at both macro- and micro-levels. At the macro-level, we can determine whether any parts are missing or whether the appropriate supports and joints are present. At the micro-level the method shows us where stress concentrations might occur and thus be avoided, or their effects minimized. This is the type of information normally found by using photoelasticity.

Material Overhead Costs

A further means of reducing material costs is to reduce the material overhead costs, which can amount to 5% to 20% of the material costs. The means of achieving this is to use fewer types of materials and fewer different purchased parts. This reduces storage requirements, handling costs, and administrative expenses. Figure 106 shows that a manufacturer was able to reduce the total number of rolled stock sections from 136 to 65 through such rationalization. To be sure, storage requirements can be reduced by using just-in-time (JIT) delivery. Since material must be stored at some stage, there is a possibility that JIT will shift the costs upstream, resulting in higher costs of delivered materials. The only way we can avoid material storage is to synchronize its processing and handling from extraction through use.

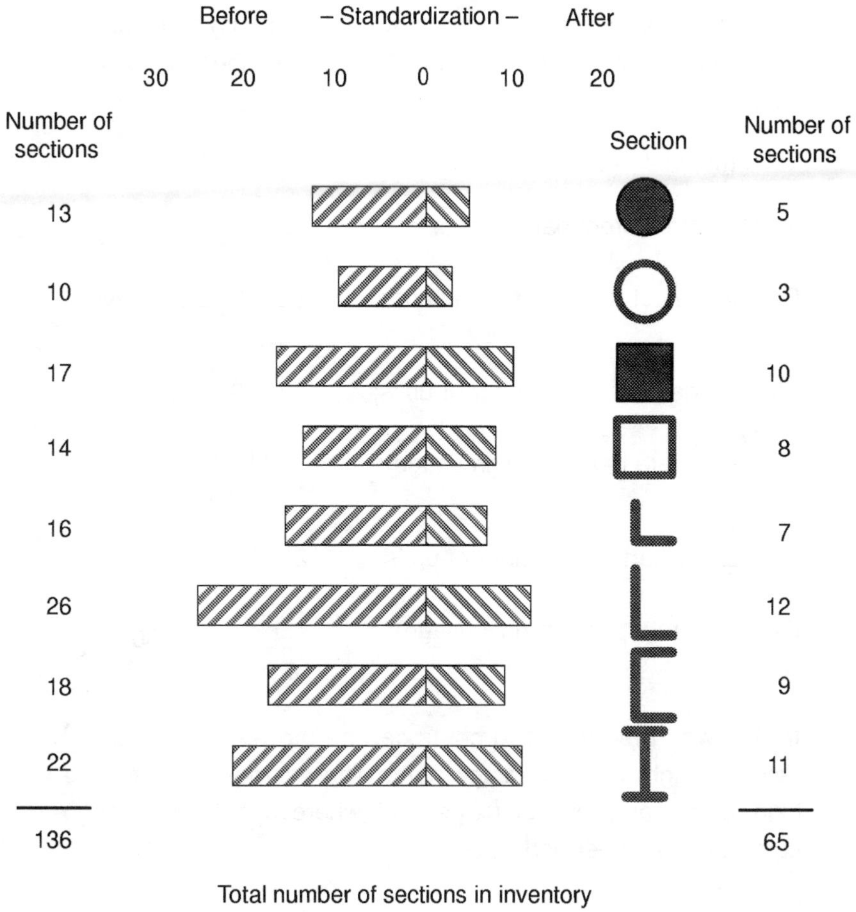

Before – Standardization – After

30 20 10 0 10 20

Number of sections	Section	Number of sections
13		5
10		3
17		10
14		8
16		7
26		12
18		9
22		11
136		65

Total number of sections in inventory

FIGURE 106. Reduction in types of material by standardization.

• • • • •

EXAMPLE
ELECTRONIC INSTRUMENTS

This example shows how the number of different parts affects the overhead costs. An electronic instruments manufacturer (Cooper and Kaplan, 1991) uses a large number (10,000) of a certain part A, and a low number (20) of another part B. (These are two of

5000 different parts in the company's inventory.) The company allocates overhead expenses based on part numbers as follows:

Total material overhead pool:	\$1,000,000
Number of different part numbers:	5000
Overhead cost for each part number:	$\frac{\$1,000,000}{5000} = \$200$
High-usage part (A), number of units:	10,000
Overhead rate per part (high usage):	$\frac{\$200}{10,000} = \$0.02$
Low-usage part (B), number of units:	20
Overhead rate per part (low usage):	$\frac{\$200}{20} = \$10.00$

In this way the overhead costs can be more fairly allocated between high- and low-usage parts. The overhead rate for the high-usage part is only \$0.02 per part, whereas for the low usage part it is \$10.00 per part.

• • • • •

EXAMPLE
ELECTRIC MOTOR HOUSING

Pahl and Beitz (1988) give the example of a welded electric motor housing. The old design used eight different plate thickness values (ranging from 2 mm to 16 mm). The redesign called for three plate thickness values (5 mm, 10 mm, 16 mm), which led to a slightly higher weight but reduced overall costs by reducing the overhead as well as the labor costs. The new design also took advantage of new numerically controlled flame cutting machines.

• • • • •

Reducing Cost per Unit Volume

After material volume reduction, the next steps to consider are those that reduce the cost per unit volume. This is achieved by using materials produced in large quantities, rather than specialty materials. A convenient means of comparing materials is to generate values of relative costs, using a common material such as mild steel as the basis. Special lists of this type can be useful, such as for simple cast parts of different materials.

In order to compare materials in terms of cost, strength, and weight, we consider the case of a bar of the same length L, carrying the same load P, made of different materials. Figure 107 shows the geometry and loading of the bar. Let

C_v = cost per unit volume of the material, \$
S = ultimate tensile strength of the material, MPa
ρ = density of the material, kg/m^3
A = cross-sectional area of bar, m^2
V = volume, m^3
C = cost (material) of bar, \$
W = weight of bar, kg
P = load, N

Table 39 shows the calculations for nine different materials. Column 2 shows their cost per unit volume, normalized to that of low-carbon steel (Kalpakjian, 1991) Column 3 shows the ultimate tensile strength (UTS), in MPa. Then the required area of the bar for a load P = 1000 N (assuming a factor of safety of 1) is shown in column 4 as

$$A = \frac{P}{S}$$

The volume of the bar is

$$V = AL = \frac{PL}{S}$$

which is also shown in column 4, if we take the length $L = 1$ m. The cost of the bar is shown in column 5 as

$$C = C_v V = C_v \frac{PL}{S}$$

which is shown in column 6 after being normalized to the value for low-carbon steel. The weight of the bar, shown in column 8, is (units are mixed)

$$W = \rho V = \rho \frac{PL}{S}$$

which uses the density values given in column 7 in kg/m^3. The weight, normalized to that of low-carbon steel, is shown in column 9.

From Table 39 we notice the following:

- The material cost for the bar to carry a given load (columns 5 and 6) is lowest for low-carbon steel and highest for the Ti alloy. The costs of nylon and Al alloy bars are also high.
- The weight of the bar to carry a given load (columns 8 and 9) is lowest for the Ti alloy and highest for gray cast iron. The weight is also relatively low for the Mg alloy, and relatively high for the nylon and Zn alloy.

The latter fact explains why Ti is finding increased use in aircraft, despite its high material (and processing) costs. We will discuss this further in the next example.

The values from column 2 (cost per unit volume), column 6 (actual cost), and column 9 (weight of bar), normalized to those for low-carbon steel, are shown in Figure 108.

The cost shown in column 5 of Table 39 is the cost per unit strength and has been suggested by Dieter for comparing different materials for strength. Let us consider two materials. In the following, the subscripts

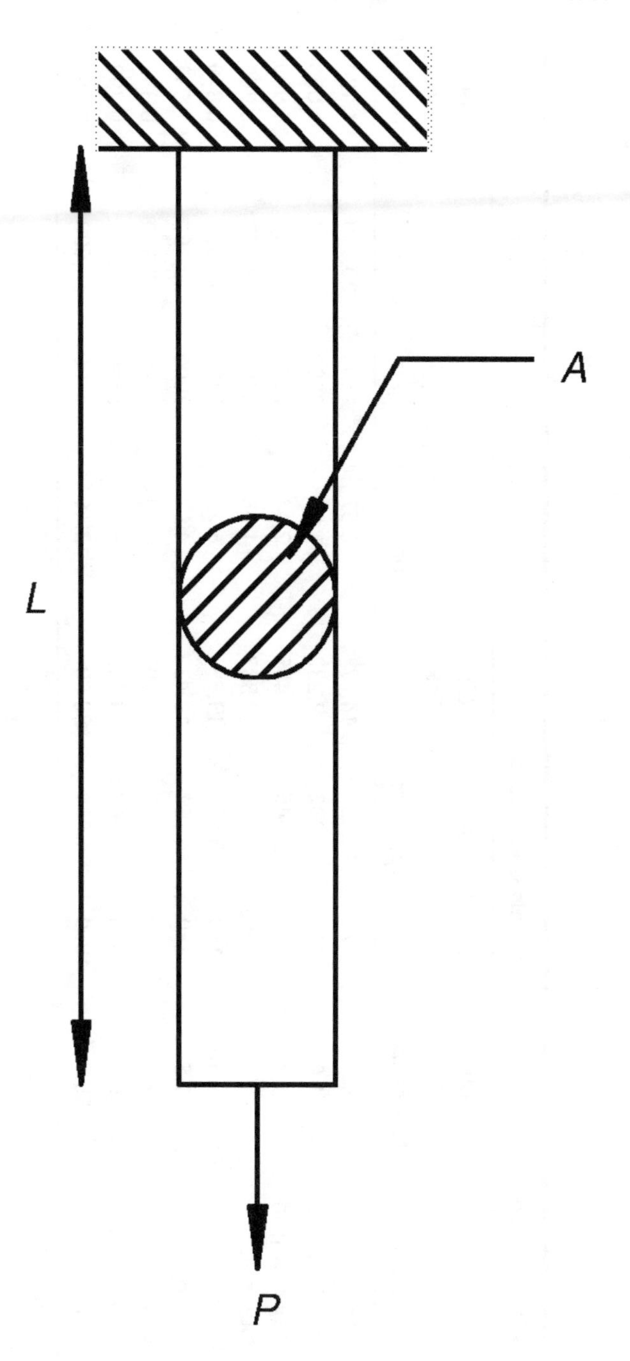

FIGURE 107. Bar under axial loading.

TABLE 39 Cost Comparison of Materials under Load

(1)	(2)	(3)	(4)	(5)	(6)	(7)	(8)	(9)
	C_v	S,	$AL = V$	C, \$	C	ρ,	W, kg	W
Material	(Normalized)	MPa	P/(3), m^3	(2) × (4)	(Normalized)	kg/m^3	(4) × (7)	(Normalized)
Nylon	1.1	75	13.33	14.67	6.57	1500	20,000	1.120
Low-C steel	1.0	448	2.232	2.232	1.00	8000	17,857	1.000
Gray cast iron	1.2	220	4.545	5.455	2.44	7500	34,091	1.909
HSLA steel	1.4	600	1.667	2.333	1.05	8200	13,667	0.765
Al alloy	2.5	180	5.556	13.89	6.22	2700	15,000	0.840
Zn alloy	2.5	240	4.167	10.42	4.67	7000	29,167	1.633
Mg alloy	3.0	300	3.333	10.00	4.48	1780	5,933	0.332
Stainless steel	5.0	600	1.667	8.333	3.73	8500	14,167	0.793
Ti alloy	30.0	1000	1.000	30.00	13.40	4600	4,600	0.258

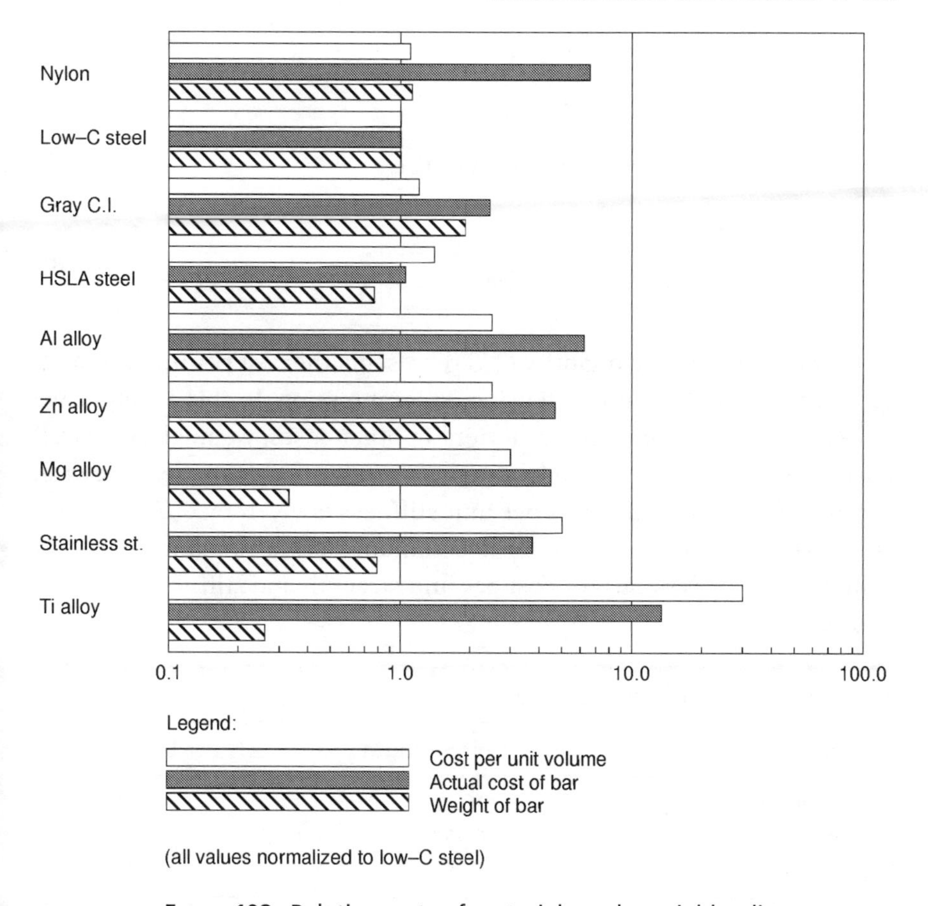

FIGURE 108. Relative costs of materials under axial loading.

1 and 2 refer to the respective materials. For the same load P and length L the costs of the two bars are

$$C_1 = C_{v1}A_1L \qquad \text{and} \qquad C_2 = C_{v2}A_2L$$

or

$$\frac{C_1}{C_2} = \frac{C_{v1}A_1}{C_{v2}A_2} \tag{45}$$

Also, since

$$P = A_1S_1 = A_2S_2 \qquad \text{or} \qquad \frac{A_1}{A_2} = \frac{S_2}{S_1}$$

we get, by substituting in Equation (45),

$$\frac{C_1}{C_2} = \frac{C_{v1}/S_1}{C_{v2}/S_2} \tag{46}$$

In equation (46) the quantity C_v/S represents the cost per unit strength for the given material under tension or compression. In this example, we have used UTS as the failure criterion. In other cases, we might very well have used the yield strength or fatigue strength, for example. Likewise, it can be shown that the cost per unit stiffness is given by C_v/E, where E is the Young's modulus. For three common cases of loading of a solid cylinder, the expressions for cost per unit strength and stiffness are given in Table 40.

TABLE 40 Minimum Cost Formulas

Loading Condition	Per Unit Strength	Per Unit Stiffness
Axial loading	C_v/S	C_v/E
Bending	$C_v/S^{2/3}$	$C_v/E^{1/2}$
Torsion	$C_v/S^{2/3}$	$C_v/G^{1/2}$

• • • • •

EXAMPLE
EFFECT OF WEIGHT ON LIFE CYCLE COST

As mentioned earlier, there are applications where the weight of a component plays a role in the operating cost of a system, e.g., for aircraft. We will consider the same materials for which the calculations were shown in Table 39. The part is in the shape of a rod, carrying a load of 1000 N. The costs of the different materials are as shown in Table 41, column 1. From the data used in Table 39, we calculate the weight, and from that the cost of the material in the part. These are shown in columns 2 and 3, respectively, in Table 41.

We shall call the cost in column 3 the initial cost of the part, C_{init}. The cumulative cost C_T after T hours of operation is given by

$$C_T = C_{init} + C_{op}WT \tag{47}$$

where

W = part weight

C_{op} = operating cost per unit weight and time

Figure 109 shows the plots of the cumulative costs for the different materials as they increase with operating life T. For these calculations we have used C_{op} = \$0.08/kg-h. From Figure 109 we notice that the rate of increase of cost with time for the high-strength materials is lower than that for low-strength materials. At the end of 500 h, the Al alloy has become more cost-effective than low-carbon steel. Since the Ti alloy part has the lowest weight, it

TABLE 41 Absolute Cost for Different Materials for Carrying a 1000-N Load

Material	(1) Material Cost, $/kg	(2) Weight, kg	(3) Part Material Cost, $
Nylon	6.45	0.0200	0.129
Low-C steel	1.10	0.0179	0.020
Gray cast iron	1.41	0.0341	0.048
HSLA steel	1.50	0.0137	0.021
Al alloy	8.15	0.0150	0.122
Zn alloy	3.14	0.0292	0.092
Mg alloy	14.83	0.0059	0.088
Stainless steel	5.18	0.0142	0.073
Ti alloy	57.39	0.0046	0.264

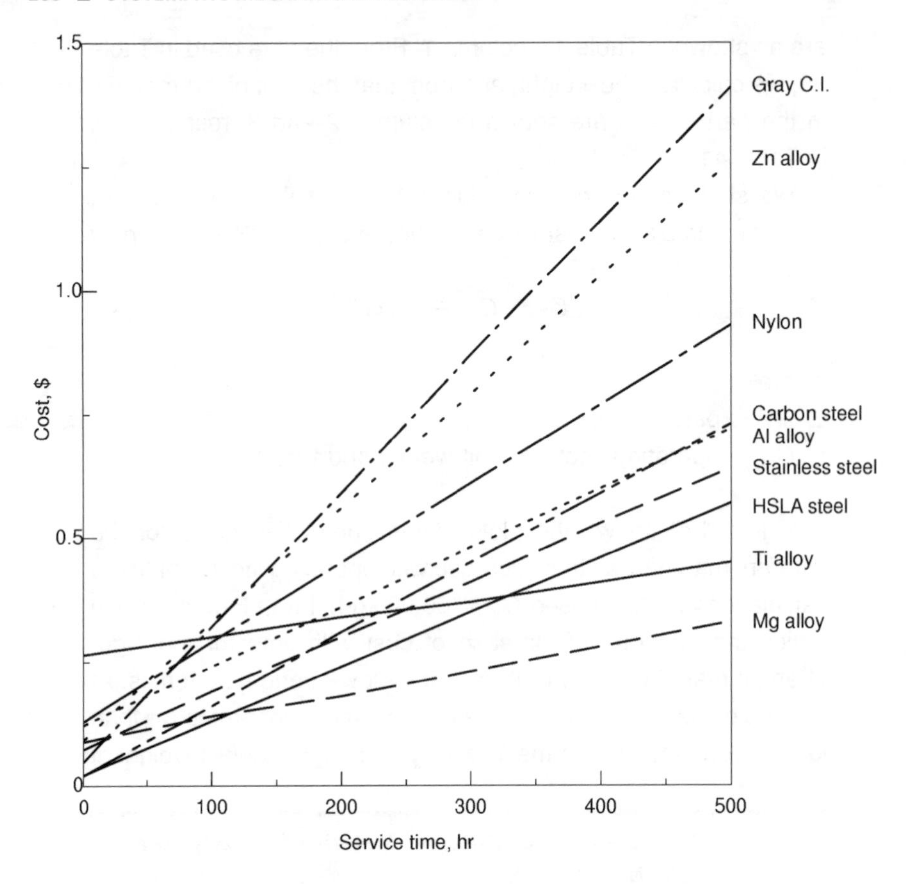

FIGURE 109. Effect of part weight on operating cost.

becomes the cheapest material in the long run. If we continue the calculations, we find that at $T = 1600$ h, the Ti alloy part also becomes cheaper than the part made out of Mg alloy.

We should note that the "initial" costs used in Figure 109 are material costs only and do not include processing costs. Were the processing costs to be included in the initial costs, the picture would look somewhat different. For example, the cost of the Ti alloy part would be relatively much higher, and would therefore require a longer service time to make it cost-effective.

• • • • •

Figure 110 shows the results for selected injection-moldable thermoplastics. These values correspond to those shown in Figure 108 primarily for metals. For each case, the rod 1 m long carries a load of 1000 N; see Figure 107. The results are based on yield strength and cost values given by Boothroyd et al. (1994).

USE OF PHYSICAL LAWS

Costs can be estimated based upon physical laws applied in the design analysis of a product. For instance, stress equations can relate load for a structural member or pressure in a pressure vessel to the amount of material. Analyses can lead not only to costs but also to their optimization. This is the most commonly used and obvious method of minimizing costs during embodiment.

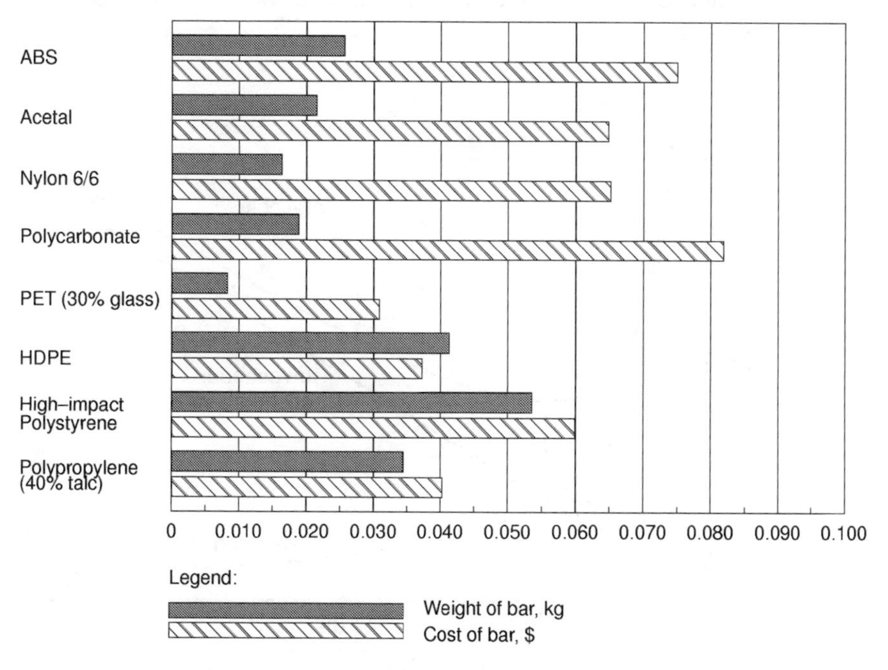

FIGURE 110. Costs of rods of molding plastics.

• • • • •

EXAMPLE
FINDING PIPE DIAMETER BY USING PHYSICAL LAWS

As a simple example, consider a pipe of constant thickness and of diameter *d*. Its annual cost is the sum of (1) operating (pumping) costs, which are proportional to $1/d^4$, and (2) investment cost proportional to *d*. The total cost is given as follows:

$$C = \frac{C_1}{d^4} + C_2 d \tag{48}$$

where C_1 and C_2 are coefficients, with units of $\$\text{-}m^4$ and $\$/m$, respectively. Figure 111 shows the plots of three cost terms in Equation (48). The value of the diameter that gives the minimum total cost can be found by setting the derivative of the cost equal to zero:

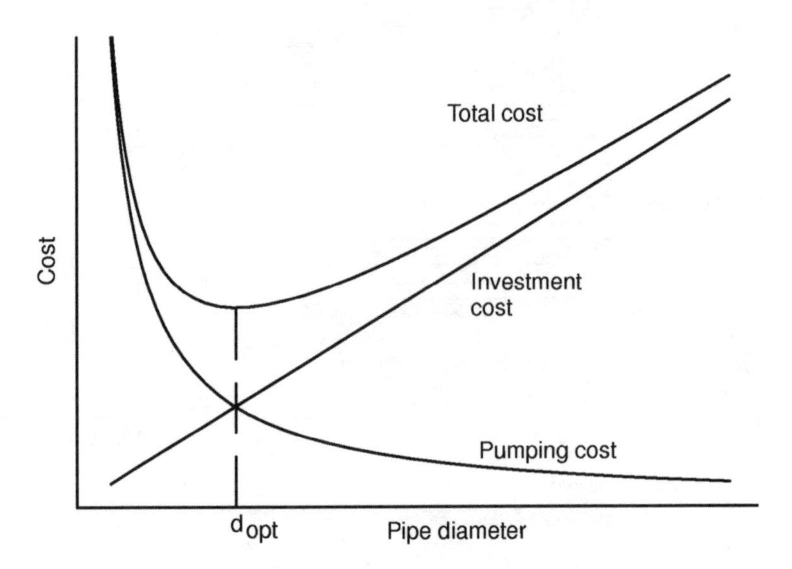

FIGURE 111. Optimum design of pipe.

$$\frac{dC}{dd} = 0 \tag{49}$$

Solving for d, we get the optimum diameter:

$$d_{opt} = \left(\frac{4C_1}{C_2}\right)^{0.2} \tag{50}$$

More details on mathematical methods of optimization are give in Appendix I.

• • • • •

SUMMARY

This chapter has provided a general treatment of engineering materials and methods of reducing material costs. The discussion is, of necessity, brief, since there are complete volumes devoted to the subject. The choice of material for a part is dictated by the requirements and other constraints—chemical environment, physical loading, quantities required, shape, manufacturing method, expertise available, and material availability. The materials discussed were metals, polymers, ceramics, and composites. Methods of reducing direct and overhead material costs were given. Costs of different materials under uniform loading were calculated.

BIBLIOGRAPHY

Askeland, D. R. *The Science and Engineering of Materials*. Boston: P.W.S.-Kent, 1994.

ASM Handbook (various volumes). Materials Park, Ohio: ASM International.

Boothroyd, G., P. Dewhurst, and W. A. Knight. *Product Design for Manufacture and Assembly*. New York: Marcel Dekker, 1994.

Cooper, R., and R. S. Kaplan. "Profit Priorities from Activity-Based Costing." *Harvard Business Review* 69: 130–135 (May–June 1991).

Dieter, G. E. *Engineering Design*. 2nd ed. New York: McGraw-Hill, 1991.

Holman, J. P. *Heat Transfer*. New York: McGraw-Hill, 1981.

Juvinall, R. C., and K. M. Marshek. *Fundamentals of Machine Component Design*. New York: Wiley, 1991.

Kalpakjian, S. *Manufacturing Processes for Engineering Materials*. Reading, Mass.: Addison-Wesley, 1991.

Kalpakjian, S. *Manufacturing Engineering and Technology*. Reading, Mass.: Addison-Wesley, 1992.

Khol, R., ed. *Machine Design 1994/5: Basics of Design Engineering*. Cleveland: Penton Publishing, 1994.

Marshek, K. M. *Design of Machine and Structural Parts*. New York: Wiley, 1987.

Matousek, R. *Engineering Design—A Systematic Approach*. London: Blackie, 1963.

Pahl, G., and W. Beitz. *Engineering Design—A Systematic Approach*. Berlin and New York: Springer Verlag, 1988.

Van Vlack, L. *Elements of Materials Science and Engineering*. 6th ed. Reading, Mass.: Addison-Wesley, 1989.

Manufacturing Processes and Economics

- **Primary Metal Shaping Processes**
  - *Example: Manufacturing Cost of a Casting*
  - *Example: Cost of Cast Parts of Different Materials*
- **Metal Deformation Processes**
  - *Example: Cost of Die-Stamped Parts of Different Materials*
- **Machining Processes**
  - *Example: Cost of Machined Parts of Different Materials*
- **Joining Processes**

- **Manufacturing Processes for Plastics**
  - *Example: Plastic Part Design for Proprinter*
  - *Example: Cost of Injection-Molded Parts of Different Plastics*
- **Manufacturing Processes for Composites**
  - *Example: A New Office Chair Design*
- **Non-Traditional Processes**
- **Dimensional Tolerance and Surface Finish**
- **Material Wastage During Processing**
- **The Make-Versus-Buy Decision**
- **Summary**
- **Bibliography**
  - *Example: Design Variants of a Bearing Pedestal*
  - *Example: Design of a Lever*

There are several excellent books on manufacturing processes; some noteworthy ones are by Kalpakjian (1991, 1992), and Groover (1995). In this chapter, only a brief account of the different processes will be given so that the designer can make informed decisions while keeping in mind the requirements, including costs. It is important to work with manufacturing experts before the design is finalized.

Manufacturing processes can be broadly classified as primary, secondary, and tertiary processes. These designations refer to processing stages *beyond* the raw material stage. Thus, molding and forging are primary processes, whereas machining and welding, which follow these, are secondary processes. Tertiary processes include heat and surface treatments of various kinds.

The manufacturing processes are largely dictated by the material selected; in fact, the choice is often narrowed down to one process for certain materials. Where alternative processes are available for a given material, the other requirements, e.g., shape, quantities required, and cost, lead to a logical choice of manufacturing processes.

Major requirements governing the choice of the manufacturing process are:

- Shape
- Material
- Quantities required
- Physical properties
- Surface finish
- Facilities available—machines, tools, fixtures
- Tolerances
- Cost

Certain processes allow the creation of more complex shapes, e.g., casting processes for metals, die stamping, and molding of plastics. Thus, these processes allow more functions to be combined into each part, which makes for easier assembly.

Specific combinations of processes and materials impose limitations on the thickness of the section that may be produced and also allow only certain relationships between the size and the section thickness. This information is shown in Figure 112 (from Schey, 1987) for a few materials and processes. The thinnest sections can be produced by cold rolling, followed by hot rolling, except for small parts, which can be die-cast to very thin shapes.

Figure 113 shows the sequence of processing for metals starting from the raw material to the finished part. The sequence of operations is not unique. For example, casting may be followed by some machining, followed by surface treatment, and again by machining. The figure identifies the primary, secondary, and tertiary processes for metals. The different processes are classified further in Figure 114.

Design of a part should be such that it takes full advantage of the particular manufacturing process. In order to reduce the manufacturing

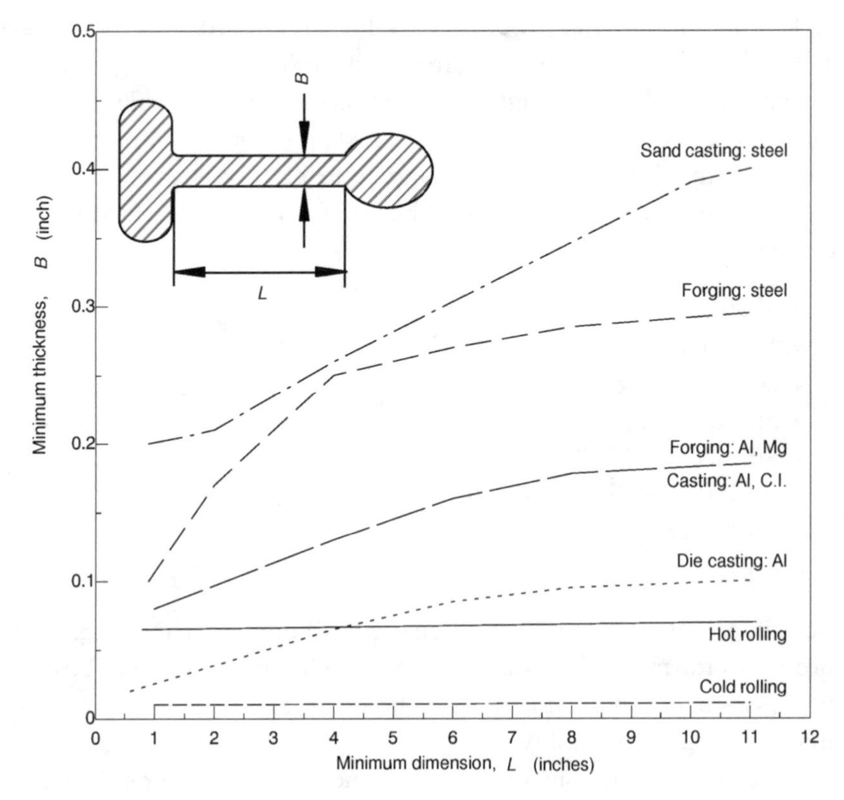

Fig. 112. Minimum size and thickness (Schey, 1987).

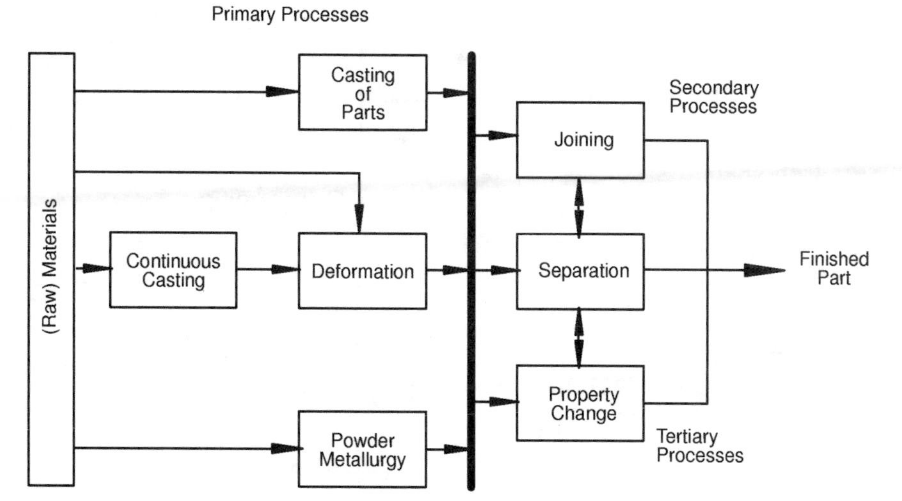

FIGURE 113. Sequence of manufacturing processes for metals.

costs, we look at the individual steps in that process and determine how costs may be reduced.

Production processes for metals can be classified as shown in the overview in Figure 114. There are over 200 production processes used by industry today, some dating back thousands of years. As stated earlier, the choices of materials, processes, and shapes are closely tied together, as was shown in Figure 83 in Chapter VII. Production processes not only change the shape but also influence properties of the materials. In fact, many of the processes are necessary to attain the required properties.

PRIMARY METAL SHAPING PROCESSES

The primary shaping processes for metals are shown in Figure 115. Casting is by far the most common process; it takes the raw material and forms it into a semifinished shape.

Cast parts are produced from a variety of metals. If there is one outstanding advantage of casting it is the production of intricate shapes, which enables integration of many functions into fewer parts. In choosing a particular casting process we keep in mind the following properties:

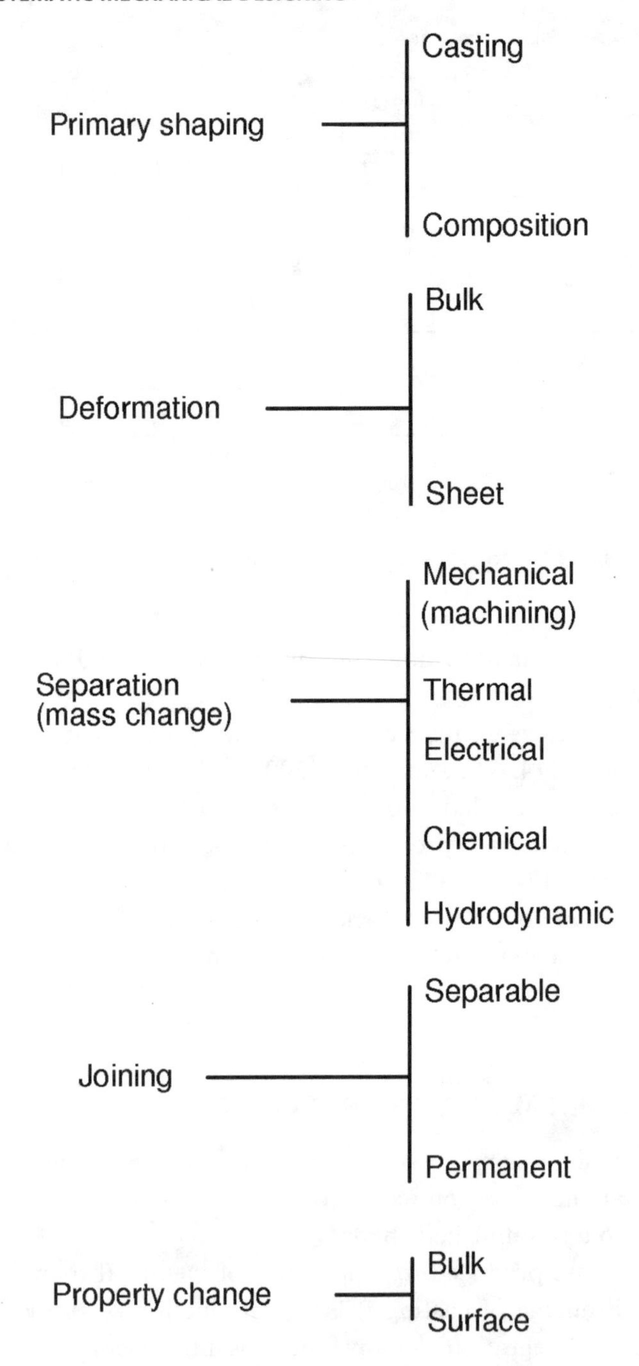

FIGURE 114. Manufacturing processes for metals.

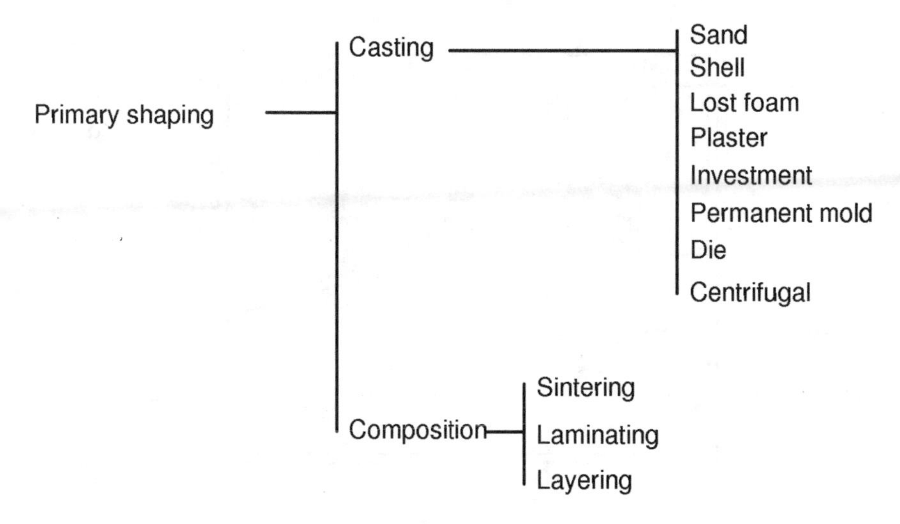

FIGURE 115. Primary shaping processes for metals.

- Weight of the casting
- Surface finish produced
- Section thickness desired (see Figure 112)
- Porosity in the casting
- Shape complexity achievable by a given process
- Dimensional accuracy desired
- Quantities to be produced and cost per unit

The different metal casting processes will be described next in brief.

SAND CASTING

Sand casting is the oldest casting method in use today. Worldwide, it accounts for the largest tonnage of castings produced. The mold is produced by placing the pattern in the box and filling the cavity with sand. Patterns are removed prior to pouring; thus typical molds are made in two or more pieces.

Figure 116 shows the details of a generic mold and casting system. The mold medium (sand, with bonding clay) is contained in the mold flask or box, which is generally in two halves. The two halves, cope and drag, join and part at the parting plane. The parting plane plays a significant role in the geometry that may be cast. The mold is produced by

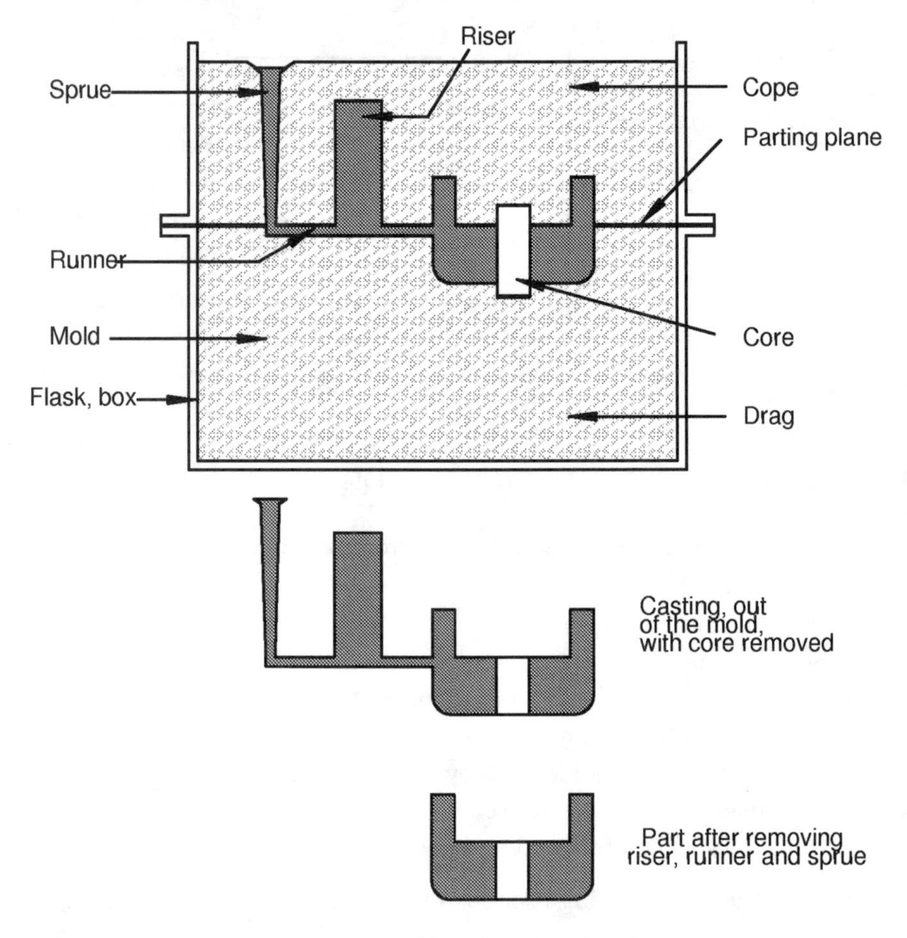

FIGURE 116. Mold and casting details.

placing the pattern, made of wood, plastic, or metal, inside the box and filling the cavity with the mold medium. The metal is carried to the cavity via the sprue and the runner. Patterns are removed prior to pouring. Cores are inserts made from sand which form the interior surfaces of the casting. A system of risers ensures filling and makes up for shrinkage.

Figure 117 shows characteristics of the sand casting process as regards the materials that are cast using this process, ranges of the casting weight, surface finish, as well as the the expected porosity in the casting, complexity of the shapes that may be cast, and the dimensional accuracy that

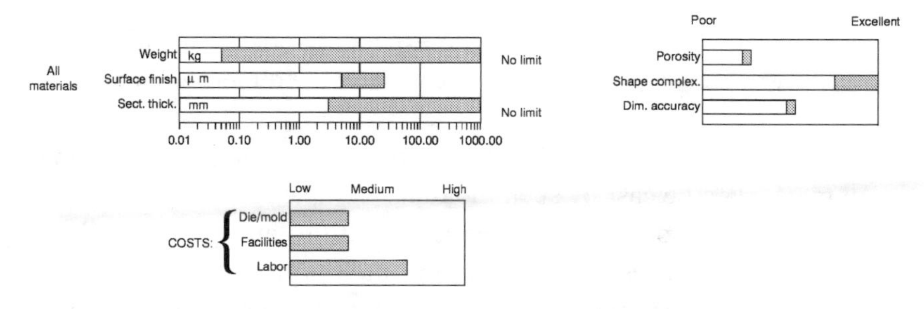

FIGURE 117. Metal casting characteristics for sand casting.

might be expected from this process. The figure also provides the relative costs for the die/mold, facilities, and labor for this casting process.

Sand casting is used to make a large variety of parts such as engine blocks and housings, machine bases, and pump housings. All metals can be sand-cast. The process is suitable for small to medium lot sizes due to its lower mold and facilities costs and medium labor costs.

SHELL MOLD CASTING

A shell mold casting is made by coating a metal pattern with a silicone parting agent, covered with sand containing a resin binder. It is heated to cure the resin. The shell is removed from the pattern after hardening. The thickness of the shell, usually 5 to 10 mm, can be controlled to provide strength and rigidity. The two halves of the shell form the mold.

Figure 118 shows characteristics of the shell mold casting process as regards the materials that are cast using this process, ranges of the casting

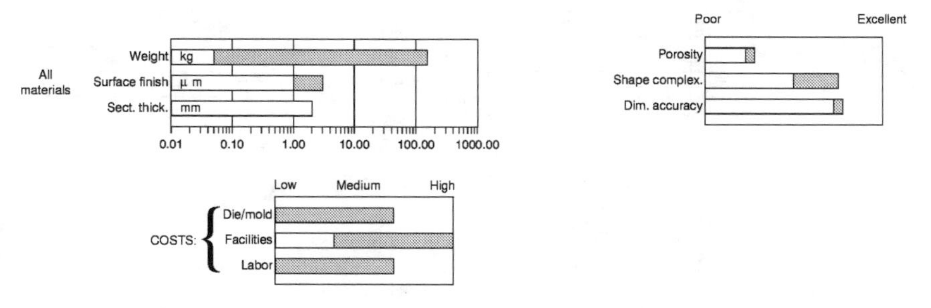

FIGURE 118. Metal casting characteristics for shell mold casting.

weight, surface finish, as well as the expected porosity in the casting, complexity of the shapes that may be cast, and the dimensional accuracy that might be expected from this process. The figure also provides the relative costs for the die/mold, facilities, and labor for this casting process.

All metals can be cast by shell molding. Compared with sand casting, shell molding is economical for higher quantities and produces shapes with smoother surfaces, sharper corners, and thinner sections. Cleaning, machining, and other finishing costs are lower. Shell molding is used for high-precision parts such as connecting rods and cylinder heads and housings.

Lost Foam Casting

Lost foam or lost pattern casting employs a polystyrene pattern that is evaporated by the heat of the molten metal as it is poured. The process starts with a heated die, which is filled with expandable polystyrene beads, which fuse and bond together to form the pattern. The pattern is placed in the mold box, which is then filled with sand and compacted. The mold, called the flask, is in one piece, with no parting line. Thus, there is much more freedom in part design, leading to function integration, with fewer fasteners and other parts. The mold and the polystyrene are inexpensive. Cleaning, machining, and other finishing costs are low. The major cost is of the initial die.

Figure 119 shows characteristics of the lost foam casting process as regards the materials that are cast using this process, ranges of the casting weight, surface finish, as well as the expected porosity in the casting,

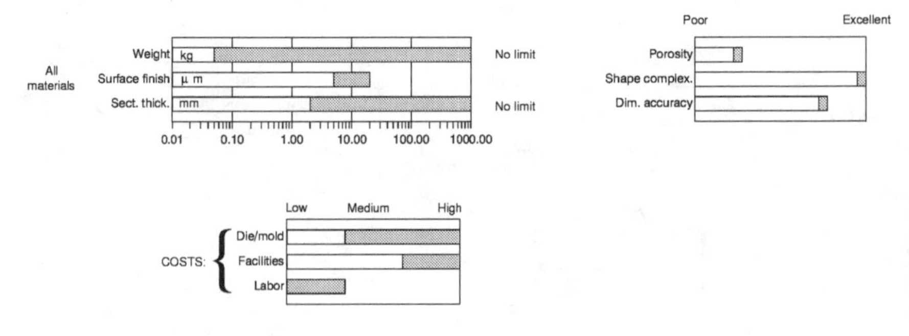

Figure 119. Metal casting characteristics for lost foam casting.

complexity of the shapes that may be cast, and the dimensional accuracy that might be expected from this process. The figure also provides the relative costs for the die/mold, facilities, and labor for this casting process.

Lost foam casting is used for automobile engine parts such as aluminum engine blocks and cylinder heads, crankshafts, manifolds, brake parts, and machine bases.

Investment Casting

Investment casting, also called the lost wax process, dates back thousands of years. The process is similar to the lost foam process. It uses either wax or polystyrene to make the pattern. Investment casting is used for parts made from ferrous or nonferrous materials, such as small components for office machines. Its advantages are a high degree of dimensional accuracy and its usefulness in obtaining complex shapes with low porosity of castings.

Figure 120 shows characteristics of the investment casting process with regard to the materials that are cast using this process, ranges of the casting weight, surface finish, as well as the expected porosity in the casting, complexity of the shapes that may be cast, and the dimensional accuracy that might be expected from this process. The figure also provides the relative costs for the die/mold, facilities, and labor for this casting process.

Plaster Mold Casting

Plaster molds are made by pouring plaster of Paris over a pattern made of nonferrous alloy or plastic. Also called precision casting,

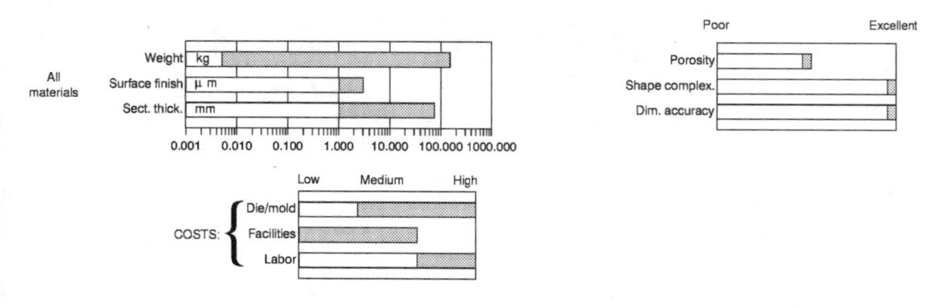

Figure 120. Metal casting characteristics for investment casting.

this process provides a high degree of surface finish, shape complexity, and dimensional accuracy. Its use is limited to nonferrous alloys due to temperature limitations of plaster. A variation of this process, the ceramic mold process, is used for high-temperature metals. Plaster mold casting is used for ornaments, fittings, valves, and small components.

Figure 121 shows characteristics of the plaster mold casting process as regards the materials that are cast using this process, ranges of the casting weight, surface finish, as well as the expected porosity in the casting, complexity of the shapes that may be cast, and the dimensional accuracy that might be expected from this process. The figure also provides the relative costs for the die/mold, facilities, and labor for this casting process.

PERMANENT MOLD CASTING

Permanent mold casting uses molds that can be reused — permanent molds — made of steel, cast iron, refractory metal alloys, or graphite. Permanent mold casting is used for casting metals with lower melting points than that of the mold material. Due to high equipment and low labor costs, this process is suitable for large production runs. Parts made by this process include gear blanks, kitchenware, and automobile engine parts. Advantages are high surface finish and close tolerances. Controlled heat dissipation produces good mechanical properties.

Figure 122 shows characteristics of the permanent mold casting process as regards the materials that are cast using this process, ranges of the casting weight, surface finish, as well as the expected porosity in the casting, complexity of the shapes that may be cast, and the dimensional

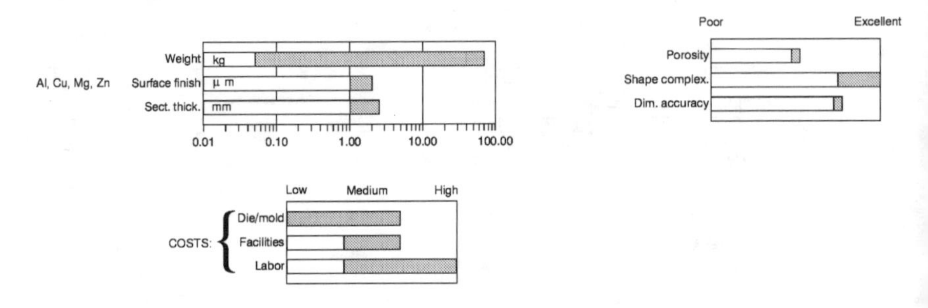

FIGURE 121. Metal casting characteristics for plaster mold casting.

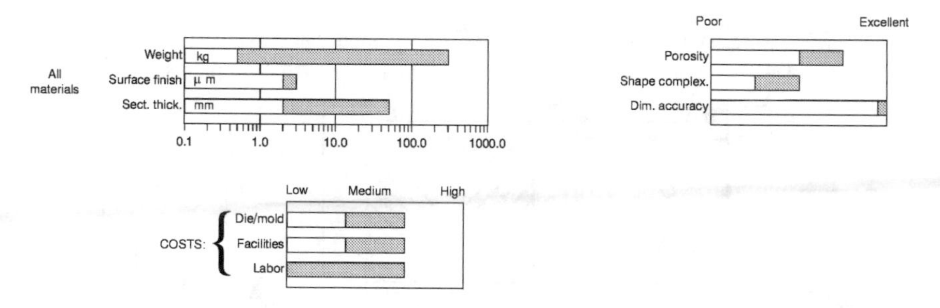

FIGURE 122. Metal casting characteristics for permanent mold casting.

accuracy that might be expected from this process. The figure also provides the relative costs for the die/mold, facilities, and labor for this casting process.

DIE CASTING

Molten metal is forced into the mold in die casting, a type of permanent mold casting process. Die casting is used for making parts for toys, appliances, automobiles, and business machines. Die casting is of two types. Hot chamber die casting is used for low-melting point alloys of zinc, tin, and lead. Pressures in hot chamber die casting range from 2000 to 5000 psi. The cold chamber process is used for higher–melting point alloys of aluminum, magnesium, and even ferrous metals. Pressures in cold die casting chambers range from 5000 to 20,000 psi. Figure 123 shows the general layout of a cold chamber die casting machine.

Figure 124 shows characteristics of the die casting process as regards the materials that are cast using this process, ranges of the casting weight, surface finish, as well as the expected porosity in the casting, complexity of the shapes that may be cast, and the dimensional accuracy that might be expected from this process. The figure also provides the relative costs for the die/mold, facilities, and labor for this casting process.

Die casting produces castings with high dimensional accuracy and low porosity and can be used where moderate shape complexities are required. Economically, this process is suitable for large production quantities. Inserts such as screws and pins can be cast in place. Die casting is used to produce a wide variety of parts such as for appliances, automobiles,

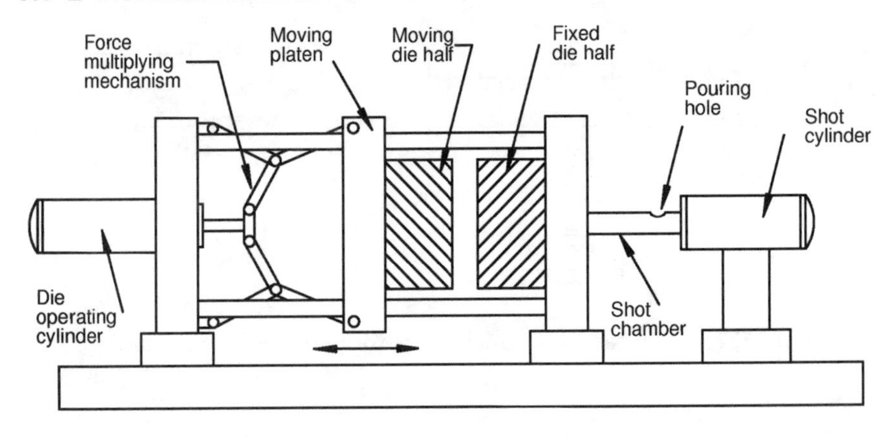

FIGURE 123. A die casting machine.

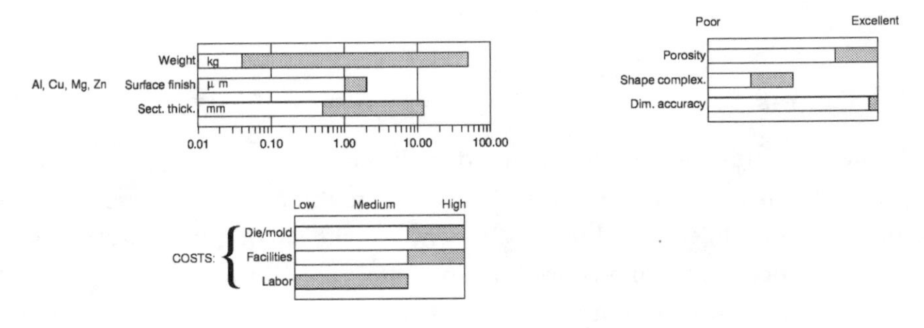

FIGURE 124. Metal casting characteristics for die casting.

electric motors, plumbing fixtures, and power tools. Alloys most commonly die-cast are of aluminum, copper (brass), magnesium, and zinc. It is a near-net-shape forming process. Die casting allows for function integration, with fewer parts and manufacturing operations.

CENTRIFUGAL CASTING

Centrifugal casting uses rotation of the mold during the casting process. Centrifugal forces produce pressure, which improves properties of the casting. Certrifugal casting is used for casting cylindrical parts such as cylinder liners, bushings, and bearing rings, and parts with axial symmetry such as wheels.

Figure 125 shows characteristics of the centrifugal casting process as regards the materials that are cast using this process, ranges of the casting weight, surface finish, as well as the expected porosity in the casting, complexity of the shapes that may be cast, and the dimensional accuracy that might be expected from this process. The figure also provides the relative costs for the die/mold, facilities, and labor for this casting process.

Cost Drivers for Metal Castings

The major cost drivers for a metal casting are:

- Pattern and cores
- Material for the part
- Postpouring processes, e.g., cleaning, machining, surface treatment
- Equipment used in the casting process

The cost charts shown in Figures 117 through 125 give an indication of the economics of a process as regards the quantities produced. A process that has low die or mold and facilities costs and higher labor costs is suitable for low production quantities. On the other hand, a process with high die or mold and facilities costs and low labor costs is justified only for large-quantity production.

Figure 126 shows the components of the manufacturing cost of steel castings (Ehrlenspiel, 1985) as a function of weight, another example of magnitude-based costing (MBC). The various components are seen to vary linearly with the logarithm of casting weight. Such a cost structure shows where the greatest potential cost savings lie. For example, for small castings there is little to be gained in trying to save on material; there

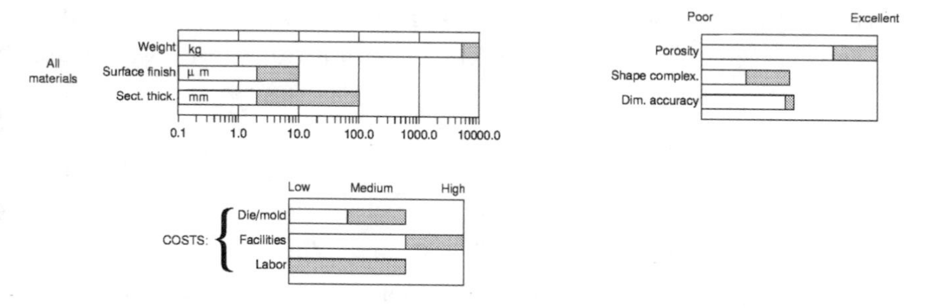

Figure 125. Metal casting characteristics for centrifugal casting.

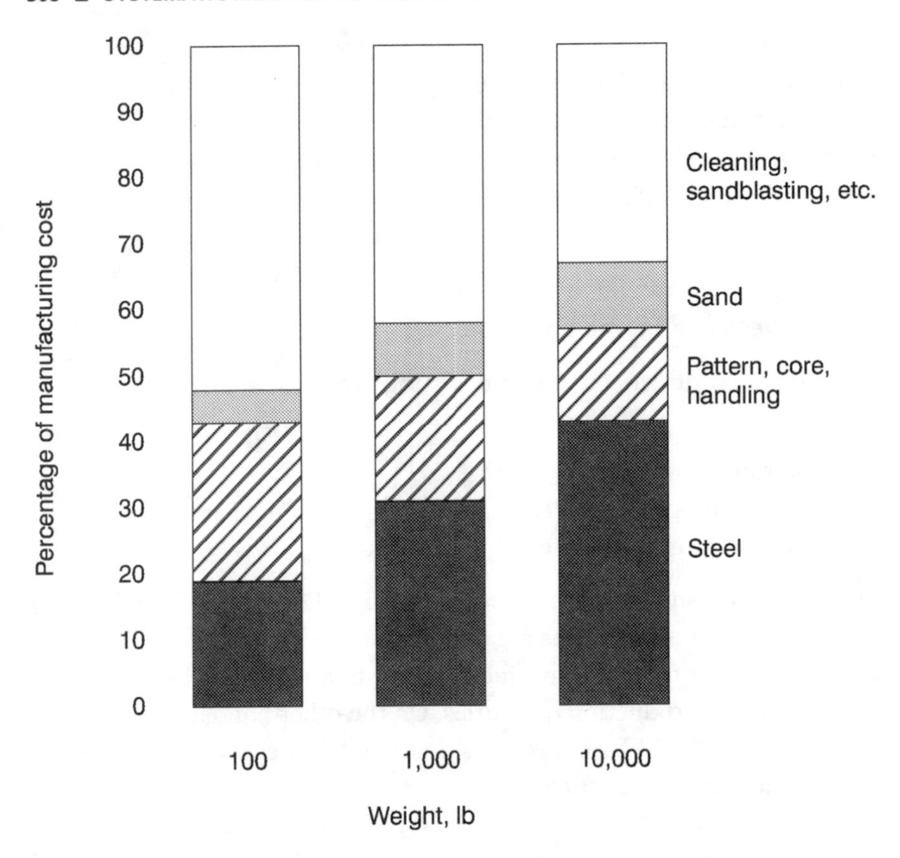

FIGURE **126.** Cost components of cast steel parts. (Ehrlenspiel, 1985)

are more fruitful possibilities—try to reduce the cost of cleaning and sandblasting, or the pattern and core costs.

The cost of castings depends on casting weight and the complexity of the casting (Pahl and Rieg, 1984). The general formula for cost per unit weight is of the form:

$$C = c_{\text{cl}} W^m \tag{51}$$

where W is the weight, c_{cl} a coefficient depending on the "complexity class" (from c_{I} through c_{X}), and the exponent $m = -0.12$. The classes range from *Complexity class I*, solid casting with no cores and recesses, to *Complexity class X*, hollow casting with complex coring. Figure 127 shows graphs of Eq. (51).

We will now address the major cost drivers for castings and see how each may be controlled to reduce costs. Figure 128 shows the implementation of some of these concepts.

Steps for Reducing Pattern Costs

- Use simple shapes, composed of flat or cylindrical surfaces.
- Aim for one-piece patterns.
- Avoid cores. Design the casting with ribs, instead of hollow shapes.
- Use tapers starting at the parting plane.

Steps for Reducing Rejects

- Cores should be well supported to avoid movement during pouring.
- Keep wall thickness uniform.
- Avoid sharp corners and junctions. They cause cracks.
- Avoid horizontal surfaces. They impede flow of gases.
- Avoid a narrow section between coupon and riser. The greater the shrinkage, the more important this is.
- Make stiffening ribs thinner than the wall being stiffened.

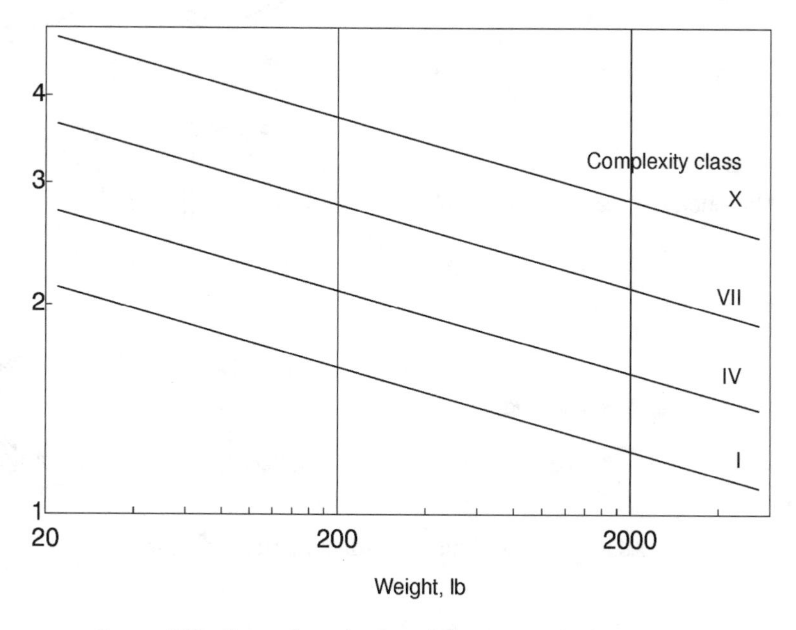

FIGURE 127. Cost of castings and their complexity.

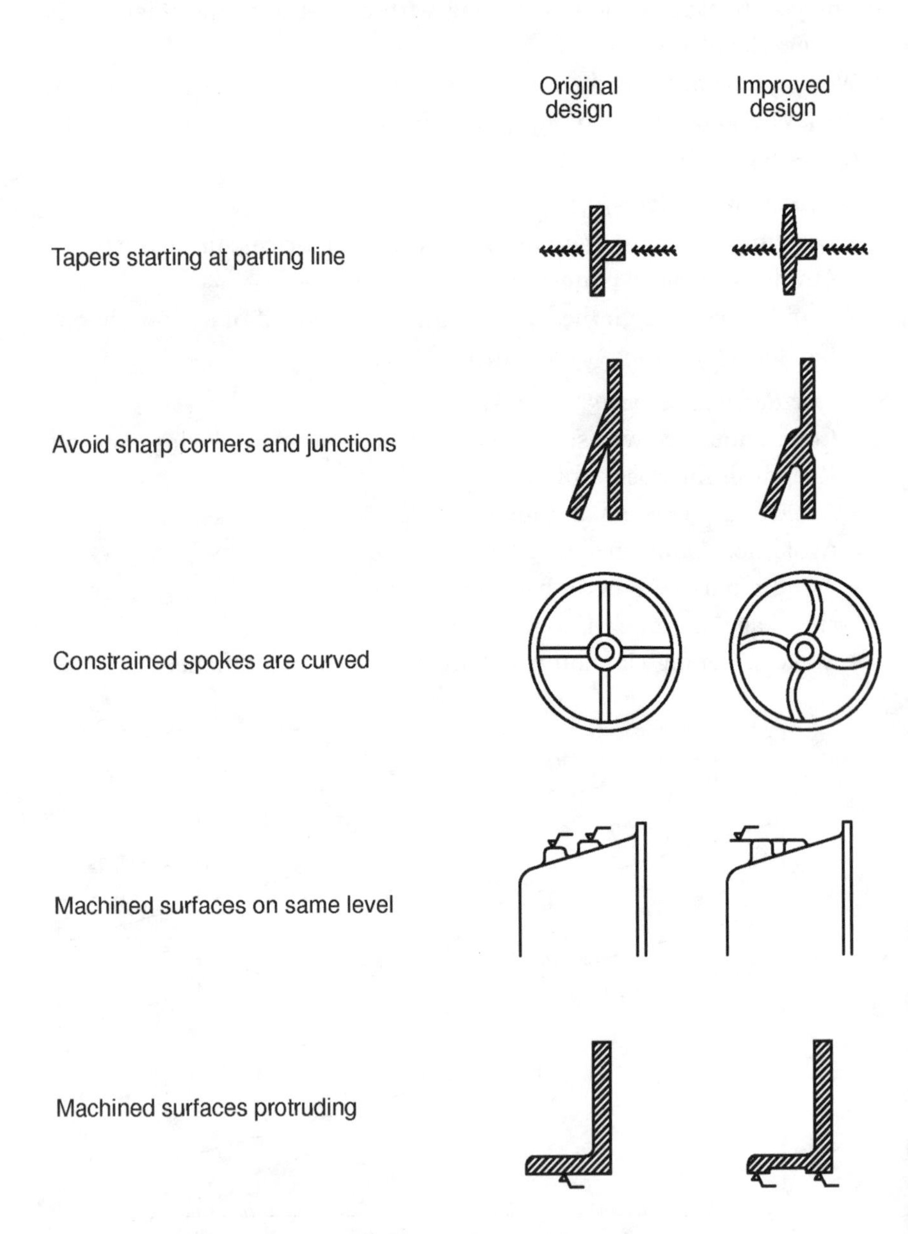

FIGURE 128. Examples of reducing metal casting costs.

- Allow shrinkage of constrained arms, ribs, and spokes by using curved shapes.

Steps for Reducing Cleaning Costs

- Avoid dead-end spaces, cavities. Provide for easily accessible reentrant corners.
- Put coupon and riser on a flat or convex surface where it can be easily chipped off.
- Surfaces to be ground should be easily accessible.

Steps for Reducing Finishing Costs

- Arrange parting plane so the flash can be easily chipped off.
- Provide easy access for tools.
- Closely located machined surfaces should be on the same level.
- Machined surfaces should stick out from as-cast surfaces.
- Avoid machining at an angle.

● ● ● ● ●

EXAMPLE
MANUFACTURING COST OF A CASTING

An application of regression analysis is shown by Pacyna et al. in their article on cost of cast parts for simple gray iron castings (1982). They base the cost model on the geometric ratios r_e, r_t, and r_v, which are based upon considering a solid of similar overall shape and same volume as the net volume of the casting, as shown in Figure 129.

$$r_e = \text{elongation ratio} = \frac{d_c}{d_s}$$

$$r_t = \text{wall thinness ratio} = \frac{s_s}{f_c}$$

$$r_v = \text{volume (bulkiness) ratio} = \frac{V_c}{V_s}$$

These ratios account for the influence of shape in the cost model.

The cost per unit $C_{M.C.I.}$ for manually cast gray cast iron can be approximated by:

$$C_{M.C.I.} = K_p K_c n^{-0.0782} v_s^{0.8179} r_e^{-0.1124} r_t^{0.1655} r_v^{0.1786} n_c^{0.0387} S_u^{0.2301} \quad (52)$$

where

K_p = a proportionality factor, \$/unit
K_c = complexity factor (0.9 to 1.4)
v_s = volume of casting material
n_c = number of cores (= 0.5 if no cores)
S_u = tensile strength

The authors also provide cost models for patterns. For a hand-made pattern, the cost equation is of the form:

$$C_{MP} = K_p v_P^{0.373} r_t^{0.572} P_C N_{CB}^{0.155} F_Q \quad (53)$$

where

v_p = enclosing volume of the pattern parts
P_C = complexity of the pattern (0.3 to 3)
N_{CB} = number of core boxes
F_Q = quality factor (from table)

• • • • •

EXAMPLE
COST OF CAST PARTS OF DIFFERENT MATERIALS

Ulrich and Eppinger (1995) show examples of unit part cost of sand cast gray cast iron and investment-cast aluminum alloy and yellow brass parts. The parts are of varying degrees of complexity. Figure 130 shows the graphs of these values as a function of quantities produced. The fixed costs consist of the one-time tooling cost. The variable costs per unit are the material cost and "processing" costs, which include machine time, direct labor, and overhead costs. The unit cost for producing one unit is dictated almost

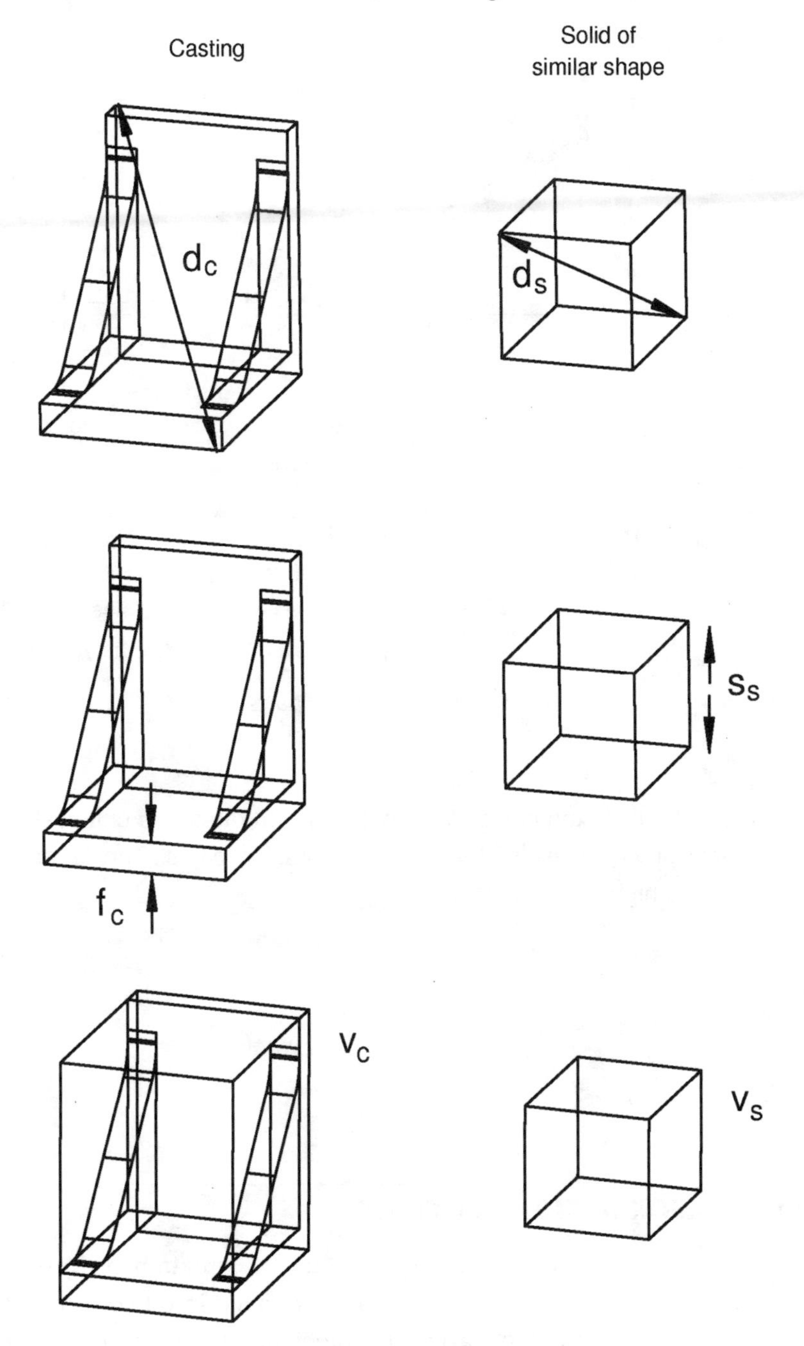

FIGURE 129. Casting geometry.

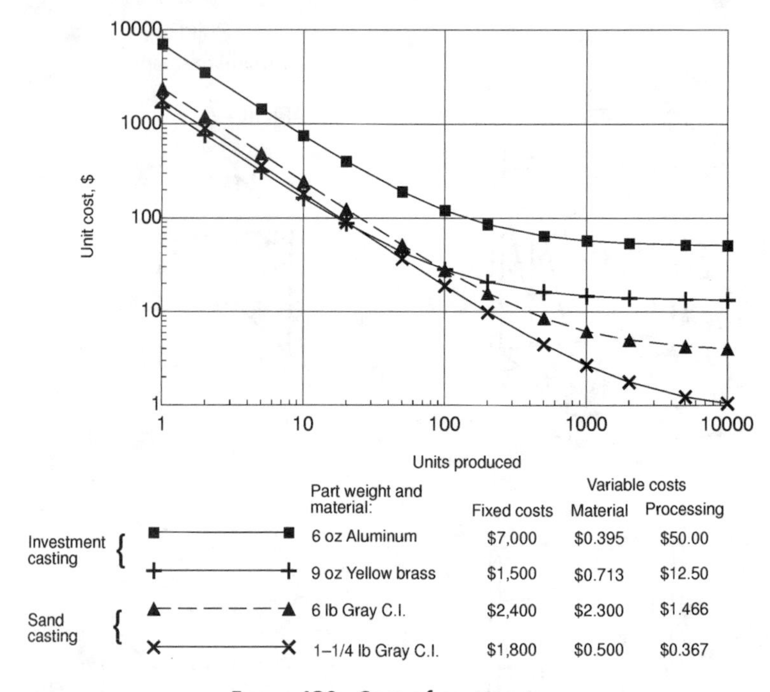

FIGURE 130. Cost of cast parts.

entirely by the fixed costs. At high production rates, the unit cost reflects primarily the material and processing costs. The latter costs are high for complex parts.

• • • • •

METAL DEFORMATION PROCESSES

Figure 131 shows the classification of metal deformation processes in more detail. These processes can be broadly divided into two categories: bulk forming, i.e., those that start with a block of raw material; and those used on metal already in sheet form.

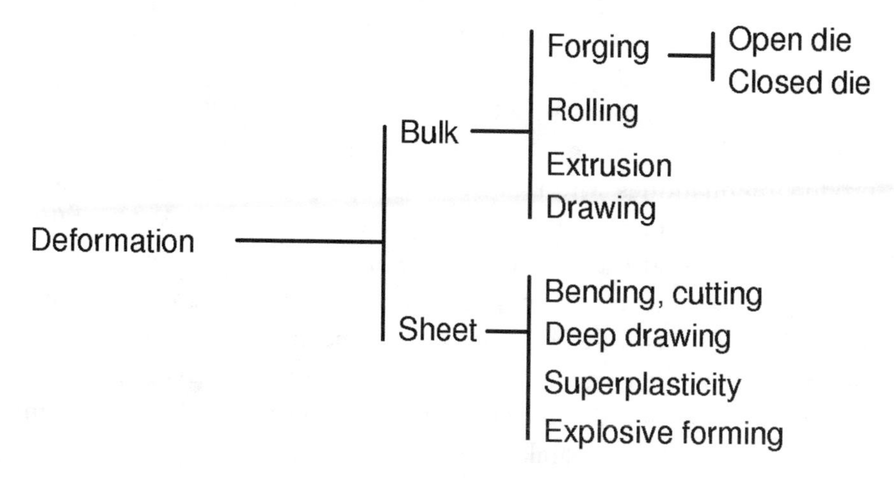

FIGURE 131. Metal deformation processes.

FORGING

Forging dates back thousands of years. Manual forging is still practiced by blacksmiths and goldsmiths. The development of the drop hammer in the early nineteenth century, and later of forging presses, made forging an important process of the Industrial Revolution. Forging produces discrete parts with high strength and toughness, particularly in the direction of material flow. For critical applications such as shafts, disks, gears, tools, connecting rods, fasteners, railroad wheels, and aircraft parts, forging is generally the preferred forming method.

Nonferrous metals and alloys, such as Al, Cu, and Mg, and their alloys, are the easiest to forge. By ease of forging we mean, for example, low force requirement, intricate shapes, and low cost. Nonferrous metals are followed, in increasing difficulty of working, by carbon steels, stainless steels, and Ti alloys. Superalloys of Fe, Co, Mo, and Ni are the most difficult to forge.

Forging may be classified by the speed of operation, i.e., drop forging versus press forging. It may also be classified by how the metal flows in the dies as open or closed die forging. Open die forging requires manual manipulation of the workpiece, and thus depends on the operator's skill. It is suitable only for one-of-a-kind or small lot sizes. Open dies are of flat or other simple shapes. They are used for producing simple shapes, but can produce parts in a wide range of sizes. The parts have very poor dimensional accuracy.

Closed die forging can be either impression die or flashless forging. Closed dies deliver good dimensional accuracy and can be used for large production quantities with repeatable accuracy. Impression die forging is the term used for dies shaped to produce the desired part. Metal flow produces a flash at the die interface, which must be trimmed off. So-called precision closed die forging can produce very high dimensional accuracy, requiring little or no subsequent machining. Flashless forging is a true closed die forging. The metal stays completely within the dies and no flash is produced. This process requires close control of the amount of metal in the starting blank.

Other forging operations include upsetting, heading, and roll and orbital forging. Figure 132 shows some of the forging operations in schematic form.

Forging may be also classified by the temperature of the workpiece. Temperatures at which the metal is worked divide the process into hot or cold forging. Cold forging produces a smooth finish and high dimensional accuracy, but can only be performed on highly ductile materials. Because of high forces required during the process, cold forging is generally limited to the production of small parts, or small deformations in large parts. Figure 133 shows a few of the many guidelines for the design of cold-forged parts.

Hot forging requires smaller forces and therefore is used for large deformations. The finish and dimensional accuracy are poorer than with cold forging—the parts thus require more finishing operations. Figure 134 shows a few of the many guidelines for the design of hot-forged parts.

The major cost drivers in forging are:

- Forging press and auxiliary equipment
- Dies
- Material for the forgings
- Postprocessing, e.g., machining, surface treatment

Figure 135 shows how the lot size influences the shape of the forged part and the amount of machining required. The part shown is half of a

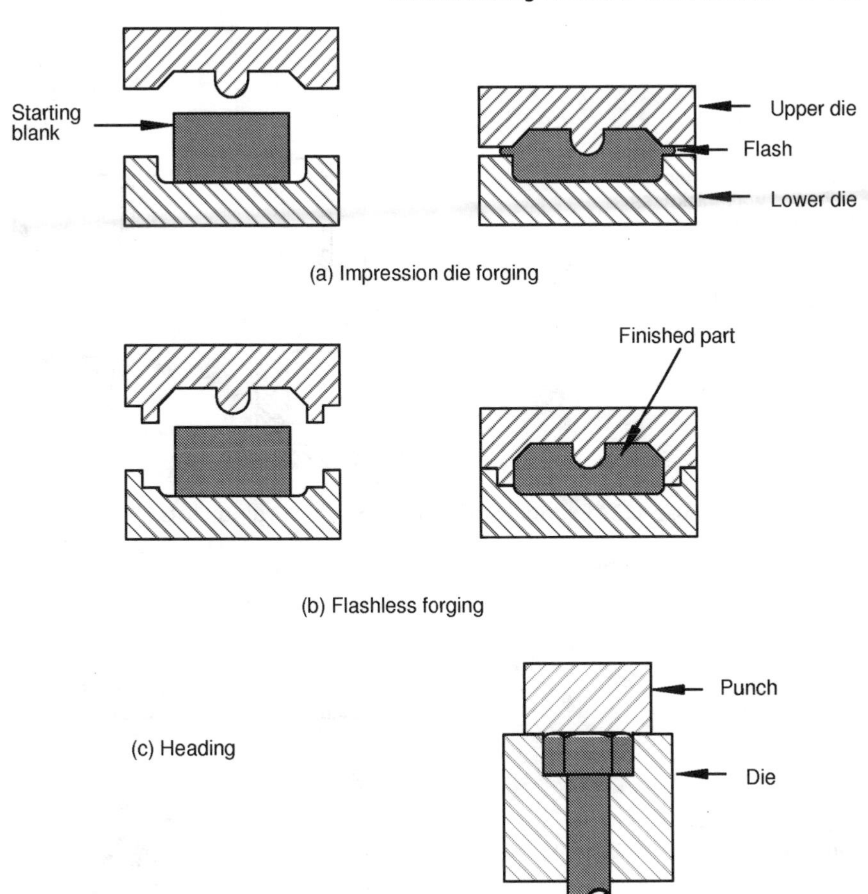

FIGURE 132. Some forging operations.

centrifugal coupling. In the upper part of the figure is shown the outline of the forged part that is most economical for the given lot size. For small lot sizes (< 150), the die is of a simple shape and a considerable amount of material is removed by machining. For large lot sizes (> 1000), the forged part is nearly of the final shape, and little machining is required. For this case the die is of a more complex shape and higher forces are required to squeeze the material into the cavities. As the lot size increases, we see increased forging cost, which is more than compensated by the reduced amount of machining.

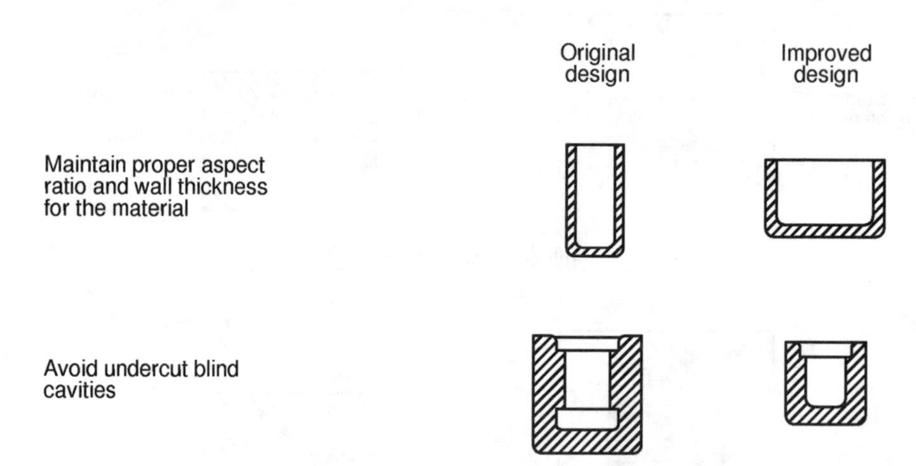

FIGURE 133. Examples of reducing cold forging costs.

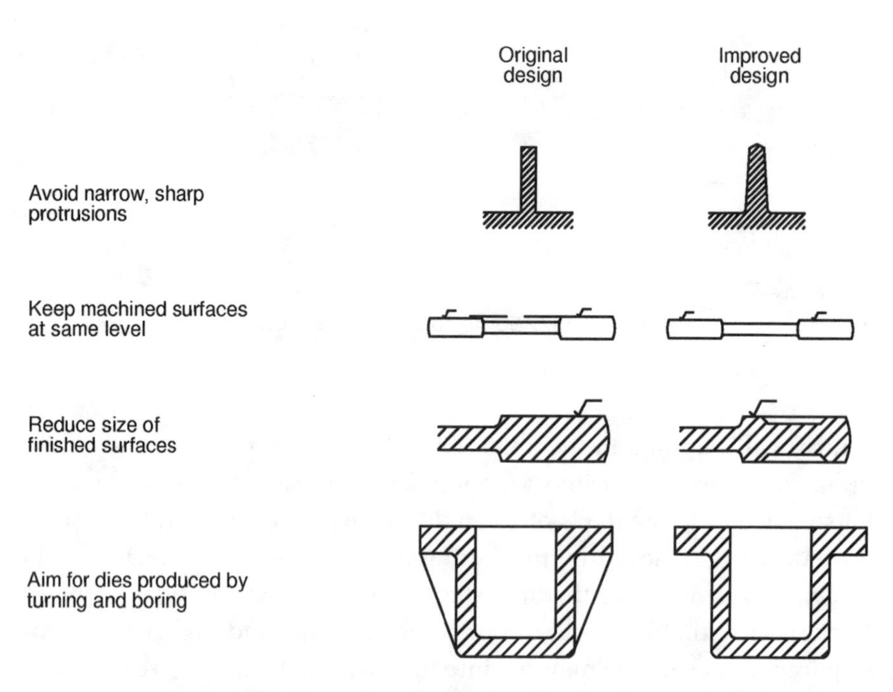

FIGURE 134. Examples of reducing hot forging costs.

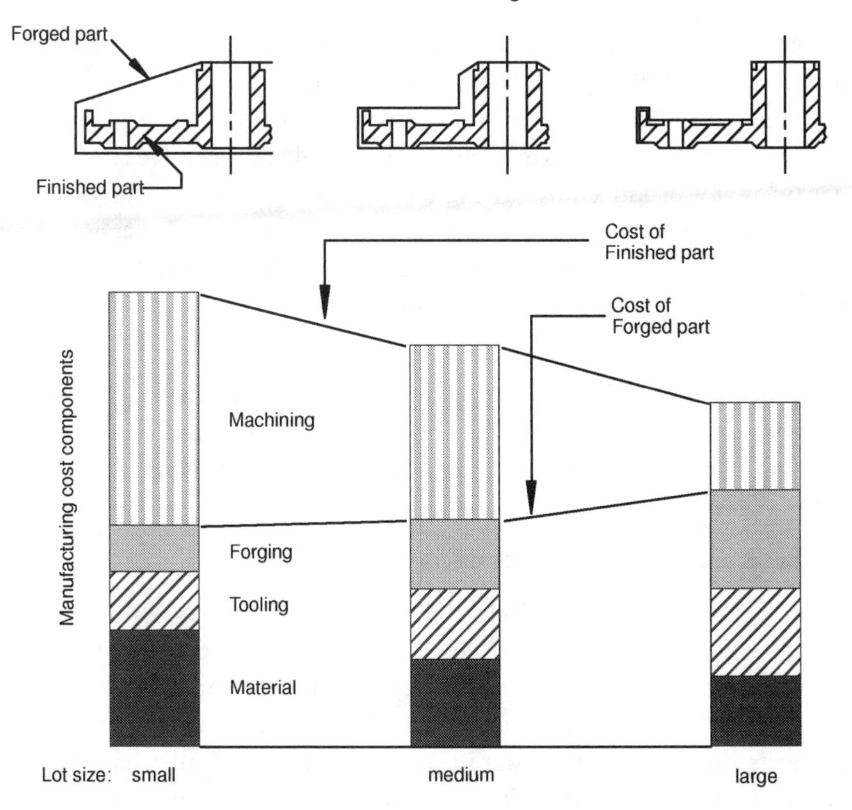

Figure 135. Cost components for a forged and machined part.

Other Bulk Deformation Processes

The other processes for bulk deformation of metals shown in Figure 131 are the following.

Rolling. The metal stock is squeezed between two rotating rolls, thus deforming it and, in most operations, reducing its cross-sectional area. The rolls can be of plain cylindrical shape, to produce plates and sheets, or they may have grooves, which produce structural profiles (angles, I-beams, etc.). Generally, several passes are required before the final shape is produced. Cold rolling is used to produce a smooth finish on sheets and bars.

Extrusion. In extrusion, a block of metal is forced through an appropriately shaped die. The process is used mainly for nonferrous metals. Tubes and architectural shapes (e.g., door and window frames) are made by extrusion. Discrete parts such as gear blanks, decorative shapes, and hardware are produced by sawing the extruded metal. Extrusion may be cold or hot, depending on the ductility of the material. In direct extrusion, a piston or ram is used to push the metal block. In hydrostatic extrusion, fluid pressure is applied directly to the metal.

Drawing. In drawing, the metal is pulled through a die, as opposed to being compressed, as in extrusion. Drawing is used for making shafts, rods, wire, welding rods, springs, and spokes. It is also an intermediate step in the production of fasteners such as bolts and rivets.

Common Sheet Metal Processes

Sheet metal stamping, bending, and cutting offer the possibility of lower weight and cost as compared with casting. Figure 136 shows a bearing pedestal that was formerly a casting replaced by a stamping with a significant (≈50%) cost savings. In the case of large parts such as covers and supports, this is especially significant, since castings must have a minimum wall thickness. Stampings, like castings, also allow more than one function to be integrated into one part.

Metal stampings are gradually being replaced by injection-molded plastic parts. The reasons for this change are lower cost and improvement in the properties of plastic materials.

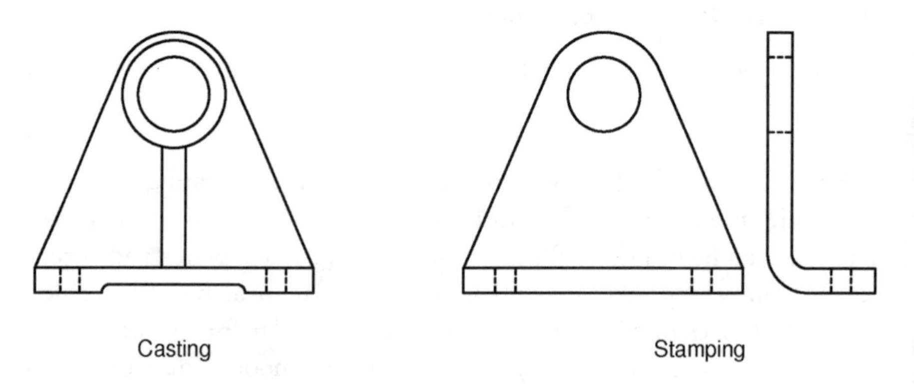

Figure 136. Cast part replaced by stamping.

• • • • •

EXAMPLE
COST OF DIE-STAMPED PARTS OF DIFFERENT MATERIALS

Ulrich and Eppinger (1995) show examples of unit cost of die-stamped parts of brass, stainless steel, copper, and galvanized steel. The parts are of varying degrees of complexity. Figure 137 shows the graphs of these values as a function of quantities produced.

The fixed costs consist of the one-time tooling cost. The variable costs per unit are the material cost and "processing" costs, which include machine time, direct labor, and overhead costs.

• • • • •

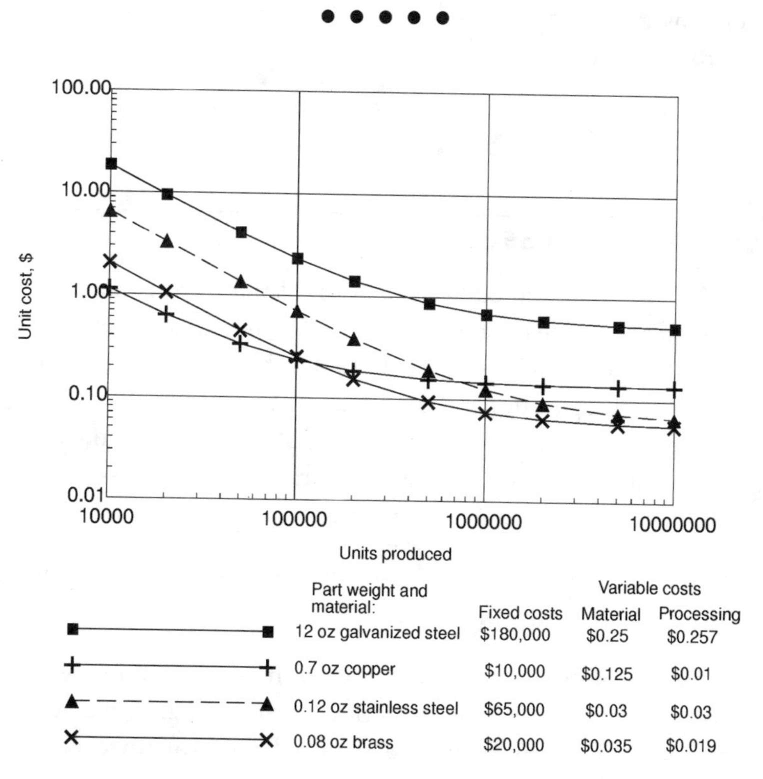

FIGURE 137. Cost of die-stamped parts.

Other Sheet Deformation Processes

Sheet metal can also be shaped by other processes, including the following:

- *Deep drawing* is used to form containers such as tanks, kitchen sinks, beverage containers, and flashlight tubes. It is usually carried out in a succession of dies (progressive dies).
- *Superplastic forming,* whereby highly ductile materials such as titanium-aluminum-vanadium and zinc-aluminum alloys can be formed into complex shapes. Actual processes used are common metal and plastic forming methods such as forging, extrusion, and compression molding.
- *Explosive forming* uses an explosive charge in a tank filled with water. The die, with the sheet above the cavity, is set at the bottom of the tank. When the explosion is set off, the pressure wave in the water deforms the metal. The die cavity is evacuated of air to help with the deformation.

MACHINING PROCESSES

Figure 138 shows a detailed classification of the various manufacturing processes that separate or remove material. When this separation occurs by mechanical means, it is embodied in a number of machining processes. While all of the separation processes are important, we will focus our attention here on machining. Some of the other processes mentioned in Figure 138 will be discussed under "Nontraditional Processes."

Machining processes are distinguished by

- The type of surface produced—cylindrical, flat, or a general shape
- The quality of finish produced

Figure 139 shows three of the most common machining processes: drilling, turning, and milling. The workpiece (or material, or blank) is acted upon by the tool to carry out the material removal. In each of these processes there is a rotation and a feed involved.

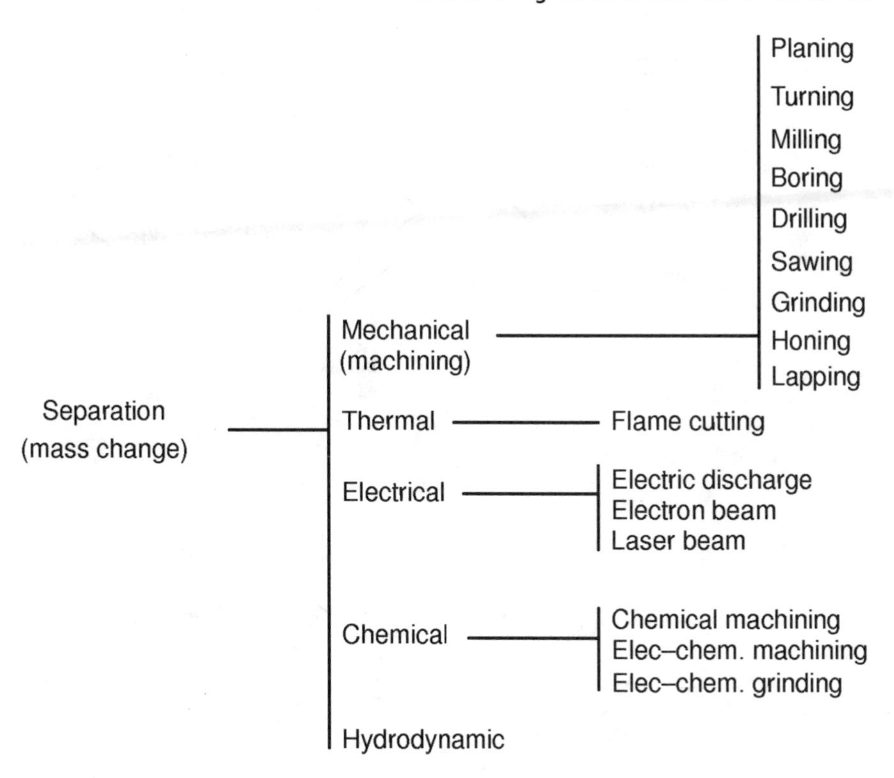

FIGURE 138. Separation processes.

- Drilling is carried out in a drill press or by a handheld drill. It produces a hole in the material. Drilling involves both the rotation and the feed on part of the tool, the drill bit. The workpiece is held fixed. The tool rotates while it is fed along its axis into the material.
- Turning is carried out in a lathe. It produces cylindrical parts. In a lathe the workpiece rotates while the tool is forced against it to remove material. Special shapes of cutting tools can be used to produce specific profiles, e.g., screw threads.
- Milling is carried out in a milling machine. It produces flat or sculptured surfaces. The workpiece is mounted on a table which can move (feed) along one, two, or three axes. The tool (milling cutter) rotates. The axis of rotation of the cutter may be horizontal (in horizontal milling machines) or vertical (in vertical milling machines). Specially shaped milling cutters are employed to produce special sections.

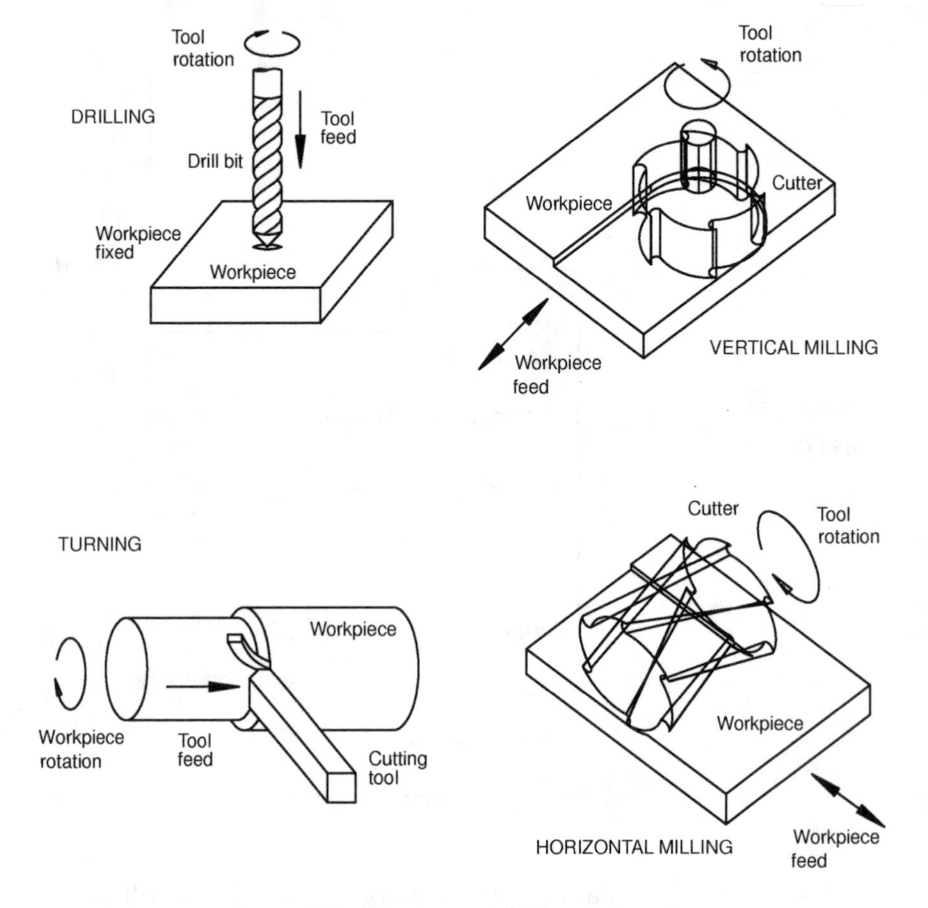

FIGURE 139. Common machining processes.

Machining processes are used, as a rule, as the finishing operations on parts that have been preprocessed by other means, such as casting, forging, or welding. However, we should also realize that the dies and molds used in casting and forging are produced by machining processes. As a point of interest, the majority of parts produced by machining operations are rotational parts.

Rules for Designing Machined Parts

We list below some of the rules the designer should keep in mind. Some of these rules were already illustrated in Figures 128 and 134. Figure 140

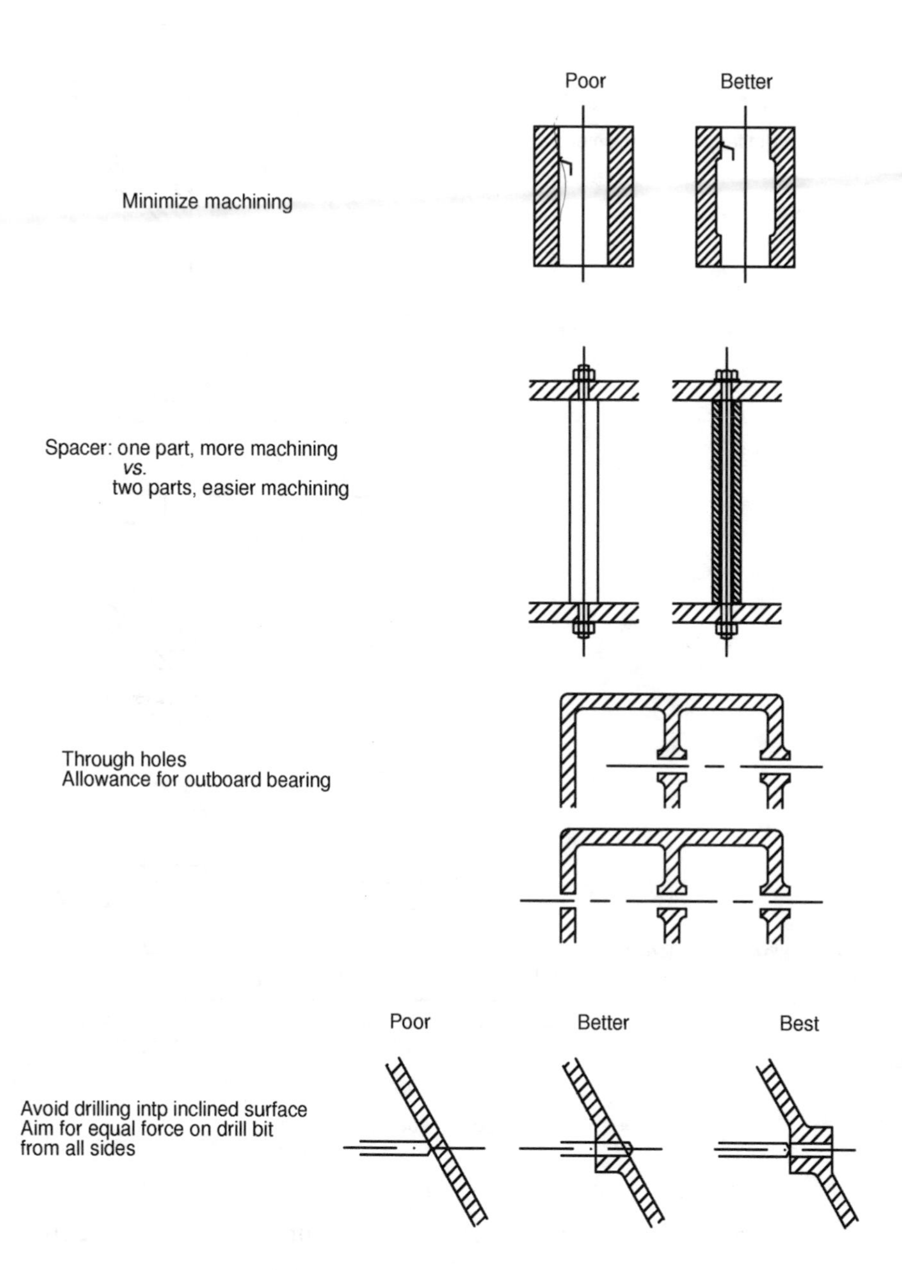

FIGURE 140. Examples of poor and good machined part design.

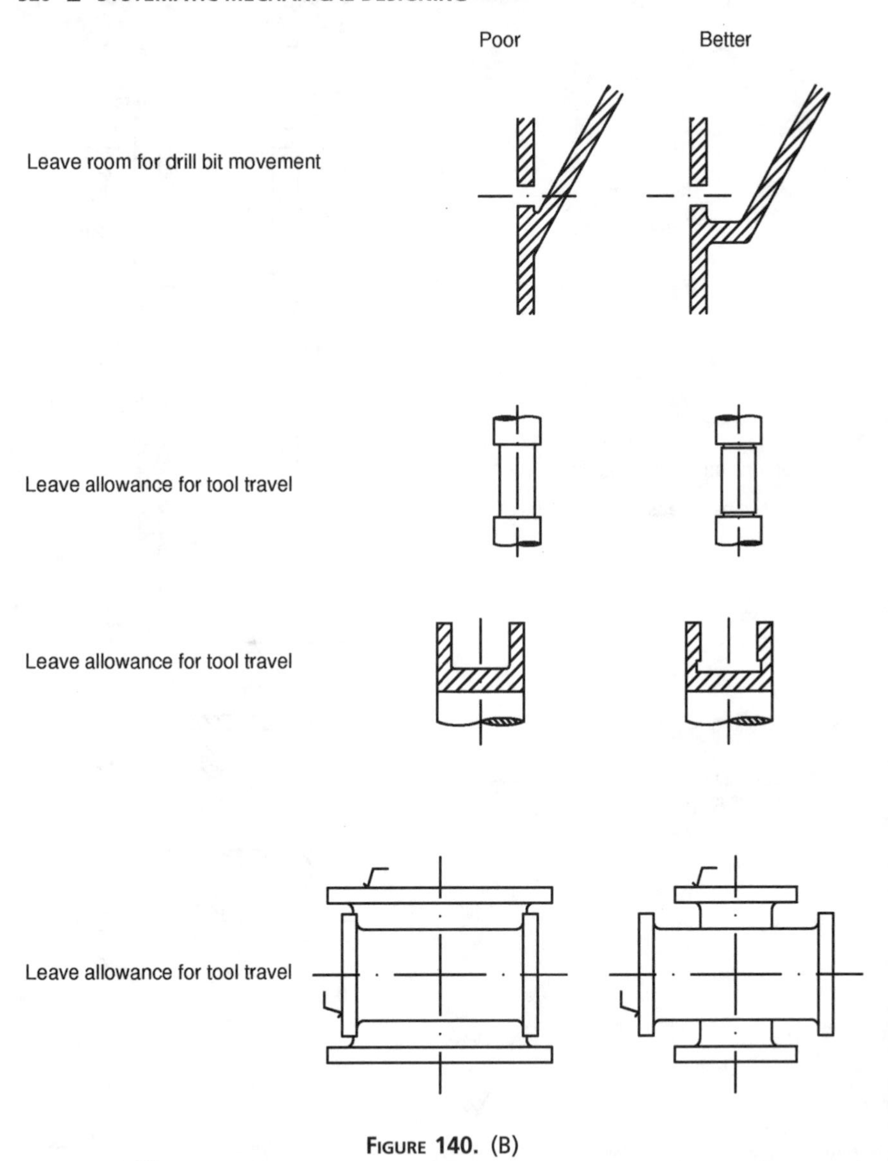

FIGURE 140. (B)

depicts more of these principles of good machining practices. These rules address the cost drivers (Ehrlenspiel, 1985) in machining:

- Amount of material machined
- Type of material machined

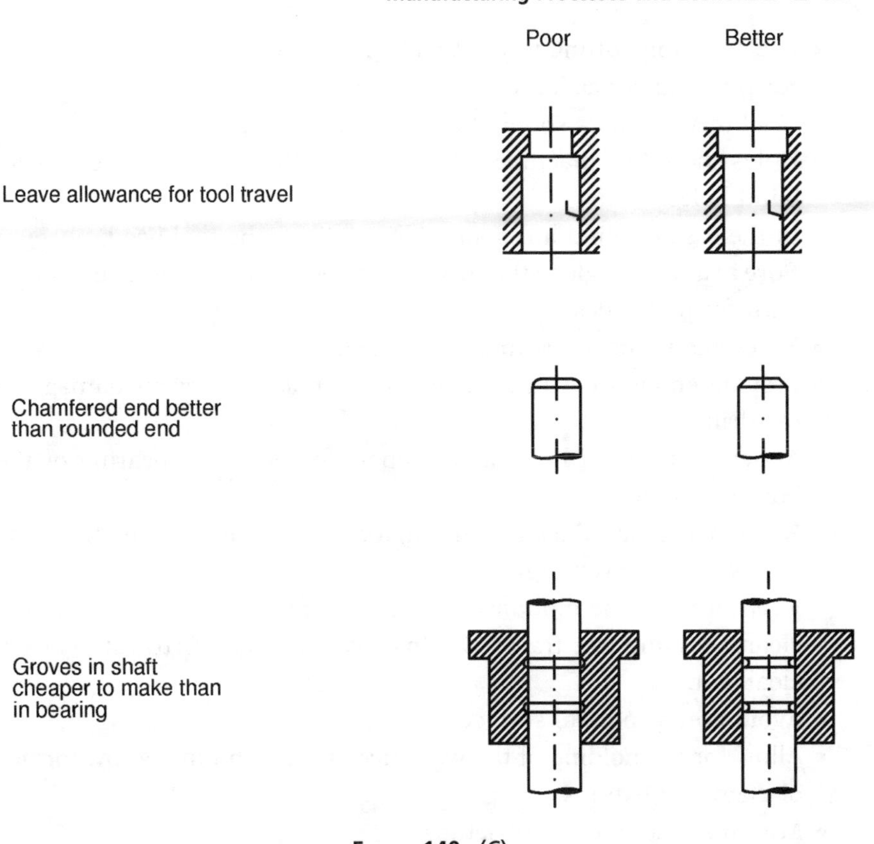

FIGURE 140. (C)

- Tool costs
- Production costs, determined by set-up, cutting, and idle times

Matousek (1963) states that two primary rules in keeping machining costs low are:

- Machining should be kept to a minimum.
- Machining should be needed only to reduce friction and to change shape to improve stress distribution—that is, to increase strength.

General rules for good machining practice and lowering machining costs include:

- Keep the amount of material removed to a minimum.
- Keep the surface quality as rough as the requirements allow.

- Use common cutting tools. Avoid special tools.
- Keep tolerances as loose as the requirements allow.
- Machined surfaces should be on the same plane.
- Holes should be machined through with the same diameter. Avoid blind holes.
- In the case of blind holes, provide space for the tool head.
- Bore at a right angle to the surface. The drill should have equal forces on it from all sides.
- Holes should be accessible to the drill.
- Provide enough clearance beyond the hole to prevent damage to the drill.
- Allow for through holes and support for outboard bearing of the boring spindle.
- Keep tool changes to a minimum, for example, by keeping the same hole size in a given region.
- Keep in mind the motions and limits of motion of the tools. Allow clearances and free travel. Cutting tools require overtravel to give a clean cut.
- Avoid noncylindrical surfaces.
- Allow for the holding of the workpiece for machining. Allow for use of clamping fixtures.
- Minimize the number of setups.
- Provide grooves in shafts, rather than in bearings.
- At shaft ends, chamfers cost less than rounded ends.
- When machining—drilling a hole—at an interface of two materials, use materials of similar hardness.
- Allow for a number of similar parts to be machined at the same time.
- Keep fillets and chamfers that lie close together of the same radius and slope, respectively.
- Holes should not be placed too near the edge—breakage or uneven deformation can result.

Cost Models for Machining Operations

Models for calculating machining costs utilize items such as cutting times, idle times, and setup times, along with the machine and labor rates per hour of use.

Dewhurst and Boothroyd (1987) describe cost models for machining and injection molding operations. For machining costs, they cite the example of replacing an assembly with a single casting. The cost C_m of machining one feature on a machine is expressed as a sum of machining costs, costs of idle time, and the cost of a new tool (cutting edge):

$$C_m = Mt_m + \frac{Q(Mt_{ct} + C_t)t_m}{t_{lt}} \tag{54}$$

where

M = machine and operator rate
t_m = total machine operating time
Q = fraction of t_m that tool is cutting
t_{lt} = tool life
t_{ct} = tool changing time
C_t = cost of providing new tool

In general, the cost of a machined part (Dieter, 1991) is expressed by the sum:

$$C_{\text{mfg}} = C_{\text{mach}} + C_{\text{tool}} + C_{\text{mat}} \tag{55}$$

where

C_{mach} = machining cost
C_{tool} = tool cost
C_{mat} = material cost

The tool cost for one part is proportional to the ratio of machining time for this part to the total tool life. The tool life depends on whether it is resharpened or is a throwaway type. The machining cost is determined by labor and machine rates:

$$C_{\text{mach}} = M(1 + \text{MOH}) + L(1 + \text{LOH}) \tag{56}$$

where

M = machine rate, \$/h
MOH = machine overhead

L = labor rate, \$/h
LOH = labor overhead

● ● ● ● ●

EXAMPLE
COST OF MACHINED PARTS OF DIFFERENT MATERIALS

Ulrich and Eppinger (1995) show examples of unit part cost of parts made of aluminum and UHMW polyethylene produced by computer-numerically controlled (CNC) machining. The parts are of varying degrees of complexity. Figure 141 shows the graphs of these values as a function of quantities produced. The fixed costs consist of the one-time setup and tooling (programming and

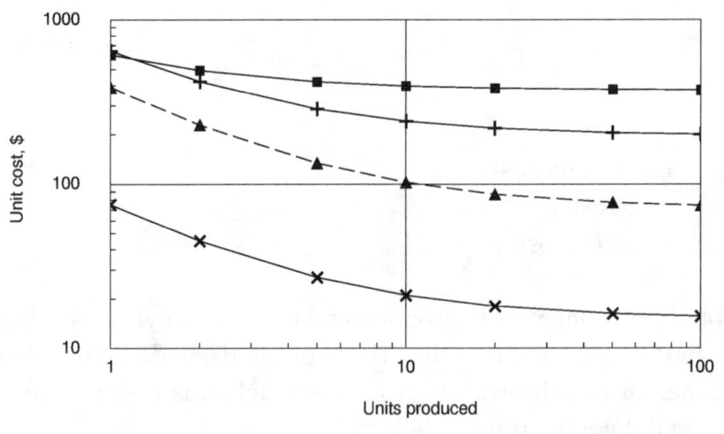

	Part weight and material:		Fixed costs	Variable costs	
				Material	Processing
■—■	3.3 lb	Aluminum	\$240	\$12	\$360
+—+	10 lb	UHMWPE	\$450	\$25	\$171
▲- -▲	4.3 lb	Aluminum	\$315	\$16	\$55
×—×	2.5 lb	Aluminum	\$60	\$9	\$6

FIGURE 141. Cost of CNC machined parts.

fixture) costs. The variable costs per unit are the material cost and "processing" costs, which include machine time, direct labor, and overhead costs.

• • • • •

ENERGY REQUIREMENTS

The energy required for machining various materials is shown in Figure 142. The chart gives an indication of the relative machining costs of these materials.

PRODUCTION QUANTITIES AND COSTS

The quantities of parts to be produced by machining operations dictate the type of equipment used for the purpose. Figure 143 shows the spectrum of

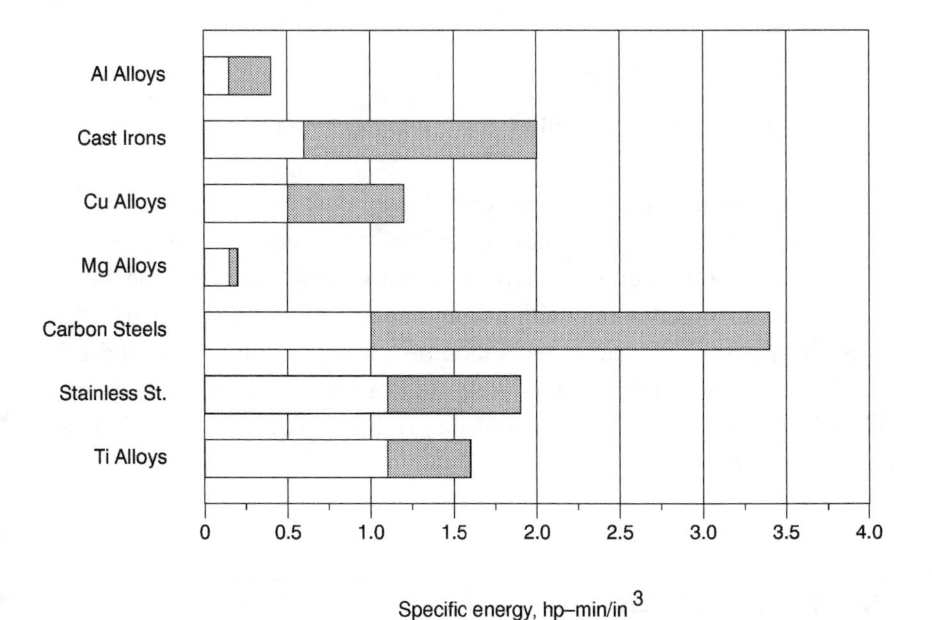

FIGURE 142. Energy requirements in machining. Shaded portions show the ranges of variation for different metals.

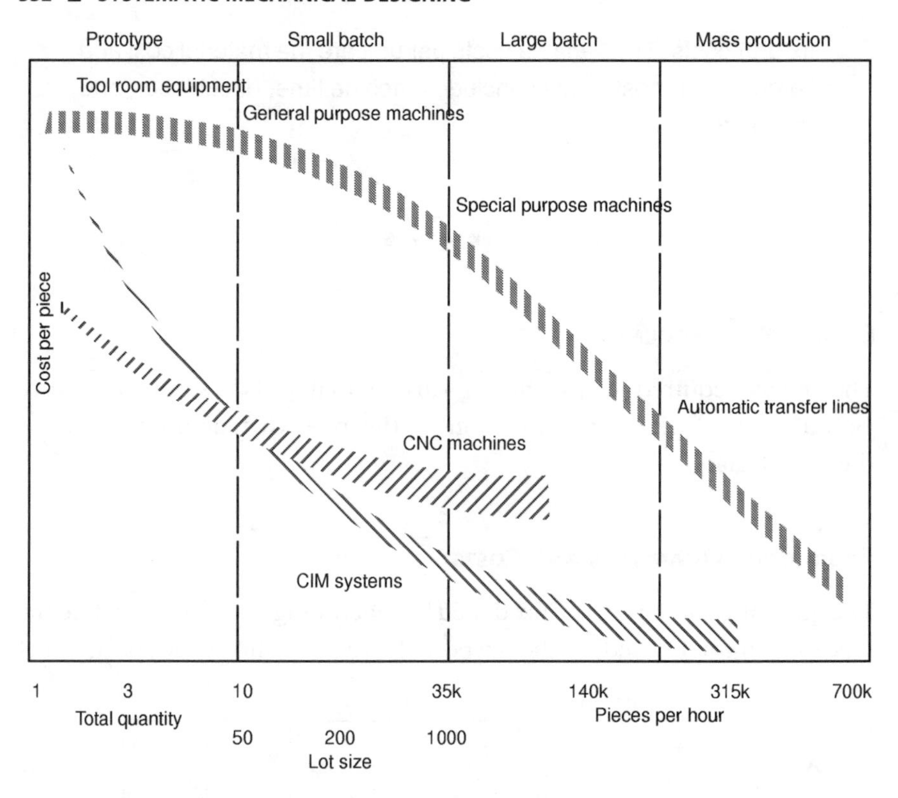

FIGURE 143. Production quantities, costs, and machine types (Curless, 1996).

types of machines, quantities produced, and cost per piece for different processes. Prototypes and small quantities are made by tool room and general-purpose machines. For production in large batches, special jigs and fixtures are used, along with numerical control. Very high production rates dictate specially designed machines. Numerically controlled (NC/CNC) machines and computer-integrated manufacturing (CIM) systems fill niches in small- to medium-volume production for significantly lower cost per piece.

JOINING PROCESSES

Since most manufactured products are made of several parts and are not built as monoliths, the parts need to be joined together. Figure 144 shows

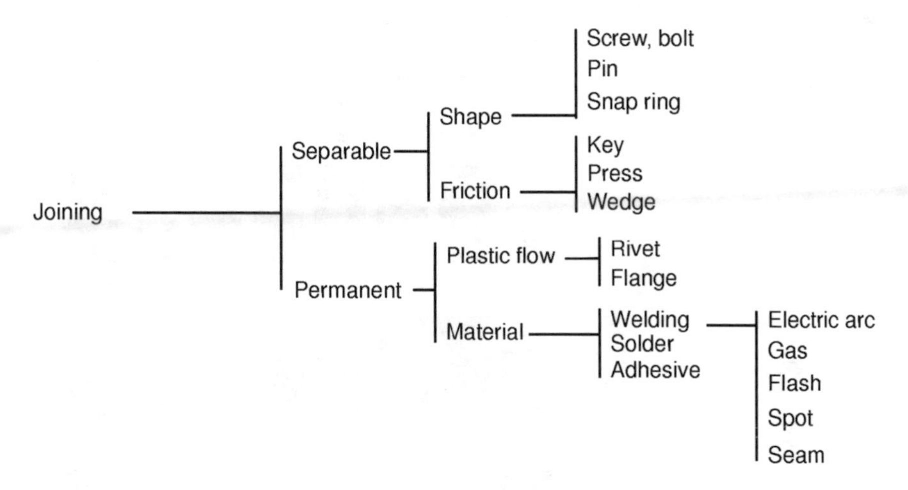

FIGURE **144.** Joining processes.

the classification of joining processes in some detail. The discussion here will be limited to fixed joints and will not include those which allow mutual motion of connected members. The latter, such as bearings and slides, are adequately covered in textbooks on machine elements.

The joining processes, as Figure 144 shows, are broadly classified by the types of joints produced: separable or permanent. Separable joints are further divided into those dependent on the shapes of the parts (screws, bolts, pins, rings) and those dependent primarily on friction (key, wedge, and press fits). In separable joints, forces play an important role.

Permanent joints depend on either the plastic flow of material, as in riveted and flange joints, or on material that itself forms the bond. Material-based joints are bonded by either a separate material (soldering; use of adhesives) or material of the same type as the joined parts (welding).

Figure 145 shows a rating of six types of joints according to different properties, including cost. A simple comparison on the basis of cost is not possible. In order to choose a joint type for an application, a matrix method, such as that shown in Table 5 of Chapter III, can be used. The type of joint used is also influenced by environmental requirements (see Chapter XI).

For sheet metal joining, the relative cost values (Ehrlenspiel, 1985) are shown in Figure 146. These are based on joining ⅛ in. steel sheets and normalized to the cost of the tack-welded joints.

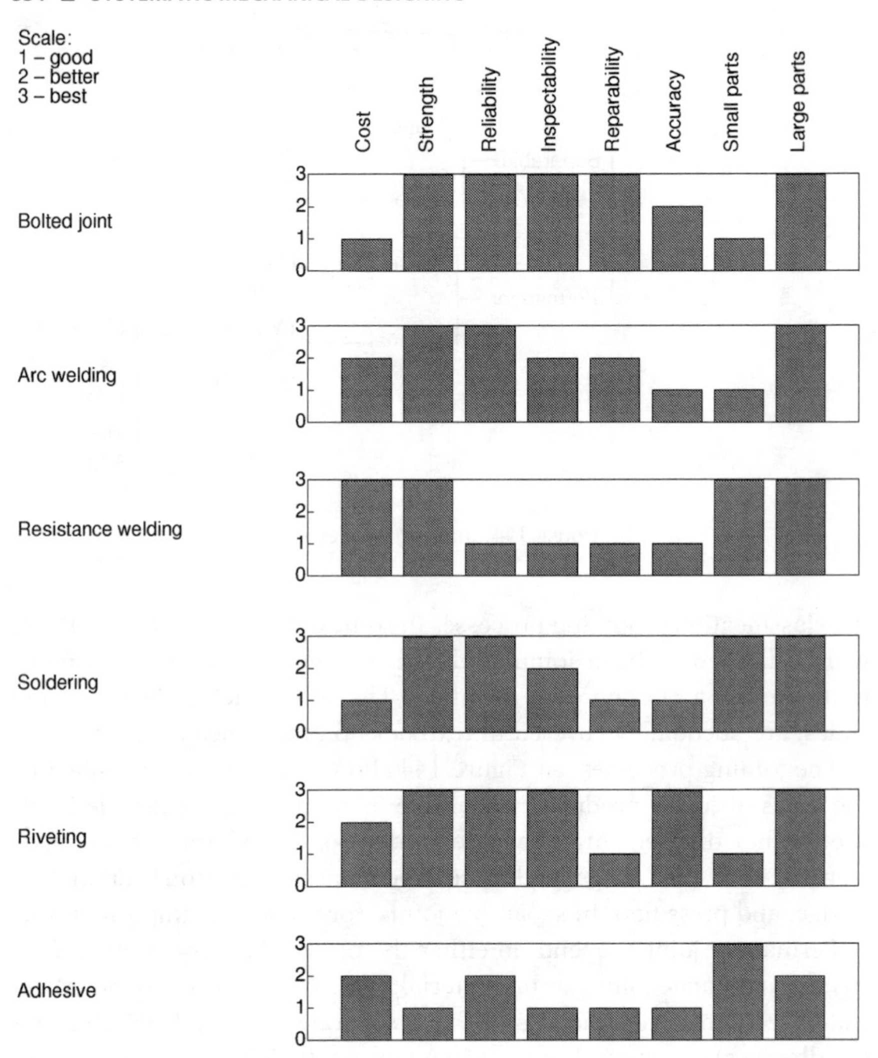

Figure 145. Comparison of different types of joints.

Cost Models for Welding

Ott and Hubka (1985) describe a typical method for calculating the manufacturing cost of weldments based on weld dimensions. They calculate time requirements for each welding operation and the complete costs using the labor and machine rates. The total welding cost $C_{W,\,tot}$ is the sum:

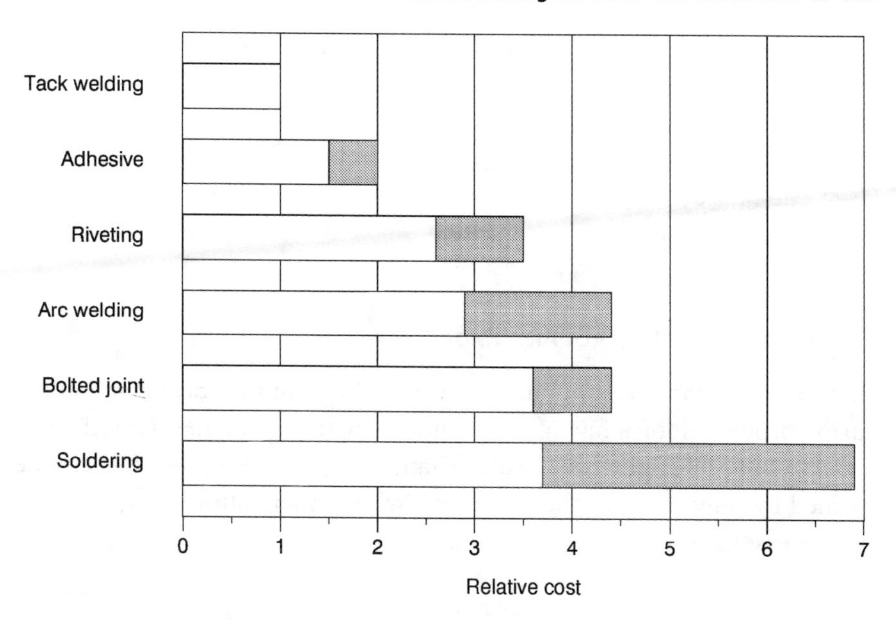

FIGURE 146. Relative cost of sheet metal joints normalized to that of tack welding. Shaded portions show the ranges of variation of different processes. (Ehrlenspiel, 1985).

$$C_{W,\,\mathrm{tot}} = C_{\mathrm{prep}} + C_W + C_{\mathrm{post}} \tag{57}$$

where

C_{prep} = preparation cost
C_W = welding cost
C_{post} = post-processing cost

The preparation cost is given by

$$C_{\mathrm{prep}} = \sum_i (C_{\mathrm{mat}} + C_{\mathrm{lab},\,d}) \tag{58}$$

where

C_{mat} = material cost
$C_{\mathrm{lab},\,d}$ = direct labor

The latter is found from the hourly rate and the setup, activity, and idle times. The welding cost in Equation (57) is the sum:

$$C_W = C_{WM} + C_{WE} + C_{WL} \tag{59}$$

where

C_{WM} = material (electrodes) cost
C_{WE} = energy cost
C_{WL} = labor cost

Compatibility of Metals and Processes

Figure 147 provides a summary of the suitability of the various manufacturing processes for a few of the common metals and alloys (Boothroyd, Dewhurst, and Knight 1994; Kalpakjian, 1992). Certain materials can be formed by only a few of the processes, while others allow a wider range of processes.

MANUFACTURING PROCESSES FOR PLASTICS

The processing of plastic materials entails some operations similar to those for other materials—e.g., machining and molding—and other processes unique to these materials. For forming simple shapes, some plastics are available in the form of sheet, rod, pipe, and plate. However, the majority of plastic products start out in a powder or pellet form, which is put through one of the processes. We will take a brief look at the important characteristics of these processes. The points to be noted with regard to the different processes are:

- Type of materials processed
- Limitations on shapes
- Range of sizes produced
- Type of shapes produced—hollow versus solid, various wall thicknesses
- Economical production volumes
- Equipment and tooling costs
- Production rates

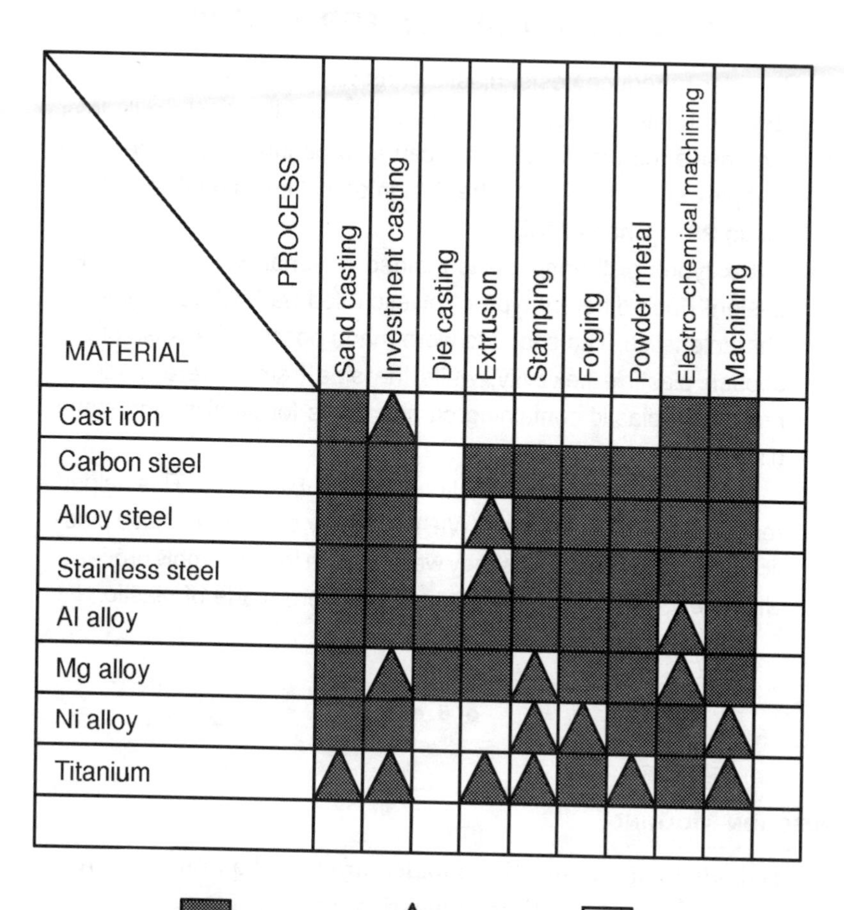

FIGURE 147. Metals and process compatibility.
(Boothroyd, Dewhurst, and Knight, 1994) and (Kalpakjian, 1992)

• • • • •

EXAMPLE
PLASTIC PART DESIGN FOR PROPRINTER

In order to reduce costs, plastic parts were incorporated into the IBM Proprinter design. Polymers are generally significantly less expensive than metals. Plastic can also be injection-molded into complex shapes, allowing the design of multifunctional parts (see Table 28 in Chapter VII).

Six types of thermoplastic molded material were used in the design. Each different type of plastic used had different material characteristics. The snap fits were made possible because of the chosen plastic's flexibility, while the side frames were made of more rigid plastic containing carbon fibers for stability and static discharge.

Plastic parts could be injection-molded at the plant. This helped reduce the work in process (WIP) inventory of parts because plastic parts could be made as they were used in the assembly process. Ultimately, 77% of the Proprinter parts were made of plastic.

• • • • •

INJECTION MOLDING

Materials that can be injection-molded are found among all three basic categories of polymers: thermoplastics, thermosets, and elastomers. Melted granules are forced out of a hot chamber into a split-die chamber. A wide variety of products are made by injection molding: toys, flashlight parts, electrical parts, and many others. Typical production volumes range from thousands to billions, often using expensive equipment and tooling. Production rates are high. Injection molding can produce intricate shapes, open hollow shapes with controlled wall thickness, and very small parts. Figure 148 shows the layout of an injection molding machine. The raw plastic is melted and fed from a heated screw feeder similar to that used in extrusion, as shown in Figure 149.

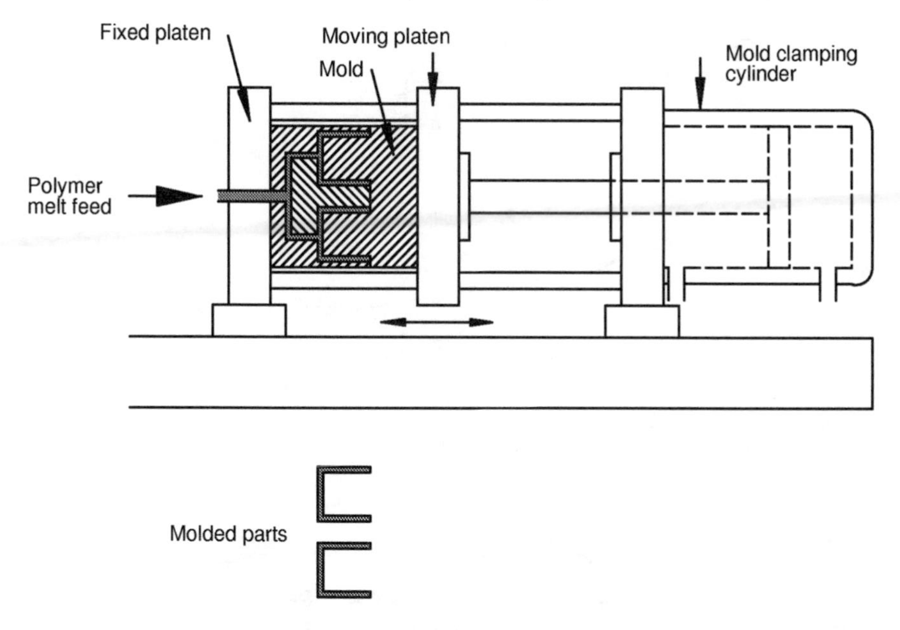

FIGURE 148. Injection molding machine.

Reaction injection molding involves two or more fluids in a cavity reacting and solidifying to produce a thermoset. Reinforcing fibers can be added. Typical products are bumpers, fenders, and other structural components.

Cost Models for Injection Molding

Dewhurst and Boothroyd (1987) provide a cost model for injection molding operations. Their injection molding cost model gives the cost C of manufacturing N components:

$$C = \left(\frac{Nt_i}{n_c}\right)(C_r + C_s) + C_n + C_b + NC_p \tag{60}$$

where

t_i = mold cycle time
n_c = number of cavities in mold base
C_r = machine rate
C_s = machine supervision rate

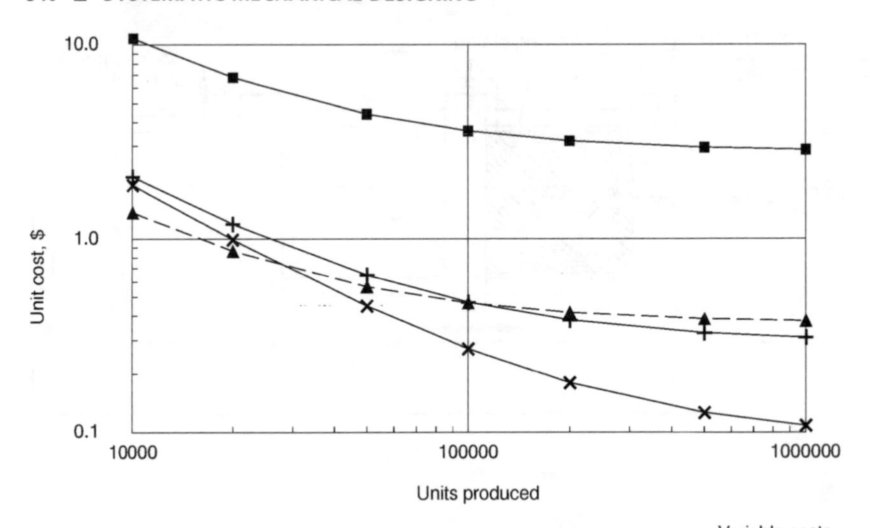

	Part weight and material:	Fixed costs	Variable costs Material	Processing
■——■	8 oz polycarbonate with brass inserts	\$80,000	\$2.38	\$0.42
+——+	0.8 oz polyphenylene oxide	\$18,000	\$0.144	\$0.146
▲- - -▲	0.35 oz steel–filled polycarbonate	\$10,000	\$0.147	\$0.218
×——×	1.6 oz low density polyethylene	\$18,000	\$0.06	\$0.03

FIGURE 149. Cost of injection-molded parts.

C_n = cost of manufacturing n_c cavities in mold base
C_b = cost of mold base
C_p = cost of polymer per component

Kiewert (1979) provides a formula for cost per part C of injection-molded thermoplastic parts in quantities greater than 200,000 per machine:

$$C = \frac{t_i(C_r + C_s)}{n_c} + C_{\text{tool}} + WC_{m,\,\text{mt}} \tag{61}$$

where

t_i = mold cycle time
C_r = machine rate
C_s = machine supervision rate
n_c = cavities in mold base

c_{tool} = tool cost
W = weight of part
$C_{m,\,\text{mt}}$ = material cost per unit weight

• • • • •

EXAMPLE
COST OF INJECTION-MOLDED PARTS OF DIFFERENT PLASTICS

Ulrich and Eppinger (1995) show examples of unit part cost of injection-molded plastic parts. The parts are of varying degrees of complexity and are made from different plastics. Figure 150 shows the graphs of these values as a function of quantities produced. The fixed costs consist of the one-time tooling cost. The variable costs per unit are the material cost and "processing" costs, which include machine time, direct labor, and overhead costs. The costs for producing one unit are dictated almost entirely

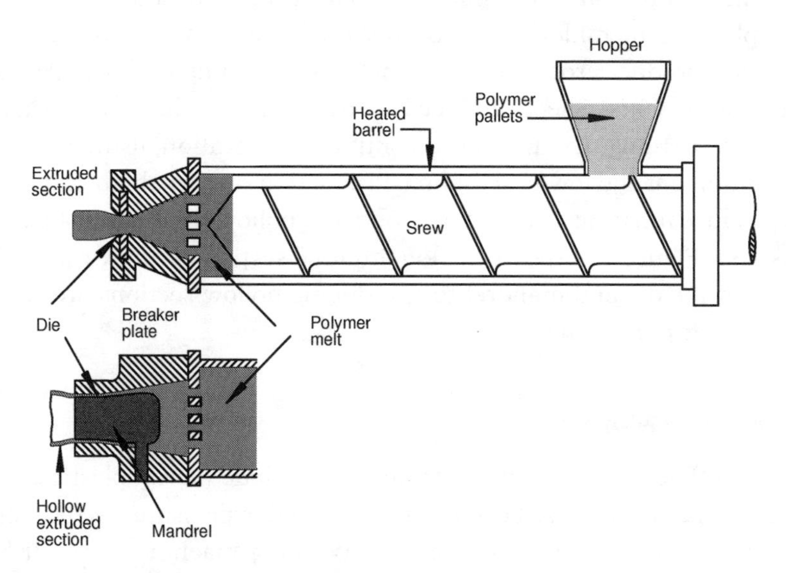

FIGURE 150. Extrusion of plastics and elastomers.

by the fixed costs. At high production rates, the unit cost reflects primarily the material and processing costs. The latter costs are high for complex parts.

• • • • •

Blow Molding

In blow molding, material is melted and blown to fill the cavity. Extrusion blow molding is used to produce pipe and tubing. Injection blow molding produces beverage bottles and other containers. Typical production volumes are from thousands into billions, and the process typically uses medium-cost equipment and tooling. Production rates are medium. Blow molding can produce open hollow shapes, and enclosed hollow shapes with large enclosed volume.

Extrusion

Materials that can be molded by extrusion include thermoplastics and elastomers. Material is fed into a heated barrel with a screw. Melting takes place in the middle of the barrel. It is then pumped through a die. Extrusion produces rod, sheet, channels, and building hardware, among other components. Sheet produced by extrusion can be further shaped by thermoforming. Extrusion is a continuous operation, using medium-cost equipment and low-cost tooling. Production rates are high. Extrusion can produce intricate shapes, both solid and hollow, with controlled wall thickness. Figure 149 shows the longitudinal section of an extruder. The details of the die and mandrel for producing hollow sections are shown in the lower part of the figure.

Compression Molding

Materials that can be compression-molded include thermoplastics, thermosets, and elastomers. Forming is done under pressure by the upper half of a die in a heated mold cavity. Washing machine parts, dishes, fittings, and electrical parts are made by compression molding. Typical production volumes range from thousands to millions, and expensive equipment and tooling are usually required. Production rates are medium. Compression molding can produce intricate shapes, and open hollow shapes with controlled wall thickness.

Rotational Molding

The materials that are molded by rotational molding are thermoplastics. Large, hollow parts are formed in molds that can rotate about two axes. Typical products are tanks and other large containers, boat hulls and appliance housings. Typical production volumes are in the hundreds and thousands, and low-cost equipment and tooling can be used. Production rates are low. Rotational molding can produce open hollow shapes, and enclosed hollow shapes with large enclosed volume.

Transfer Molding

Materials molded by transfer molding include thermosets and elastomers. In a variant of compression molding, a plunger is used to force the elastomer or uncured thermoset into the mold cavity. Typical products are electrical parts. Typical production volumes are from thousands to millions, and expensive equipment and tooling are typically required. Production rates are medium. Transfer molding can produce intricate shapes, open hollow shapes with controlled wall thickness, and very small parts.

Casting

Materials that can be cast include thermoplastics and thermosets. Large parts such as wheels, gears, and plates are cast. Flexible molds can be used to produce intricate shapes. Typical production volumes are in the tens and hundreds, and low-cost equipment and tooling can usually be used. Production rates are low. Casting can produce intricate shapes with controlled wall thickness.

Thermoforming

The materials formed by thermoforming are thermoplastics, including ABS, acrylic, polyethylene, polypropylene, polystyrene, and PVC. The process is used for display of products ("blister" or "bubble" packaging), medium-sized items such as office machine covers, toys, and light fixtures; and large items such as skylights (acrylic), bathtubs, boat hulls, shower stalls, and refrigerator liners (ABS). The thermoplastic in sheet form is heated and then formed by vacuum, pressure, or mechanical thermoforming.

MANUFACTURING PROCESSES FOR COMPOSITES

Manufacturing of composite material parts involves bringing the components together. The separate material components—polymers, ceramics, metals—are produced by their own appropriate techniques. The manufacturing process then depends on what the components are and the geometry required. Metals and ceramics for composites are generally in powder form. Reinforced plastics are produced by molding (vacuum or pressure), pultrusion, and filament winding.

● ● ● ● ●

EXAMPLE
A NEW OFFICE CHAIR DESIGN

McDowell, in his presentation "EVO—A New Chair Design" (1995), describes the development of a new plastic material–based office chair. The new chair has 50% fewer parts (36, down from 50 to 60) and 30% lower weight (40 lb, down from 60 lb) than a conventional chair. Most significant is the design of the C-shaped main spring. It is made of laminated epoxy/fiberglass, with an HDPE surface for lubricity. The new design integrates three parts (back support pan, the spine, and the seat support pan) into one part. The design group worked closely with the injection molder in developing the design. Among other novel approaches developed was the use of Be-Cu core inserts to aid in heat removal, which lowered the cycle time from 2 minutes to 70 seconds. The spring was subjected to 160 kpsi in a fatigue test and was deemed to have an infinite fatigue life. The unibody seat spine in the chair is made of 33% glass-filled nylon. It successfully withstood 2 million cycles in a fatigue test. Tooling was reduced from four molds to one large mold with an insert.

• • • • •

NON-TRADITIONAL PROCESSES

Certain manufacturing processes make use of physical processes other than those described above. These processes have been developed to serve special needs and, as might be expected, are more expensive per unit mass of material removed than the conventional processes. Typical applications of these processes are for:

- Holes of very small diameters
- Holes with high length-to-diameter ratio
- Noncircular holes
- Pockets, blind holes, die impressions
- Narrow, irregular cuts

Nontraditional processes may be divided according to the physical effects used: mechanical, chemical, electrochemical, and thermoelectric. We will briefly describe some of these processes.

ULTRASONIC MACHINING

Ultrasonic machining removes material by using the ultrasonic vibration (20 kHz) of a tool in a slurry containing abrasive particles. The cut formed is in the shape of the tool. The process is well suited to cutting hard materials such as glass and ceramics, and less well suited for metals and plastics.

Water Jet Cutting

Water jet cutting, also called hydrodynamic machining, utilizes cutting fluid at high pressure (to 60,000 psi) and high speed. It is used for cutting plastics, cardboard, textiles, and leather, among other materials. A variation of this process is abrasive water jet cutting, in which abrasive particles are added to the fluid after it exits the nozzle. This makes the process suitable for cutting metals. These processes cause no thermal damage to the materials. A related process is abrasive (gas) jet machining.

Electrochemical Machining

In electrochemical machining, the tool is made a cathode and the workpiece the anode. A flow of electrolyte between the two carries away material from the latter. It is the opposite of electroplating. It is suitable for conductive materials, particularly hard materials such as titanium and superalloys. Related processes are electrochemical deburring and grinding. This process has the highest removal rate of all nontraditional processes, requires expensive equipment, and has high tooling costs. It is suitable for medium to high production rates.

Electric Discharge Machining

The electric discharge machining setup looks similar to that for electrochemical machining; however, the physical process is entirely different. Electric discharge occurs between the anode and the cathode, raising the local temperature to a very high value. The material (from both the anode and the cathode) melts and is carried away by the electrolyte. Surface quality of the finished surface depends on the frequency and discharge current—a low current and high frequency combination produces smooth surfaces. The process requires expensive equipment and has high tooling costs. A related process is wire electric discharge machining, in which a wire electrode is fed through the workpiece, much like a band saw.

Chemical Machining

The term *chemical machining* includes the processes of chemical milling, blanking and engraving, and photochemical machining. Chemical milling involves the following steps. (1) A mask of a nonreactive material is applied to the cleaned surface of the blank. (2) The areas of the mask under which material is to be removed are cut away, using a template. (3) An etching solution is applied, eating away the material. (4) The mask is removed and the part cleaned. The reaction leaves an undercut. Up to ½ in. of material can be removed, at rates up to 0.0004 in. per minute. The process carries low cost and is suitable for low production rates. Figure 151 provides a schematic representation of these processes.

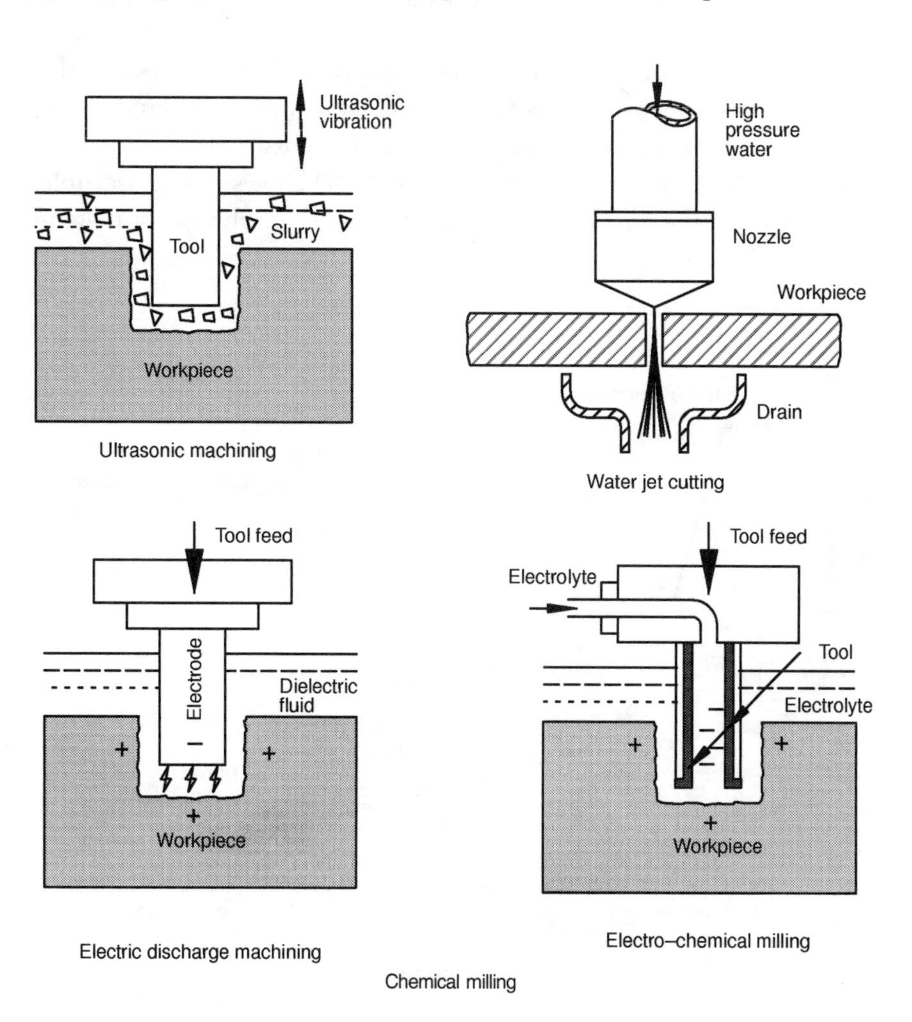

Figure 151. Some non-traditional processes.

DIMENSIONAL TOLERANCE AND SURFACE FINISH

Dimensions specify the geometry of a part. Since a dimension cannot be held exactly in a manufacturing process, a variation or tolerance is specified, if necessary. Each manufacturing process produces a "nominal" tolerance. The final drawings should indicate only those tolerances that are absolutely necessary but are not implied by the material and manufacturing process. Close tolerances are needed, for example, when parts must fit properly.

As any experienced designer knows, smooth surface finishes and tight tolerances are expensive. It is therefore obvious that we should not ask for more than is needed to satisfy the requirements.

Figure 152 (Ehrlenspiel, 1985; Trucks, 1987) shows how closer tolerances increase the cost of machining operations. Each type of machine

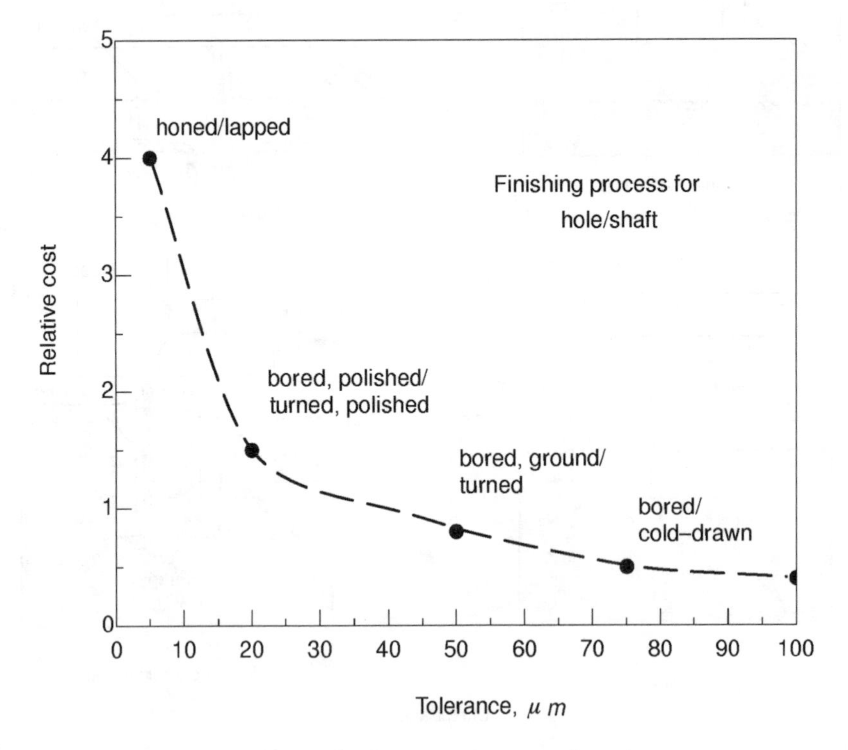

FIGURE 152. Effect of tolerance on costs of mating parts. (Ehrlenspiel, 1985) and (Trucks, 1987)

has limits of dimensional tolerance and surface finish; even within the same machine type there are differences. Figure 153 shows tolerance capabilities of common machining processes (Kalpakjian, 1992). Given the differences, it is necessary for the manufacturing department to decide on the optimum sequence of operations to achieve the required finish.

In machining, the cost of achieving a given finish goes up with the workpiece size. The finer machining operations contribute significantly only to production costs when roughing operations costs are small, since material costs are unaffected. Figure 154 shows the increase in costs of achieving finer surface finishes by machining. The general rule of thumb, as with dimensional tolerances, is: Each halving of surface roughness becomes twice as costly.

Hayes et al. (1988) give the example of the design and manufacture of a single-card hard-disk drive for a PC. The design was produced by one company and then passed over to another for manufacturing. The manufacturing company found that the gap tolerance specified as 0.002 in. between two pieces in an optical encoder could not be maintained at high production rates. The design group could not justify the close tolerance; they were able to relax it to 0.010 in., with no adverse effect on performance. With the new tolerance, the drive could be manufactured more economically.

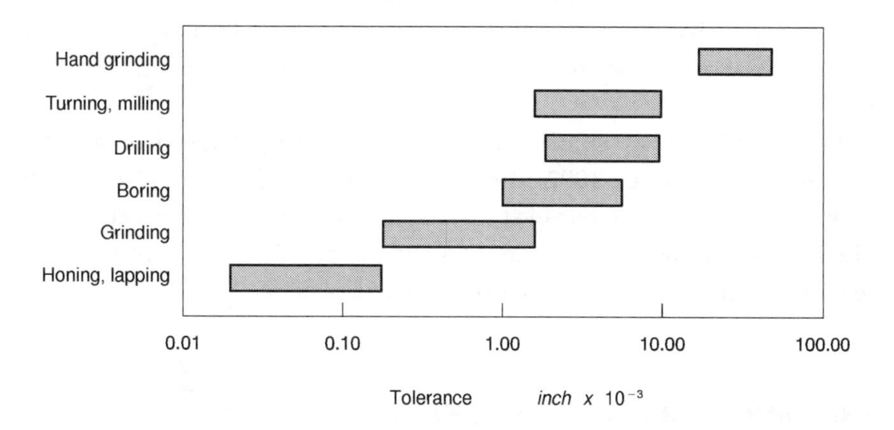

FIGURE 153. Tolerance limits for machining processes.

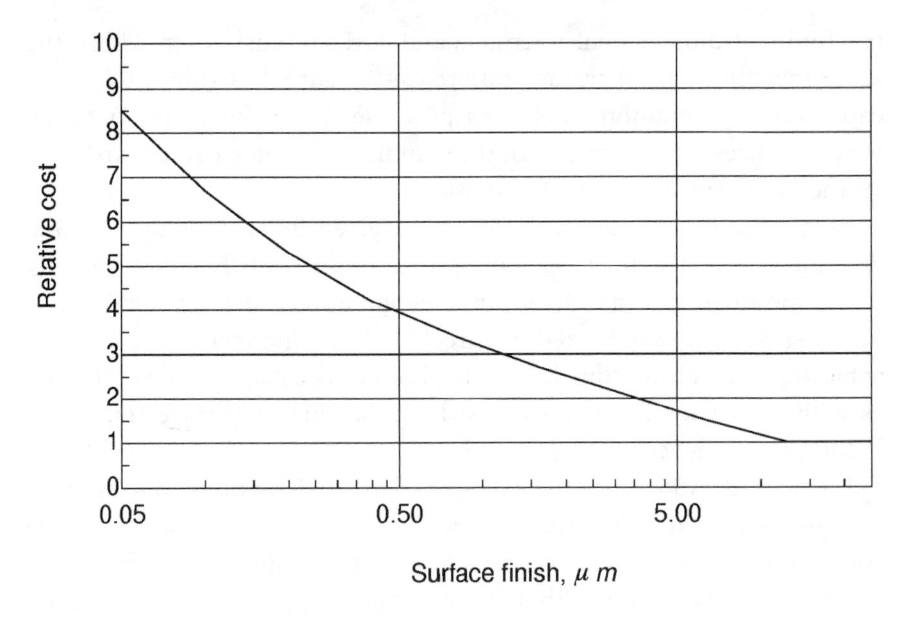

FIGURE 154. Effect of surface finish on costs.

MATERIAL WASTAGE DURING PROCESSING

Among the factors to be kept in mind in selecting materials and manufacturing processes is the amount of wastage during processing. Processes that produce near-net shapes may be more cost-effective, even while using a more expensive raw material. The wastage produced is important from a purely economic standpoint, as well as from an environmental perspective.

Figure 155 shows typical proportions of scrap produced in selected processes (Kalpakjian, 1992; Michaels and Wood, 1989). We also need to look at how easy it is to recycle the scrap. For example, nonferrous metals are, in general, easier to recycle than ferrous metals. Use of optimization techniques can lead to a reduction in waste produced.

THE MAKE-VERSUS-BUY DECISION

One of the most crucial decisions during product development is whether to make a part or subassembly for the product in-house or to purchase

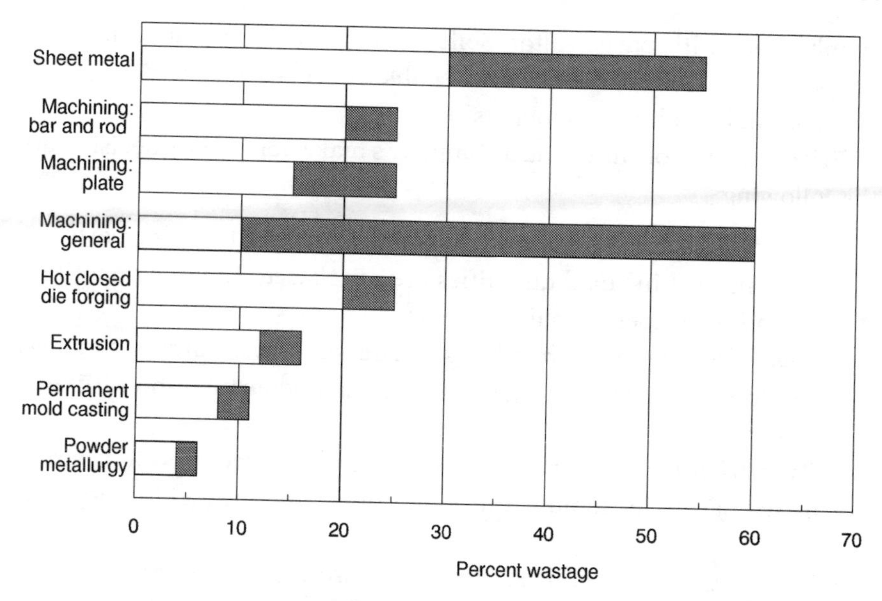

FIGURE 155. Material wastage in different processes. Shaded portions show the ranges of variation for different processes. (Kalpakjian, 1992) and (Michaels and Wood, 1989)

it. All manufacturing companies purchase materials for their products. The degree to which the purchased material has been processed can vary from none—extracted from nature—to parts and subassemblies ready to be assembled. Typical examples of material purchases are plastic molding compounds, plastic and metal extrusions, metal rolled sections, sheet metal, and standard parts such as fasteners. There are few totally integrated manufacturers today after the model of the Ford Motor Company in its early days, which purchased raw materials—ore, coal, rubber, etc.—and shipped out finished automobiles.

In making such a make-versus-buy decision, cost is obviously one of the most important considerations. There are, however, other factors involved, such as whether the item is a standard item, whether the company has the capabilities to make the item, or what quantities are required. Often there are vendors with specialized knowledge and equipment who can supply made-to-order items of a higher quality and at a lower cost than a company that uses those items in its products. Large industries such as automobiles and computers have spawned smaller industries that supply them with parts. Automobile manufacturers in particular have long-term alliances with a few but reliable vendors who can deliver low-cost and high-quality components on time.

Briefly, the factors affecting a company's make-versus-buy decision are the following:

- *Standard parts and production quantities*. Standard, catalog parts and parts needed in small quantities are purchased. In large quantities, it can be cheaper to make the part.
- *Technology*. If the part requires some unique and sensitive technology developed by the company, it is a prime candidate for in-house manufacture.
- *Production facilities and expertise*. A company needs both expertise on the part of its manufacturing personnel and the facilities to make a part.
- *Vendor expertise and reliability*. These are essential prerequisites to a decision to outsource.

In the final analysis, make-versus-buy is often a strategic decision on the part of a company. The technological advantage as a rule goes to the company which does the manufacturing. Thus if a company contracts out the manufacturing, it must keep close control of its technology, particularly if proprietary materials and manufacturing processes are involved.

We consider an example of a machined part made of UHMWPE, similar to one of the parts previously used in Figure 141. The part is needed in a lot size of 100. The company's own manufacturing costs are shown in Table 42. The table shows that the total manufacturing cost per unit to the company is $217.30 in a lot of 100 units. The company has the facilities and personnel to make the part in-house. Let us say a vendor quotes $200.00 for the same part. Should the company purchase the part or make it?

A simple argument that would decide this issue is to assume that the facilities which would have been used for this part would stay idle if the company decided to buy the part. The labor overhead cost of $54.50 and machine costs of $121.00 would continue to accrue. Thus, in reality, the purchased part would cost $200 + $54.50 + $121 = $375.50. Thus it is cheaper for the company to make the part. However, if the facilities could be used for other purposes, it may be advantageous to buy the part.

TABLE 42 Manufacturing Costs for Sleeve

Labor Costs	$	$
Setup Labor	132.00	
Tooling (programming)	48.00	
Total fixed costs (100 units)	180.00	
Fixed labor cost per unit	1.80	
Labor—processing cost per unit	20.00	
Total direct labor cost per unit		21.80
Other Costs:		
Material	20.00	
Labor overhead @ 250%	54.50	
Machine (processing) cost	121.00	
Total cost per unit		217.30

SUMMARY

This chapter has provided a general treatment of manufacturing methods for engineering materials. The discussion is, of necessity, brief, since there are complete volumes devoted to the subject. The choice of the manufacturing method for a part is closely tied to the material and constrained by the requirements—shape, quantities required, physical properties, surface finish, facilities available, and tolerances. Manufacturing processes for metals and polymers were discussed.

BIBLIOGRAPHY

Boothroyd, G., P. Dewhurst, and W. A. Knight. "Selection of Materials and Processes for Component Parts." *Proceedings of the 1992 NSF Design and Manufacturing Systems Conference* 255–263. Dearborn, Mich.: Society of Manufacturing Engineers.

Boothroyd, G., P. Dewhurst, and W. A. Knight. *Product Design for Manufacture and Assembly*. New York: Marcel Dekker, 1994. Classifies manufacturing processes as primary, secondary, and tertiary processes.

Cubberly, W., and R. Bakerjian, eds. *Tool and Manufacturing Engineers Handbook*. Dearborn, Mich.: SME Press, 1988.

Curless, R., Cincinnati Milacron, Inc., 1996 (private communication).

Dewhurst, P., and G. Boothroyd. "Early Cost Estimating in Design." *J. of Manufacturing Systems* 7(3): 183–191 (1987).

Dieter, G. E. *Engineering Design*. 2nd ed. New York: McGraw-Hill, 1991.

Ehrlenspiel, K. *Kostengünstig Konstruieren (Cost-Effective Designing)*. Berlin and New York: Springer Verlag, 1985. Data for Figure 126 are from this book.

Groover, M. P. *Fundamentals of Modern Manufacturing*. Englewood Cliffs, N.J.: Prentice Hall, 1995.

Hayes, R. H., S. C. Wheelwright, and K. B. Clark. *Dynamic Manufacturing*. New York: Free Press, 1988.

Kalpakjian, S. *Manufacturing Processes for Engineering Materials*. Reading, Mass.: Addison-Wesley, 1991.

Kalpakjian, S. *Manufacturing Engineering and Technology*. Reading, Mass.: Addison-Wesley, 1992.

Kiewert, A. Systematische Erarbeitung von Hilfsmitteln zum Kostengünstigen Konstruieren (Systematic Working on Help Media for Cost-Effective Designing). Dissertation, Munich Technical University, 1979.

Matousek, R. *Engineering Design—A Systematic Approach*. London: Blackie, 1963.

McDowell, K. A. "EVO—A New Chair Design." *Proceedings of the 1995 National Design Engineering Conference,* Norwalk, Conn.: 219–220. Reed Exhibition Companies.

Michaels, J. V., and W. P. Wood. *Design to Cost*. New York: J. Wiley, 1989.

Ott, H. H., and V. Hubka. "Vorausberechnung der Herstellkosten von Schweisskonstruktionen" (Computation of Manufacturing Costs of Weldments). *Proc. ICED 85*, 1: 478–487. Zurich: Heurista, 1985.

Pacyna, H., A. Hillebrand, and A. Rutz. *Kostenfrüherkennung für Gussteile (Early Cost Estimation for Cast Parts). VDI Berichte Konstrukteure senken Herstellkosten—Methoden und Hilfen*, No. 457. Düsseldorf, VDI Verlag, 1982.

Pahl, G., and F. Rieg. *Kostenwachstumsgesetze für Baureihen (Cost Growth Laws for Size Ranges)*. Munich: C. Hanser Verlag, 1984. Shows the dependency of cost of castings on casting weight and the complexity of the casting.

Schey, J. A. *Introduction to Manufacturing Processes*. 2nd ed. New York: McGraw-Hill, 1987.

Trucks, H. *Designing for Economical Production*. 2nd ed. Dearborn, Mich.: Society, of Manufacturing Engineers, 1987.

Ulrich, K. T., and S. D. Eppinger. *Product Design and Development*. New York: McGraw-Hill, 1995.

• • • • •

EXAMPLE
DESIGN VARIANTS OF A BEARING PEDESTAL[1]

Figure 156 shows the original design of a bearing pedestal for a chemical industry application. The unit is made out of plates which are cut and welded together. The welded housing is then machined in a borer to produce the finished unit. The pedestal contains an oil tank which is integral with the housing. The chart in Figure 156 shows that:

- The total manufacturing cost breaks down into 10% for material, 43% for the production of the welded housing (prepara-

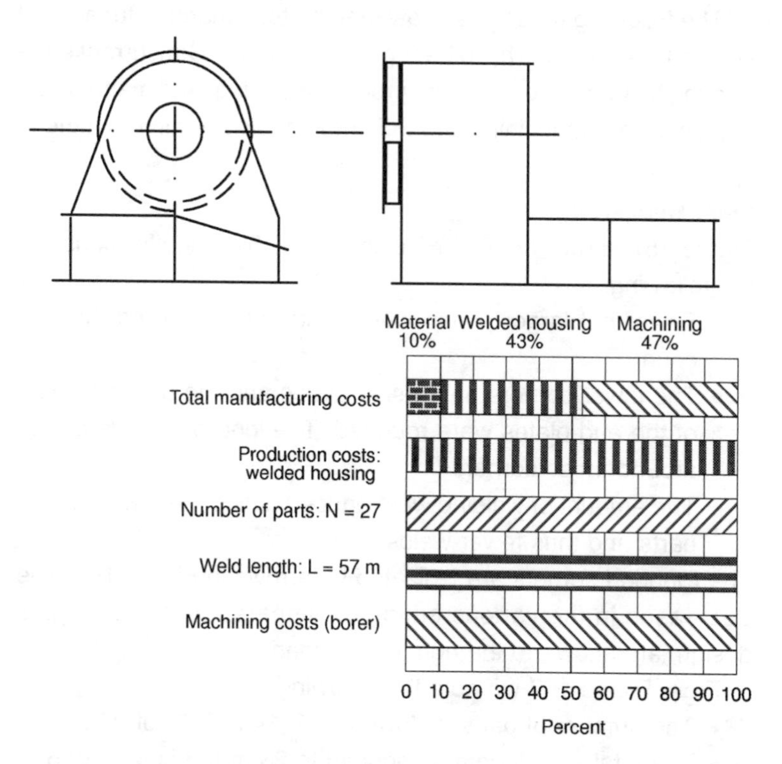

FIGURE 156. Original design of bearing pedestal.

[1]Ehrlenspiel, K. *Kostengünstig Konstruieren (Cost-Effective Designing)*. Berlin and New York: Springer Verlag, 1985.

tion, handling, fixtures, electrodes, etc.), and the remaining 47% for machining in the borer.

- There are 27 parts (plates) in this design.
- Total weld length in this design is 57 m.

In order to compare the costs and other features of this first design with those of the subsequent redesigns, all of the entities in the chart in Figure 156 have been given the baseline value of 100%. Note that the item "production costs: welded housing" shown as 100% in the second bar is the same quantity shown as 43% of the total manufacturing cost in the top bar. Likewise, the item "machining costs (borer)" shown as 100% in the last bar is the same quantity shown as 47% of the total manufacturing cost in the top bar.

The following redesigns show that the total manufacturing cost was reduced by 41% by following the DFC rules. Concurrent engineering by teams from design, process planning, and manufacturing (especially machining and welding) enabled this improvement.

First Redesign

Figure 157 shows the first redesign. The chief modifications are the following:

- Fewer but larger plates were used, thus reducing the part count.
- The lower corners of the side plates and the upper corners of the end plates were rounded. The long plates were bent around these corners.
- The amount of welding was reduced, since there were fewer parts and thus fewer welds.

The rounded corners are indicated by arrows in Figure 157. The decreases in the various entities, compared with the original design, are shown to the right of the chart.

From Figure 157 we note the following:

- The number of parts is down to 14 (14 is 52% of 27).
- The total weld length is reduced to 23 m (23 is 40% of 57).
- Due to the above steps, the production cost of the welded housing is now 70% of that of the original design.
- Machining costs are somewhat lower (87%).

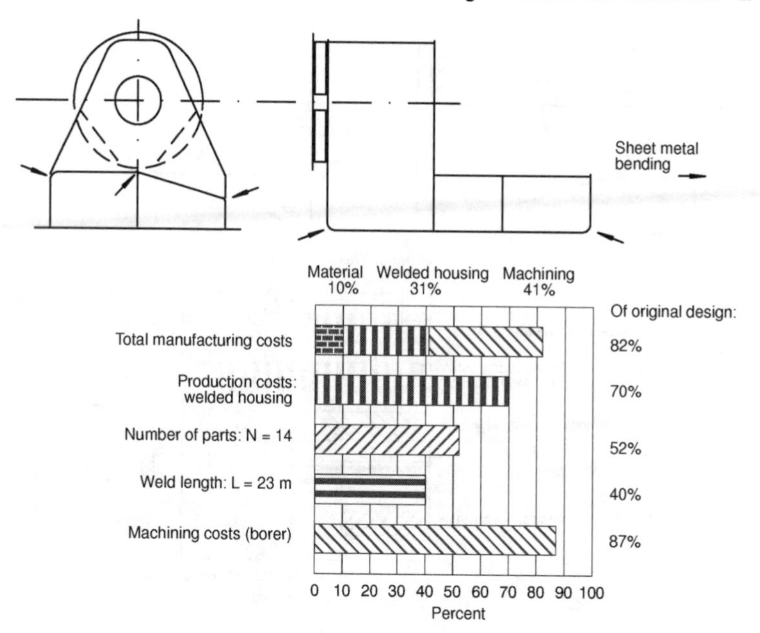

FIGURE 157. First redesign of the pedestal.

SECOND REDESIGN

Figure 158 shows the second redesign, where machining on a turret lathe was substituted for the boring operation. The oil tank, instead of being integral with the housing, was welded on to it after the machining. Without the oil tank, the unit is small enough to be set up on the lathe. This is an instance of using function separation to reduce costs, in this case for a small–lot size item. We note that despite the increase in the number of parts and the welded length, the overall cost is significantly lower than that of the first two designs.

Figure 159 shows a side-by-side comparison of the manufacturing costs of the three designs. We see that the greatest reduction that has taken place is in the machining cost. A turret lathe was used in the second redesign instead of a borer. The higher cost could be attributed to the slower throughput of the latter.

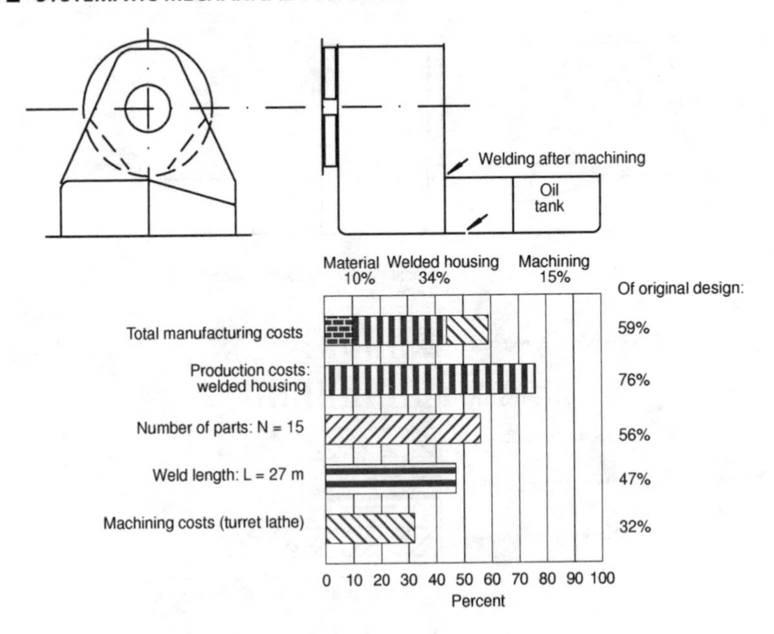

FIGURE 158. Second redesign of the pedestal.

• • • • •

EXAMPLE DESIGN OF A LEVER

A part is to be designed to serve as one element of a brake system for an industrial tractor. The brake system design takes place by following the steps shown in Figure 12 in Chapter III. This starts with the overall requirements for the system. The overall function is divided into simpler subfunctions and shown in the form of a function structure. For each of the subfunctions, we seek solutions by applying different physical effects.

CONCEPTUAL DESIGN

We now look at the design at the part level (see Figure 83 in Chapter VII). The pertinent function is stated in solution-neutral terms as

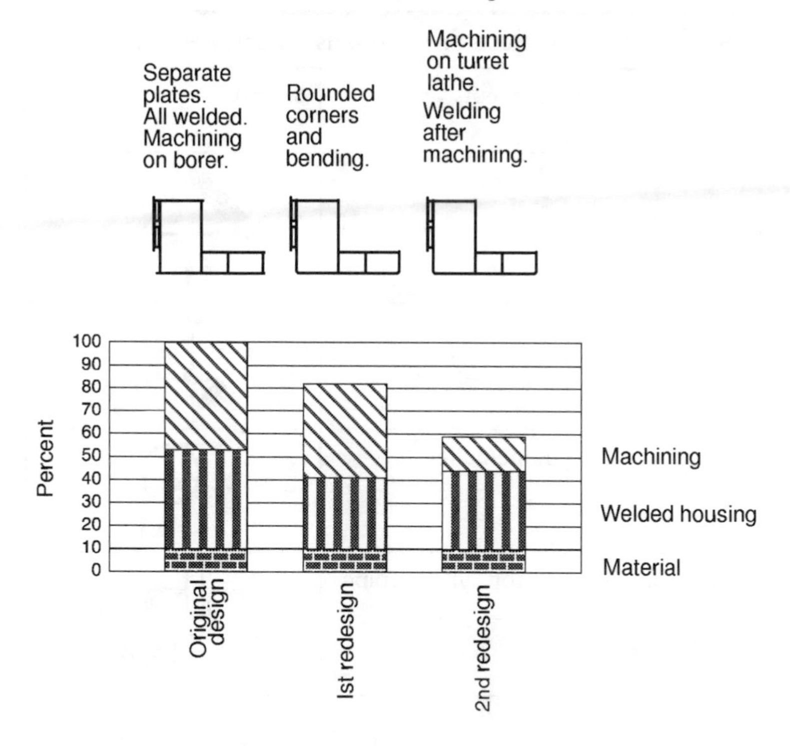

FIGURE 159. Total manufacturing costs of three pedestal designs.

Function: *Change the direction of a force by 90°*

The requirements on this part are:

Mechanical loading: 20 lb force maximum, occasional impact
Chemical environment: Exposed to rain and snow
Thermal environment: –30° to +120°F
Weight: Less than 1/2 lb
Quantities required: 1000
Manufacturing cost: As low as possible
Maintenance cost: None
Operating cost: None

The part is exposed to the elements. It is not exposed to the high temperatures from the engine.

The physical effects and the corresponding concepts for fulfilling this function are given in Table 43. By using an evaluation proce-

TABLE 43 Physical Effects and Concepts for Lever Function

Physical Effects	Concepts	Sketch
Friction	Wedge	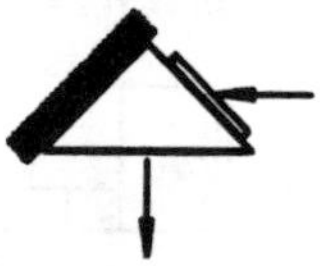
Lever	Lever	
	Rope/pulley	
Hydraulics	Piston/cylinder/pipe	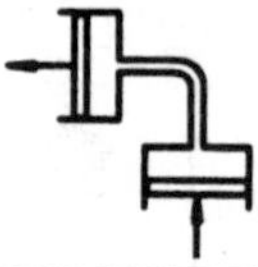

dure that employs criteria from the requirements list, we choose a bell-crank lever as the concept upon which to further develop the design, although alternative forms are possible; see Figure 104. The rough shape and the shapes of the active surfaces and the interfaces—the holes for pins and rods that support the lever and transfer forces to it—are set down (see Figure 84, Chapter VII).

We proceed next with the embodiment—the main point of this example. A lever is a basic and common mechanical element, occurring in many shapes and uses, such as handles, bars, and cranks. Levers are produced from many different materials and by different processes to fulfill the varied functions in many different environments.

Embodiment Design

Initial Material Selection

The first step in embodiment is an initial selection of the material, which is achieved by using the procedure illustrated by Figure 97 in Chapter VII, i.e., rejecting the unsuited materials. The materials passing this test for the lever are cast iron or steel, forged steel, an aluminum alloy, or a plastic. Certainly, other choices are available, e.g., ductile iron.

For a closer evaluation of these materials, we use the weighted-values method (Johnson, 1978). The following evaluation criteria are chosen: (1) tensile strength, (2) resistance to chemical attack, (3) resistance to temperature effects, (4) material cost per unit strength, (5) manufacturing cost, and (6) weight. Each of these criteria i is assigned a weight w_i (with $\Sigma\ w_i = 1.0$). The weights w_i are found by the method of binary weighting. Each of the criteria is compared with each of the others in turn. Of each pair, the criterion that is thought to be more important is given the value 1 and the other the value 0. The number of 1s for each criterion is then normalized to the total number to yield the weight.

Each of the materials j is assigned a value v_{ij} for each criterion on a scale of 1 to 10. The more favorably the criterion is fulfilled, the higher is the value. For example, the greater the weight, the lower is the corresponding v_{6j}. These data and the calculated results are shown in Table 44.

Two types of overall results are shown: the unweighted sums $\Sigma\ v_{ij}$, and the weighted sums $\Sigma\ w_i v_{ij}$. These sums rank the materials

TABLE 44 Initial Material Selection for Lever

Criterion	Weight w_i	Cast Iron v_{i1}	Steel v_{i2}	Aluminum Alloy v_{i3}	Plastic v_{i4}
Tensile strength	0.333	5	10	8	3
Resistance, chemical	0.267	10	9	10	10
Resistance, thermal	0.067	7	10	5	2
Material cost	0.133	4	10	5	3
Manufacturing cost	0.133	3	6	8	10
Weight	0.067	6	5	8	10
Sum $\sum v_{ij}$		35	50	44	38
Weighted sum $\sum w_i v_{ij}$		6.13	8.87	7.93	6.20

in the order of their value. The results show that steel is the most suitable material for this application. However, this result does not account for the manufacturing cost per piece. More detailed calculations will have to be made to make the final selection.

The operating cost in this application was taken to be negligible. However, let us say that the lever is to be used in an aircraft or some other vehicle. It is easy to see that its weight would be weighted higher in Table 44. Thus, an aluminum alloy or a plastic might turn out to be the most suitable material.

We will next take the top-rated material, steel, through the next steps, given in Figures 85 and 86 in Chapter VII, for choosing among different manufacturing processes and shapes.

Processes and Shapes. The chief processes available for forming steel are casting, forging, welding, and machining, and combinations of these processes. The active surfaces are at the connection points—(1) the pivot point and (2) the two force application points. We now consider some of the factors affecting the process selection and shape design.

Loading. The two arms of the lever are subjected to bending from the ordinary loads, as well as from any unintentional side loads. The I-cross section puts the material to the fullest use in bending. Hollow cross sections may also be used if space constraints permit.

Processes. Such cross sections can be easily produced in steel by casting and welding, although rectangular or oval sections are cheaper to produce and are therefore used in conventional lever designs. Figure 160 shows sketches of a lever produced in both conventional and I-section designs. The lever produced as a forging is made in two separate pieces, which are then joined at the pivot point.

Quantities Produced. Figure 161 compares the costs of manufacturing the lever from steel by three different processes: casting, drop forging, and welding using the conventional designs. The cost structures for the three production methods are given in Table 45. The pattern and mold, in the case of casting, and the dies, in

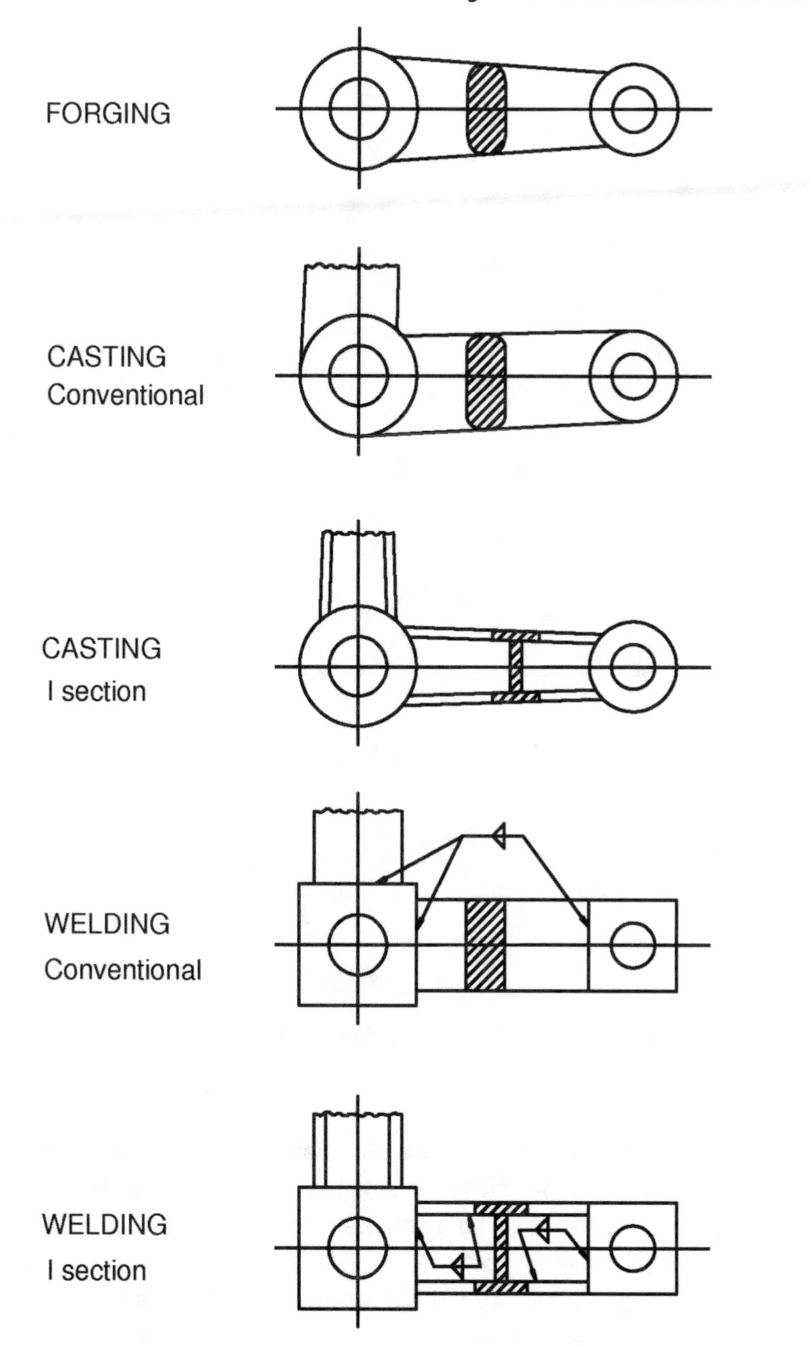

FIGURE 160. Processes and shapes for the lever.

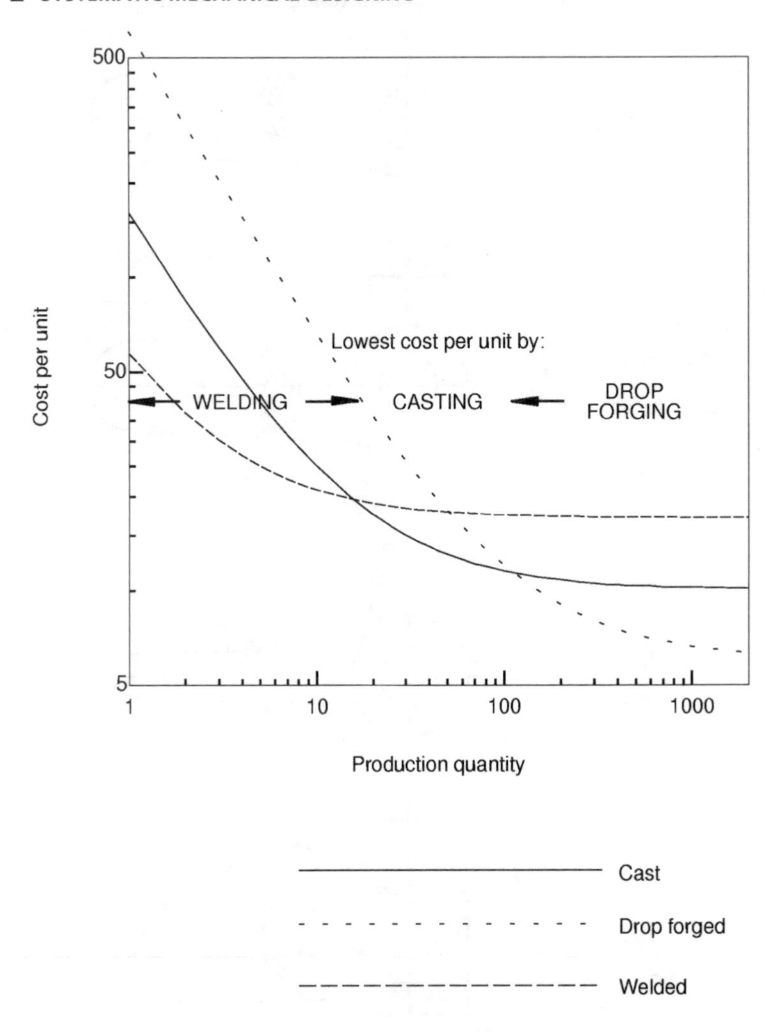

FIGURE 161. Lever costs for different production methods.

the case of drop forging, have high initial costs, which are reflected in the high cost per piece for small lot sizes produced by these methods. The cost of the jig for welding is lower; this, combined with the higher labor cost, yields the lowest cost for small lot sizes and the highest cost for large quantities. For the present case, if the part is to be produced in a lot size of 1000, the most cost-effective process is drop forging.

TABLE 45 Cost Structures for Three Production Methods for Lever

	Cast	Drop-Forged	Welded
Pattern, die, or jig	150	600	40
Material	1	1	2
Labor	5	3	13
Machine	4	2	2
Material + labor + machining	10	6	17

Summary

The example showed a systematized approach to part design. Design of parts forms the most important portion of embodiment design. It is a complex process because (1) a number of different factors must be considered simultaneously and (2) each action influences other aspects of the problem. Three main aspects of embodiment are the selection of shapes, materials, and manufacturing processes. Since the number of other factors is so large, the design process has to be iterative in nature. The systematic design method, which forms the underlying basis of the approach used in this example, allows an unprejudiced search for the best design. The systematization of part design has important ramifications for computer-based design methods, e.g., those incorporating expert systems and feature-based modeling. See, for example, Wierda (1991).

Bibliography

Domhoff, J. E., S. F. De Fosse, and D. E. Greer. "Getting a Grip on Plastic Switches." *Mechanical Engineering* 116(10): 77–78 (1994). Describes the design and optimization of a plastic handle for an electric switch. It is an example of substitution of materials: the earlier metal handle was redesigned as a two-piece injection-molded plastic part, which is claimed to be as durable as metal and has a more ergonomic shape. The load-carrying part is made of nylon 6/6 with 60% glass fibers, along with a bright-red part for high visibility, of a polycarbonate and PBT mixture.

Johnson, R. C. *Mechanical Design Synthesis*. Huntington, N.Y.: Krieger Publishing, 1978. Describes the weighted-values method.

Wierda, L. S. "Linking Design, Process Planning and Cost Information by Feature-based Modelling." *J. Engineering Design* 2(1): 3–20 (1991). Discusses feature-based design and the relationship of features to process planning, manufacturability, and costs.

Standardizing to Reduce Costs

Standardization reduces product costs by minimizing the design costs for parts and assemblies. It also reduces the time to market and improves reliability. Designers may resist standardization in the belief that it restricts their freedom. However, lack of standardization would lead to chaos. Imagine having to design a nut, bolt, or washer each time it was needed!

Standardization implies not just the use and application of standards, but rather any process that is agreed upon by the parties involved to be a time- and cost-saving measure. It is therefore very dependent on when such an agreement is reached and the state of technology at that point in time.

STANDARDIZATION STEPS AND RESULTING SAVINGS

Standardization can be realized at different stages of the design process, as enumerated in the following.

Standardization of Solution Principles

Standardized solution principles represent the lowest level of standardization. Solution principles are the representations of the application of physical principles in sketch form. (See the examples "A Temperature Measurement Device" and "A Variable Speed Transmission" in Chapter III.) The use of well-established solution principles in the overall solution leads to savings in development costs. Compatibilty of solutions must be observed.

Standardization of Geometries

The standardization of details such as shaft chamfers, fillet radii, or slots is a well-accepted practice. There are savings in design, since these details do not have to be determined each time. There are savings in production, since fewer tools and fewer different production processes are required.

Standardization of Parts

Using standard parts leads to savings in design, since existing designs can be used repeatedly. Parts can be combined in modular form in different assemblies. There is a savings in manufacturing, since larger quantities of standard parts can be made in large lot sizes. Better quality results from using standard parts, since production processes are better understood and mastered. Standardization of parts will be discussed in greater detail in this chapter.

Standardization of Modules

Standardized modules represent the highest level of standardization and the best means of cost savings, for the same reasons standardized parts offer cost savings. Interfaces between modules should also be standardized.

USE OF STANDARDS IN DESIGN

The use and application of standards must be weighed against their disadvantages. Ehrlenspiel (1985) quotes the case of AC motor design at Siemens. From 1955 to 1977, as the part count decreased by 12%, the number of standards to be considered—the operational standards, design guidelines, and so forth—rose by 25% in the same time period. The time required for searching and evaluating technical rules and requirements can take up to 10% to 20% of the total design time (according to a 1980 figure). There are even cases where the costs of proving and checking that standards have been adhered to exceed the manufacturing costs.

There is a hierarchy of standards, from international (e.g., ISO) to national (ANSI, etc.) to company standards, as shown in Figure 162. In the category of company standards are standards pertaining to specific products and more general standards that refer to raw materials, standard parts, design standards, and other company guidelines. Product standards are a must for companies that manufacture modular products and/or products in size ranges, e.g., gear drives, engines, pumps, or machine

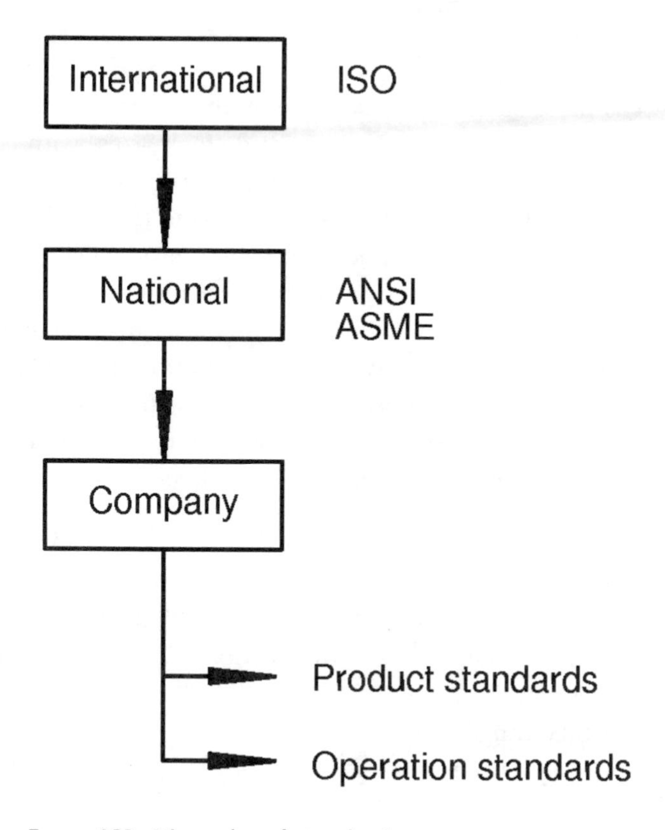

FIGURE 162. Hierarchy of standards.

tools. General standards reduce the number of different parts and suppliers and the demands on storage facilities.

COMPANY STANDARDS

A company's operation standards are set up to cover such items as the following:

- Materials
- Semifinished parts
- Standard parts
- Purchased parts

- Production processes
- Computational methods
- Drawing standards

The most outstanding example of an operation standard is reduction in the number of different parts and stock sizes, with consequent savings in different types of costs (see Figure 106 in Chapter VIII).

Product standards help save costs and time and increase reliability in every phase of product development. A product standard may be one that governs the overall design of a product, or it may be an operation standard applied to a product. Product standards may be as simple as defining part groups, or they may cover size ranges and modular designs. A product standard sets the materials, design types, dimensions, tolerances, processing, performance data, made versus purchased parts, etc.

In short, product standards include items such as:

- Performance standards
- Throughput standards
- Size standards (e.g., engine size: 2.5 liters)
- Part groupings (see the section "Group Technology" later in this chapter)
- Size ranges
- Modular designs

Product standards offer the following advantages:

- Reduced design costs, since design details are already present
- Reduced manufacturing costs, since product planning documentation already exists, production and assembly procedures are known, and production equipment exists on which personnel are experienced
- Fewer new parts, and thus lower associated costs
- Fewer different parts to be purchased and stored
- Savings in preparing brochures, advertisements, and price lists
- Reduced project engineering and quotation work, due to use of standard parts and modular designs

To be sure, there are disadvantages to having product standards, including the following:

- Product design may not be optimal for the given task
- The customer's unique requirements may be unfulfillable
- The company has less flexibility for quick response to market changes

Check on Use of Standards

During design, the following checks can be made to ensure that the use of standards helps rather than hinders the design objective:

- Check whether the use of a standard solution fulfills the desired function.
- Does the design follow dimensional standards?
- Does the design follow safety standards?

Use of standards further helps reduce costs and increase product reliability as follows:

- Use of standard ergonomic data
- By following test and inspection procedures according to standards
- Use of standard tolerances, limits, and fits
- Use of internal standards to minimize costs
- Use of standard symbols for installation, operation and maintenance

STANDARDIZATION AND THE NUMBER OF PARTS

Perhaps the greatest effect of standardization is at the part level. We will next discuss the effect of part design and quantities on costs.

Increasing the Part Quantities

The following hierarchy of design decisions can be observed in order to use fewer parts but in larger quantities to keep cost down:

- Use of purchased parts that have been produced in large lot sizes
- Use of standard parts
- Design of products in size ranges to reduce product development costs by avoiding special designs when the same function needs to be fulfilled

- Use of modular designs to permit the recurring use of the same parts and assemblies
- Use of standardized part groups
- Use of the same parts in different products
- Use of the same part many times in a given product

Cost of a New Part

The introduction of a new part requires activities in a number of departments in a company. Thus, the cost of a new part entails not only the manufacturing cost or purchase cost (if it is a purchased part), but other costs as well:

- Part introduction costs
- Support and maintenance costs
- Alteration costs
- Liquidation costs

A comparison of part development costs was shown in Figure 69 in Chapter VI.

Standard Parts

Standard parts are available "off the shelf." Typical everyday examples include nuts and bolts, but also a variety of parts such as power transmission elements. Advantages of using standard parts are:

- Savings in development costs
- Ready availability
- Produced in large quantities, therefore low cost
- Little or no need for inventory

Same Parts

A manufacturer can use identical parts in multiple locations in the same product or in different products. Advantages of using same parts are:

- Reduction in batch-level costs
- Produced in larger quantities, therefore low cost

- Higher reliability, since design and manufacturing processes are better established and more robust
- Less need for inventory, therefore low overhead

Examples of this concept are as follows:

- The design of doors of the Boeing model 777 airplane. A commonality of 96% in the parts for different doors has been achieved.
- In automobiles, the use of an identical design for doors in a sedan and a station wagon, and the use of the same wheels on both the right and left sides, have reduced the total number of parts.

Similar Parts

If identical parts cannot be used, the next best approach is to use parts that are similar in most respects, differing only in a few features. Advantages of using similar parts are:

- Reduction in batch-level costs (less so than for same parts)
- Ability to use group technology effectively to reduce design and manufacturing costs
- More robust designs, since technology is better established

Care must be used, however, during assembly so that wrong parts are not used. There is more discussion of this subject under "Manufacturing Considerations in Design" in Chapter VII.

Fewer Parts

The total number of parts has one of the most far-reaching effects on costs; see Figures 90 and 92 in Chapter VII. Advantages of using fewer parts are:

- Less handling
- Less storage space
- Fewer assembly steps, hence less time and cost
- Quicker disassembly for repair, maintenance, and recycling
- Lower overhead costs due to less record keeping (inventory lists, accounting costs)

EFFECT ON RELIABILITY

The use of fewer parts overall in a product plays a role not only in manufacturing costs but also in reliability of the product and yield of the manufacturing process. The latter depends on the number of manufacturing operations as well as parts. The dependence of final product quality on part quality can be shown as follows. Let P_i be the probability that each part or operation will be acceptable. Then, if there are n parts or operations, the probability that the final product will be acceptable, or the yield Y, is

$$Y = \prod_{i=1}^{n} P_i \tag{62}$$

If the probability that each part or operation will be acceptable is the same—i.e., if $P = P_i$—then Equation (62) becomes $Y = P^n$. Figure 163 shows a plot of this function for different probability values. We see that if the acceptability of each part is 90% and there are 20 parts, the probabil-

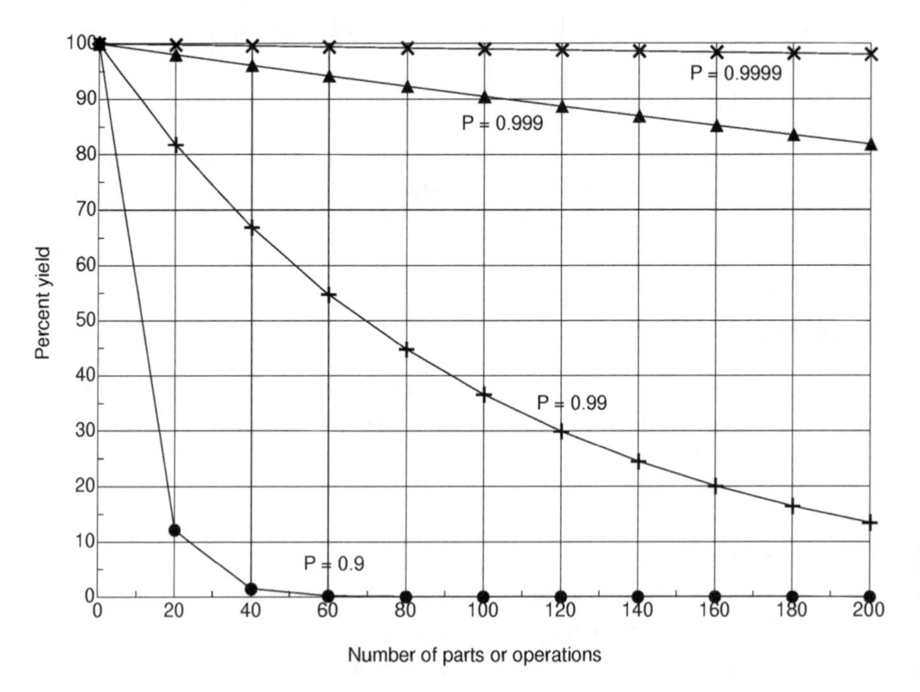

FIGURE 163. Effect on yield of the number of parts.

ity of acceptance of the whole product is only 12%. If there are 200 parts or operations, the acceptability of each must be 99.99% if the whole is to have a yield of 98%.

GROUP TECHNOLOGY

One means of reducing the number of parts and types of different parts is by using group technology (GT), a coding and classification system that enables parts to be grouped in families. An alphanumeric code is used to define the features of a part. A properly prepared and used group technology database leads to lower costs by controlling the proliferation of parts, allowing the designer to search for similar parts that may be easily redesigned, or even the same part that unbeknownst already exists. It is unfortunate that few companies use this powerful technique for rationalizing the design process and lowering costs.

The benefits of GT are most significant in manufacturing, where the grouping of machines and processes and production control and scheduling can be rationalized and optimized for families of parts. Data from industry show reductions of the order of 40% in manufacturing costs and similar savings in annual operation costs of production units when group technology is applied.

Classification of parts can be done according to their design or their manufacturing properties, which are, of course, related:

1. Design properties refer to the parts' functions, shapes, dimensions, tolerances, finishes, etc.
2. Manufacturing properties refer to the manufacturing processes, tools and equipment required, production quantities, production line attributes, etc.

STANDARD NUMBERING SERIES

Standard numbering series (SNS) have been in use in Europe for standardizing product sizes, throughputs, etc. (Pahl and Beitz, 1988). The step size in the series, designated by the letter R, can range from rough (R5) through fine (R40). The number n in the series designation Rn denotes the number of steps per decade in the particular series, as shown in Table 46. Each number in a given series is obtained by multiplying the preceding number by $10^{1/n}$. The exact numbers are rounded off for actual use, as shown by the numbers in parentheses in the R10 column. The increments provided by these geometric series have been found to be easier to comprehend than would be those of arithmetic series designed for the same purpose.

A company can use SNS for standardization and consolidation of its product line. Suppose a manufacturer of electric motors has to schedule production to satisfy the demand in sizes and quantities (for a given time period) shown in Figure 164a. This represents a total count of 333 in 37 different sizes. Such a production schedule leads to higher costs due to the small number of motors produced of each size. The manufacturer therefore consolidates the number of different sizes produced to four, by using the R5 series distribution. There are now more quantities of fewer sizes of motors being produced, as shown in Figure 164b. This more rational production schedule leads to lower costs. (The disadvantage of producing fewer sizes is that it does not satisfy the market as closely as before; less variety is offered at a lower cost.)

At the startup of a new product line, a manufacturer can keep costs down by using a rough series such as R5 or R10. As experience is gained and the market expands, the company can offer a greater variety using a finer series such as R20 or R40.

DESIGN IN SIZE RANGES

Machines produced in size ranges differ from each other only in throughput parameters and size. They have the same overall function and general arrangement of parts and generally use the same materials and manufacturing processes.

Manufacture in size ranges lowers costs, achieves shorter time to market, and results in higher quality and reliability. Lower costs are achieved by savings in design and manufacturing of similar parts and assemblies, as opposed to entirely new units. Fewer different types of parts need to

TABLE 46 Standard Numbering Series for n = 5, 10, 20, and 40

R5	R10	R20	R40
1.00	1.00	1.00	1.00
1.58	1.26	1.12	1.06
			1.12
		1.26	1.19
	(1.25*)		1.26
	1.58	1.41	1.33
			1.41
		1.58	1.50
	(1.6*)		1.58
2.51	2.00	1.78	1.68
			1.78
		2.00	1.88
	(2*)		2.00
	2.51	2.24	2.11
			2.24
		2.51	2.37
	(2.5*)		2.51
3.98	3.16	2.82	2.66
			2.82
		3.16	2.99
	(3.15*)		3.16
	3.98	3.55	3.35
			3.55
		3.98	3.76
	(4*)		3.98

(continued)

TABLE 46 Standard Numbering Series for n = 5, 10, 20, and 40 *(continued)*

6.31	5.01	4.47	4.22
			4.47
		5.01	4.73
	(5*)		5.01
	6.31	5.62	5.31
			5.62
		6.31	5.96
	(6.3*)		6.31
10.00	7.94	7.08	6.68
			7.08
		7.94	7.50
	(8*)		7.94
	10.00	8.91	8.41
			8.91
		10.00	9.44
	(10*)		10.00

*Rounded-off value.

be made, purchased, and stored, and the lot sizes are larger. Shorter time to market comes from savings in time (up to 50% or more of that for a new design) for design and production planning. Higher quality and reliability of products result from the shorter learning curve and the greater experience gained in producing similar units. The design process and cost estimation are supported by the use of such aids as similarity principles.

APPLICATION OF SIMILARITY PRINCIPLES

Similarity principles and laws have been developed for the generalization of analyses in a number of fields, most notably fluid mechanics and heat

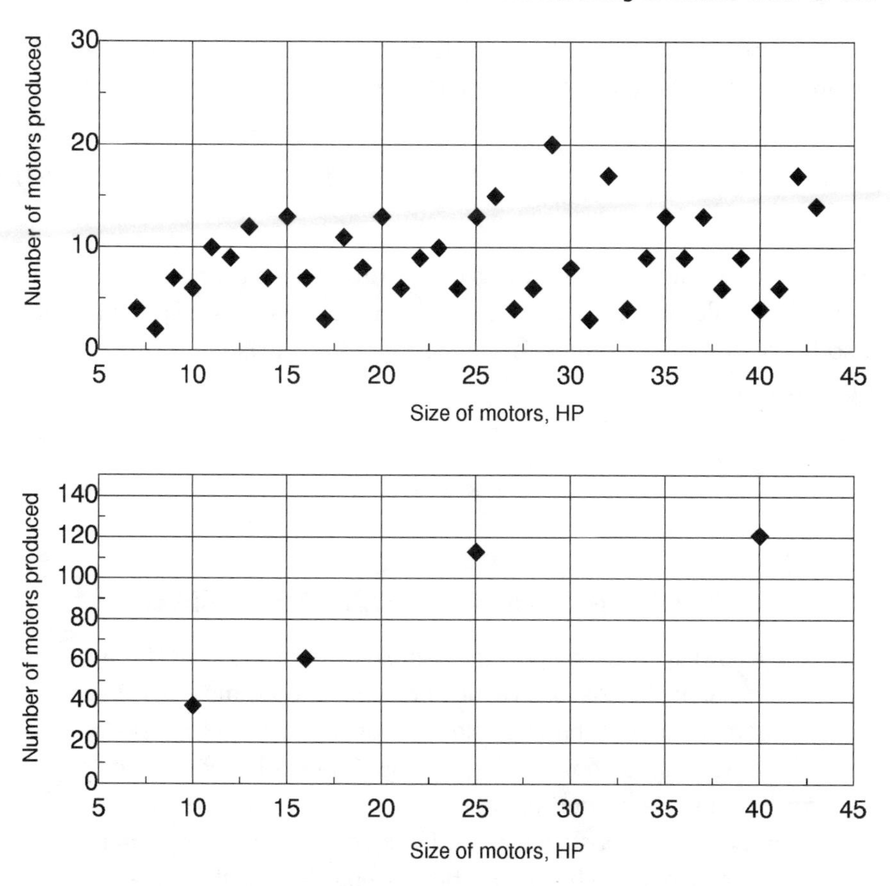

FIGURE 164. Reduced product variety for lower cost.

transfer. Similarity principles form an important basis for the design of products to be made in size ranges and are also used extensively in predicting their costs. If a design has been built in one size, then the cost of designs in other sizes can be predicted by this method more accurately than using weight or material alone. The premise is that the linear dimension ratio of two products, $\phi_l = l_2/l_1$, is the primary parameter in relating their costs. For a linear size ratio ϕ_l of two parts, their area ratio is ϕ_l^2 and the volume ratio, ϕ_l^3.

Product costs can in general be broken down into components that change with ϕ_l^i, when $i = 0, 1, 2, 3$: fixed costs ($i = 0$), costs varying with linear size ($i = 1$), and costs varying with area and volume ($i = 2$ and i

= 3), respectively. Thus, the manufacturing cost ratio relating the cost of product 2 to the cost of product 1 is given by

$$\frac{C_{\mathrm{mfg}2}}{C_{\mathrm{mfg}1}} = \phi_{\mathrm{mfg}} = \sum_{i=0}^{3} a_i \phi_l^i \tag{63}$$

where ϕ_{mfg} is the manufacturing cost ratio. The parameters a_i can in general be calculated by using regression analysis. The exponents may also have noninteger values for certain types of parts, as shown later in this chapter.

● ● ● ● ●

EXAMPLE
COST OF A SIMILAR WELDED HOUSING

The following is based upon an example by Pahl and Rieg (1984b) of calculating the cost of a welded housing. The example will show the computation of the coefficients a_i of Equation (63). A sketch of the housing is shown in Figure 165. The manufacturing costs for the unit are divided up into material, joining, machining, and other processing costs as shown in Table 47. These costs are for a certain size of the housing, which will be used as the base part for estimating costs of housings of other sizes.

The costs in Table 47 are also divided according to the power of ϕ_l to which they are proportional. The fixed or constant costs are those shown proportional to ϕ_l^0. The other allocations are as follows:

Material costs are proportional to volume, thus proportional entirely to ϕ_l^3.

Flame cutting involves \$5 fixed costs and \$20 proportional to the length cut, thus ϕ_l^1.

Chamfering costs are assumed proportional to length.

Welding includes clamping costs of \$20, proportional to the length, and material costs of \$75 proportional to volume; thus ϕ_l^3.

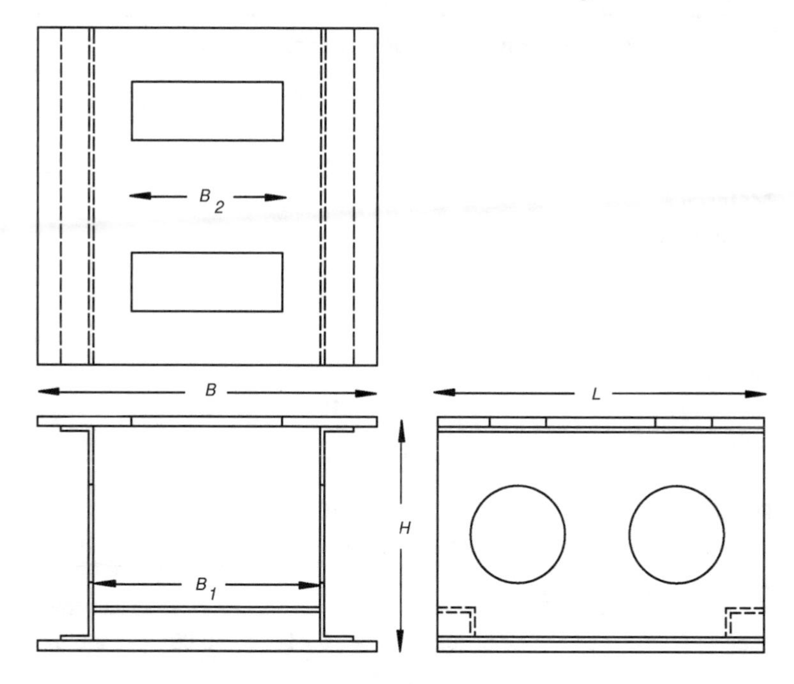

FIGURE 165. Welded housing.

Annealing is a bulk operation; thus, its costs are proportional to ϕ_l^3.

Sandblasting costs consist of those proportional to the surface area, ϕ_l^2, and the volume, ϕ_l^3.

Marking costs are proportional to the length marked, thus ϕ_l^1.

The three types of machining operations all involve some fixed setup costs. In addition:

Milling costs of $35 are proportional to the surface milled, thus to ϕ_l^2.

Drilling and boring costs are proportional to the length.

The total costs proportional to each of the ϕ_l^i terms are in the row marked "sum." The total cost of the part C_{mfg0} is given in the row marked "grand total."

The manufacturing cost of a housing of a different size is given [see Equation (63)] by:

TABLE 47 Manufacturing Cost Components for Housing

		Proportional to			
Cost		ϕ_l^0	ϕ_l^1	ϕ_l^2	ϕ_l^3
Material					200
Flame cutting		5	20		
Chamfering			10		
Welding			20		75
Annealing					30
Sandblasting				10	15
Marking			15		
Milling		10		35	
Boring		25	35		
Drilling		5	10		
Sum		45	110	45	320
Grand total	520				
a_i		0.09	0.21	0.09	0.62

$$C_{mfg1} = (a_0\phi_l^0 + a_1\phi_l^1 + a_2\phi_l^2 + a_3\phi_l^3)C_{mfg0} \tag{64}$$

In order to determine the coefficients a_i, we set $\phi_l = 0$. Thus, $C_{mfg0} = \$520$ and

$$\begin{aligned} a_0C_{mfg0} &= \$45 \quad \text{or} \quad a_0 = 0.09 \\ a_1C_{mfg0} &= \$110 \quad \text{or} \quad a_1 = 0.21 \\ a_2C_{mfg0} &= \$45 \quad \text{or} \quad a_2 = 0.09 \\ a_3C_{mfg0} &= \$320 \quad \text{or} \quad a_3 = 0.62 \end{aligned}$$

These are given in the last row of Table 47. Thus, the manufacturing cost of a housing of a different size is found by substituting these values in Equation (64):

$$C_{mfg1} = \$(45\phi_l^0 + 110\phi_l^1 + 45\phi_l^2 + 320\phi_l^3) \tag{65}$$

• • • • •

Machined Parts

In machining operations the similarity parameters for operation, idle, and setup times have been found to be of the form $\phi_t = \phi_l^n$. Ehrlenspiel and Fischer (1982, 1983) show that the setup costs for machined parts have been found to be proportional to ϕ_l^s, where s (for gears) varies from 0.14 for 2- to 8-inch diameters and 0.56 for 8- to 40-inch diameters to 1.8 for 40- to 60-inch diameters. The setup cost per unit for product 2 in terms of cost for product 1 is given by

$$C_{\text{fix2}} = C_{\text{fix1}} \phi_l^s \tag{66}$$

The production costs for machined parts are affected by actual cutting time, idle time, etc., and are found to be proportional to the surface area (approximately to ϕ_l^p); that is,

$$C_{\text{pr2}} = C_{\text{pr1}} \phi_l^p \tag{67}$$

where the exponent p varies from 1.8 to 2.2. Likewise, the material and heat treatment costs for similar parts are proportional to volume and have been found to vary as:

$$C_{\text{mat2}} = C_{\text{mat1}} \phi_l^m \tag{68}$$

where the exponent m varies from 2.4 for small gears to 3 for large sizes.

A combination of the previous equations yields the combined effects of product size and production rates. The manufacturing costs per unit for product 2 in terms of costs for product 1 are given by

$$C_{\text{mfg2}} = \frac{C_{\text{fix1}}}{n} \phi_l^s + C_{\text{pr1}} \phi_l^p + C_{\text{mat1}} \phi_l^m \tag{69}$$

where the nominal values of the exponents are $s = 0.5$, $p = 2$, and $m = 3$. This is a more specialized form of Equation (63).

For cylindrical objects in general—gears, drums, bearings, etc.—the ratios of diameters d and widths b affect the costs. For spur gears, the torque ratio ϕ_T is related to the diameter ratio ϕ_d and the width ratio ϕ_b, and thus to the linear size ratio ϕ_l, by

$$\phi_T = \phi_d^2 \phi_b = \phi_l^3 \tag{70}$$

Figure 166 shows Equation (69) in a nondimensional form, i.e., the manufacturing cost per unit has been normalized. We see that as the linear dimension ratio ϕ_l increases, the material costs become dominant, whereas the setup costs become a negligible part of the manufacturing cost. Decreasing values of ϕ_l show the opposite effect. Parts a and b of the figure show the same data: Figure 166a is a log–log plot, and Figure 166b is on linear scales.

• • • • •

EXAMPLE
ESTIMATING THE COST OF A SIMILAR GEAR

As an example, consider a machined part such as a gear or gear coupling. A company has produced the following steel gear:

Diameter $d_1 = 8$ in. Weight $W_1 = 40$ lb

The manufacturing cost per unit, $C_{mfg1} = \$210$, is made up of

Setup cost $C_{fix1} = \$110$

Production cost per unit $C_{pr1} = \$70$

Material cost per unit $C_{mat1} = \$30$

The company wishes to estimate the costs for a similar part of diameter $d_2 = 32$ in. We find that the linear size ratio $\phi_l = d_2/d_1 = 4$. Therefore, $\phi_l^{0.5} = 2$, $\phi_l^2 = 16$, and $\phi_l^3 = 64$. For the new part, the data are shown in Table 48.

Manufacturing costs per unit for the new part are \$220 + \$1120 + \$1920 = \$3260, for production in single quantities. Thus, for the larger part, material costs dominate.

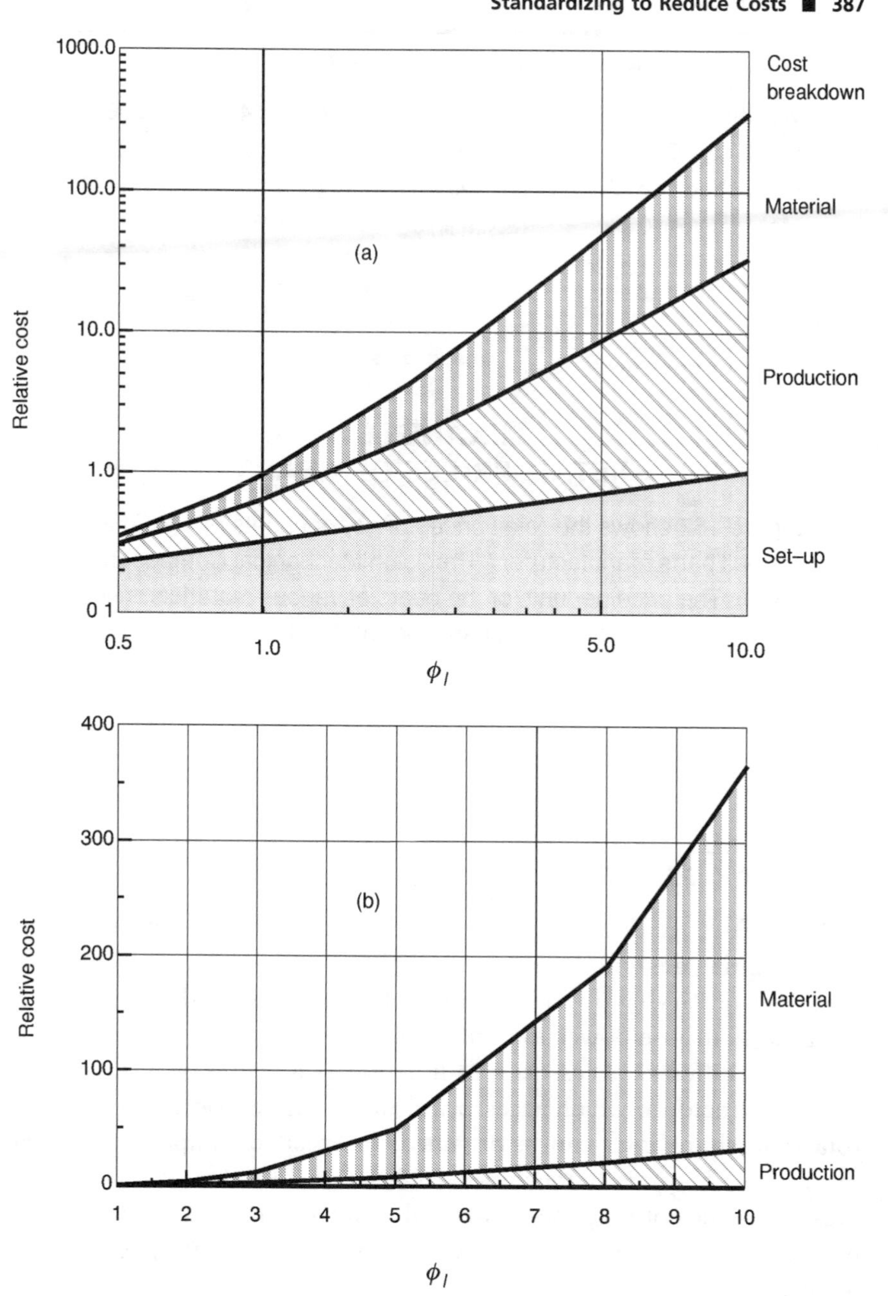

Figure 166. Relative cost as a function of linear dimension ratio.

TABLE 48 Weight and Costs for the New Gear

Weight W_2	$W_1\phi_l^3$	$= 40 \times 64$	$= 2560$ lb
Setup cost C_{fix2}	$(C_{fix1})(\phi_l^{0.5})$	$= 110 \times 2$	$= \$220$
Production cost C_{pr2}	$(C_{pr1})(\phi_l^2)$	$= 70 \times 16$	$= \$1120$
Material cost C_{mat2}	$(C_{mat1})(\phi_l^3)$	$= 30 \times 64$	$= \$1920$

• • • • •

EXAMPLE
RETAIL PRICES OF GEARS

Figure 167 shows the retail prices of gears (in lots of 1 to 9) from a small parts manufacturer ("Precision Mechanical Components," 1988). The price per unit for the gears of various materials is seen to be proportional to the square root of the diameter, i.e., $d^{0.5}$.

• • • • •

MODULAR DESIGN

Unlike size-range design, where the overall function remains the same, in a modular design parts and assemblies are combined to produce machines of varying overall functions. Examples of modular design are (1) a multipurpose power tool in which the same motor and housing is used with different accessories to function as a drill, saw, grinder, etc., and (2) a tractor to which accessories such as mower deck, snowblower, rotary brush, or tiller can be attached. In order to compare a modular machine with specialty machines, each serving a different function, we must consider not only the cost but also other attributes such as convenience to the user, time taken to change accessories, and space requirements.

The savings in cost arise as follows. The user needs to buy only one of the basic unit, e.g., the drive. The manufacturer reduces costs by producing a larger number of units. There is also a corresponding increase in

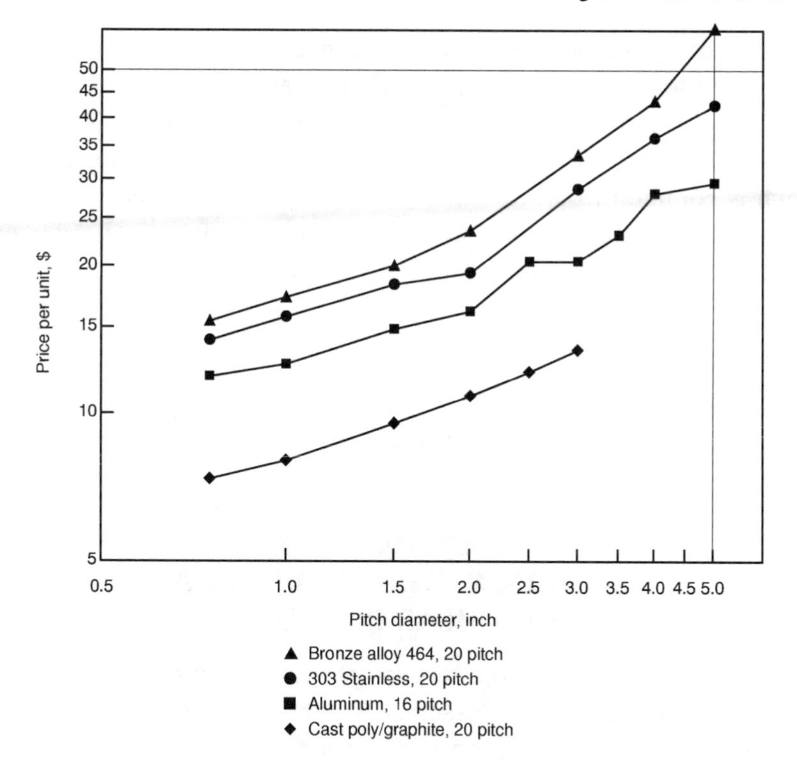

FIGURE 167. Trends in gear prices with diameter and materials.

quality and a shortening of supply times for the units. The manufacturer gains customers who return for a greater variety of spare parts and, potentially, for accessories.

There are disadvantages to modular equipment. Each setup is generally not as optimum a design as a specialty machine for a given function. For example, a combination drill/screwdriver might produce too low a torque and too high a speed as a screwdriver, and vice versa as a drill. A single-unit snowblower or lawn mower needs less space in a given place than a tractor with attachments, for the same capacity. The time taken to change attachments can be a disadvantage, particularly in a commercial setting. When redesigning the main or accessory units the manufacturer must maintain the same design of interfaces in order to keep the interchangeability between old and new units.

The primary decisions in the design of modular products are regarding which functions must be incorporated in different modules, which functions may be combined, and which must be kept separate. The overall function is subdivided into three subfunctions: (1) the main function, which is common to all variants; (2) the application function, which is different for each application; and (3) the interface function used to couple the main function with each of the application functions.

• • • • •

EXAMPLE
FUNCTION STRUCTURE FOR A MODULAR PRODUCT

Figure 168 shows such a function structure for the case of the tractor with different accessories. The names of corresponding modules are indicated along with the functions. The main functions "store fuel," "convert energy type," "connect," and "convert energy," which are embodied in the fuel tank, engine, clutch, and transmission, respectively, are contained in the main body of the tractor. There are three accessories that may be used with the tractor: mower, snowblower, and tiller. The "connect" function, embodied by a coupling, is divided between the tractor and the accessory. The function structure of each accessory is similar in form but has different final application functions.

As a rule the functions which will be used most often should be combined and incorporated into one module. On the other hand, functions used less frequently should be separated, to keep the cost and complexity of the main module low. Thus, one of the important entries during problem definition is an estimate of the frequency of use of each of the accessories. One should strive for as few subfunctions as possible and only those which recur the most often. Functions with the highest use should be incorporated in the same module as the main function.

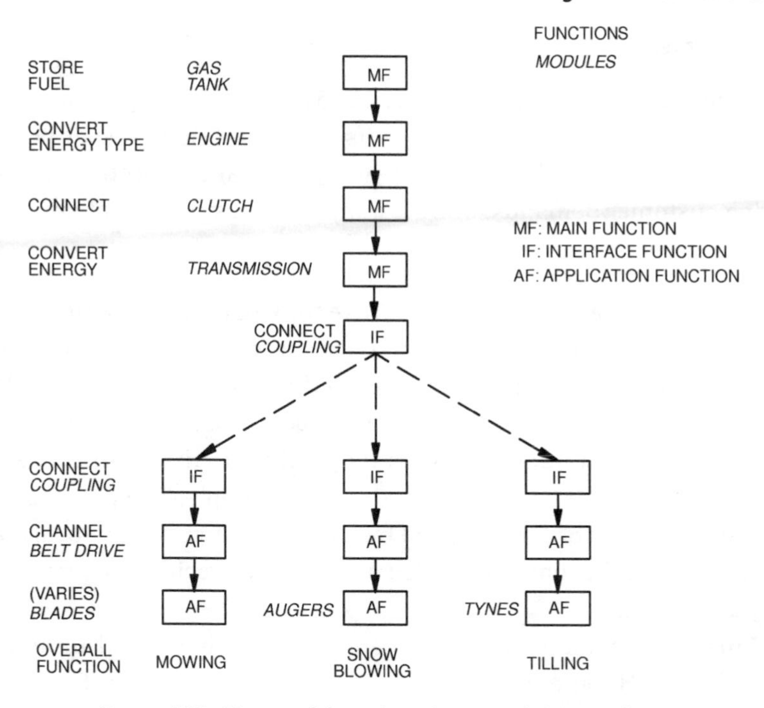

FIGURE 168. Types of functions in a modular product.

• • • • •

EXAMPLE
HYDRAULIC SYSTEM MODULAR DESIGN

Hayes et al. (1988) give the example of an equipment manufacturer who redesigned a hydraulic system. The company used value engineering and modular design to reduce the number of components from 63 to 9 and time to market from 15 months to 5 months. It found that it could offer a range of products and at the same time reduce the complexity of the manufacturing operation by reducing the number of standard parts and modules. For example, the number of different cooling systems used in the hydraulic systems could be reduced by standardizing and restricting the available sizes. (See Figure 106 in Chapter VIII for another example.)

• • • • •

Postponement

Postponement (Davis and Sasser, 1995) refers to keeping a product generic and customizing it as late in its manufacturing cycle as possible. Advantages of this policy are lower inventories and quicker response to varying customer demands. Some examples of postponement are the following:

- An apparel manufacturer used to dye the wool and then knit sweaters. Changing customer tastes would leave it with large unsold stocks of certain colors. The company changed to knitting natural wool and then dyeing the sweaters to different colors as demand dictated.
- A printer manufacturer ships its product to distribution centers around the world, but without the power supply. The power supply, including the cord, is installed by the distributors to suit local conditions. Even the manuals, which are country-specific, are printed and packed locally.
- Furniture manufacturers frequently ship unassembled units to be assembled by the distributor or even by the customer. This significantly reduces shipping costs.
- Refrigerators are made with doors that can be changed on site from left-opening to right-opening, or vice versa.

SUMMARY

Standardization may be considered a time- and cost-saving operation—one need not design the same part over and over again; repetitive problems are solved once and for all. Standardization of a part represents the current state of technology. In general, standardization promotes rationalization and quality assurance in all phases of life. The chapter has described the types of standardization, use of standards, group technology, and numbering systems. Standardization is the starting point for modular design and design of products in size ranges—both cost-saving procedures. Similarity principles were also discussed. These are used in the design of products in size ranges and estimating costs of similar parts.

BIBLIOGRAPHY

Davis, T., and M. Sasser. "Postponing Product Differentiation." *Mechanical Engineering* 117(11): 105–107 (1995). The discussion on postponement is based on this paper.

DeVries, M. F., S. M. Harvey, and V. A. Tipnis. *Group Technology: An Overview and Bibliography*. Cincinnati: Machinability Data Center, 1976. A short but adequate presentation of the subject.

Ehrlenspiel, K. *Kostengünstig Konstruieren (Cost-Effective Designing)*. Berlin and New York: Springer Verlag, 1985. Provides extensive discussion of the use of similarity principles in cost estimation.

Ehrlenspiel, K., and D. Fischer. "Relativkosten von Stirnrädern in Einzel- und Kleinserien- Fertigung" (Relative Costs of Gears in Single and Short Run Production). *FVA Forschungsvorhaben* 61(116, 146). Frankfurt: FVA, 1982 and 1983. Provides polynomials for estimating costs of gears.

Hayes, R. H., S. C. Wheelwright, and K. B. Clark. *Dynamic Manufacturing*. New York: Free Press, 1988.

Pahl, G., and W. Beitz. *Engineering Design—A Systematic Approach*. Berlin and New York: Springer Verlag, 1988. The discussion of the standard numbering series and design in size ranges is from this book.

Pahl, G., and F. Rieg. "Relativkostendiagramme für Zukaufteile—Approximationspolynome Helfen bei der Kostenabschätzung von Fremdgelieferten Teilen" (Relative Cost Diagrams for Purchased Parts). *Konstruktion* 36: 1–6 (1984a).

Pahl, G., and F. Rieg. *Kostenwachstumsgesetze für Baureihen (Cost Growth Laws for Size Ranges)*. Munich: C. Hanser Verlag, 1984b. Provides polynomials for obtaining relative costs of commonly purchased parts, e.g., bearings, screws.

Precision Mechanical Components Catalog B8. East Rockaway, N.Y.: W. M. Berg, Inc., 1988.

Design and Manufacturing for the Environment

DFE (design for environment), also called "green design" or "environmentally friendly design," is one of the greatest challenges facing engineers today but also holds out the greatest potential benefits to society. It addresses environmental problems at the design stage, where the potential for impact is the greatest. Decisions made at the design stage affect all of the life phases of a product—manufacturing, transportation, operation, maintenance, and disposal. The traditional field of environmental engineering, on the other hand, deals with waste and pollution after the fact—after wastes are generated—and searches for methods of mitigating the environmental damage.

The need to develop products that have minimal damaging effects on the environment has become increasingly more evident. Products are important in fulfilling human needs, but the side effects of pollution and the depletion of natural resources must also be of concern to designers and manufacturers.

This chapter summarizes and highlights the opportunities for the engineer and also the limitations on what an engineer can do. After all, the engineer's work is dictated by management, is driven by market forces, and is circumscribed by various regulations. How does one design and manufacture products that have either a less harmful or a more beneficial impact on the environment? Are such products more expensive? Do they give lower performance? Does the engineer look at the specific product, or the total system? What is the impact of design decisions on waste at the preconsumer, use, and postconsumer stages?

DFE—design for environment—may be defined as "the design of products to minimize undesirable impacts on the natural environment." There are subsets of DFE, such as design for recycling (DFRec) and design for remanufacturing (DFRem).

THE MATERIALS FLOW

The mix of raw materials used in the United States has changed over the years. The picture today is dominated by mined metals and nonmetals, plastics, artificial fibers, and petrochemicals, while use of materials produced in agriculture and forestry has decreased (see Figure 169). The trends in materials used also show an increasing variety, including chemicals, structural materials, and a number of "engineered" materials, such as conductive plastics and high-temperature superconductors. The use of new material processing methods has led to increases in efficiencies—lighter beverage cans, buildings, and information transmission cables, to name a few.

Modern economies are material-intensive. The United States extracts 10 tons of material (not including air, water, stone, gravel, and sand) per year per person, of which only 6% is embodied in durable goods; the rest constitutes waste, according to a report by the U.S. Congress, Office of Technology Assessment (OTA, 1992). The overall picture of a product's life cycle and the materials flow associated with it are shown in Figure 170. This figure also illustrates three terms frequently used in DFE literature: *reuse, remanufacturing*, and *recycling*.

In a product's life cycle, the design stage is the most critical as far as many of the impacts on and of the product are concerned. The decisions on materials and manufacturing processes are made at this stage. These decisions determine the total material flow, thus affecting both upstream (preconsumer) and downstream (postconsumer) impacts.

Although the most visible environmental impact of products is municipal solid waste (MSW), it is one of the smallest parts of the total waste generated during a product's life cycle. The difference between perception and reality is shown in Figure 171. The figure shows solid wastes produced annually, as defined by the Resource Conservation and Recovery Act (RCRA), from material extraction to processing, manufacturing, transport, use, and, finally, disposal (OTA, 1992). The total solid waste produced in 1988 was 11.7 billion tons, of which 0.7 billion tons was hazardous waste, as shown in Figure 171a. In Figure 171b, the nonhazardous waste is shown divided up into its various components by source. We see that MSW is only 2% of total nontoxic waste. An example of the public's perception of waste is that at fast-food outlets it consists mainly of the containers. Actually, 80% of the waste occurs "behind the counter," unseen by the public. This is one part of the wastes produced in manufacturing, which, as shown in Figure 171, constitute about one-half of the

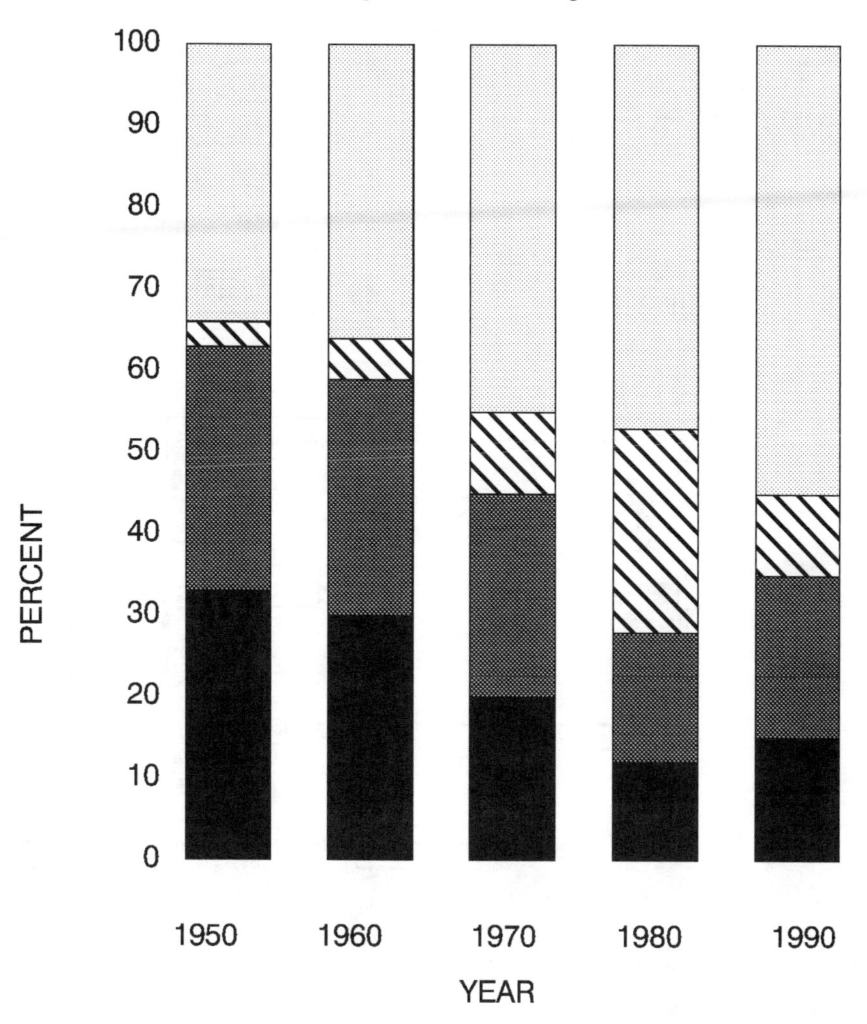

FIGURE 169. Trends in raw material consumption in the United States.

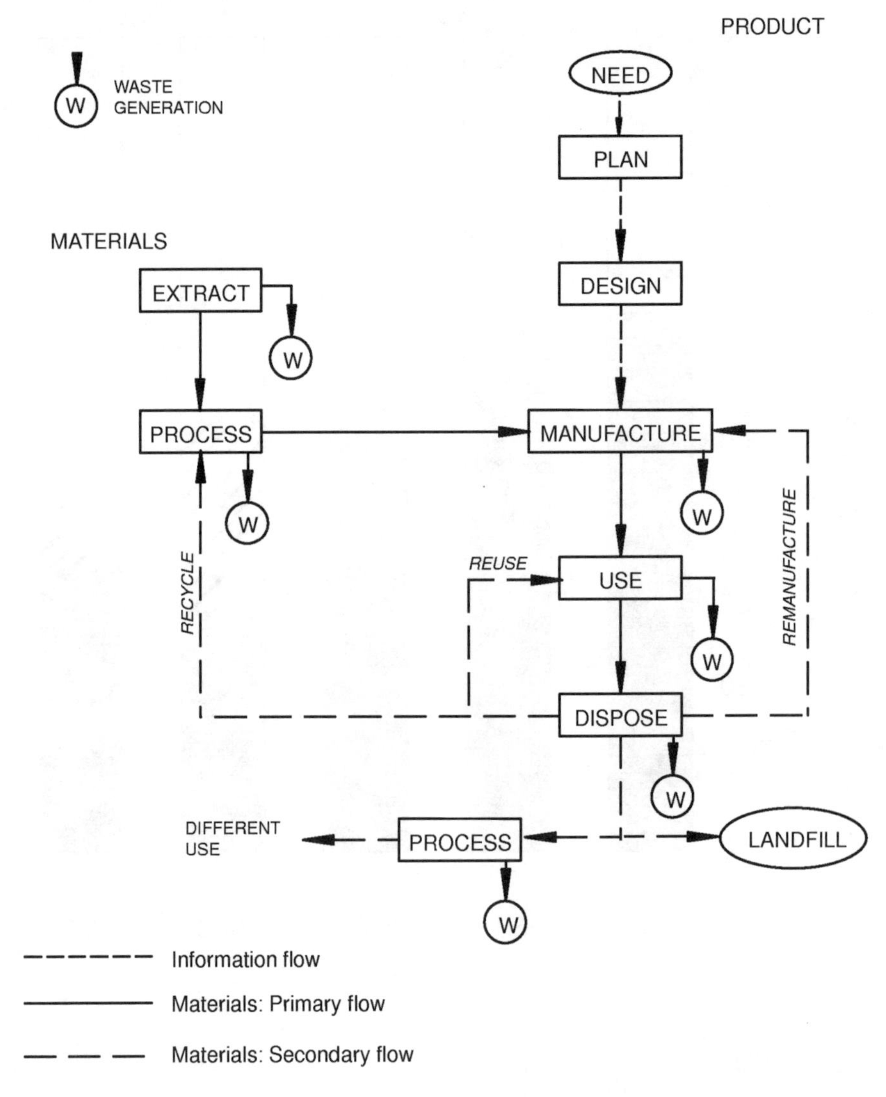

FIGURE 170. Product life cycle and materials flow.

total waste stream and thus hold the potential for significant reductions through proper design and manufacturing practices.

In order to be effective, DFE must affect all stages of the materials cycle. The U.S. Office of Technology Assessment (OTA) has defined "green design" as addressing two goals: preventing waste, and optimal use of

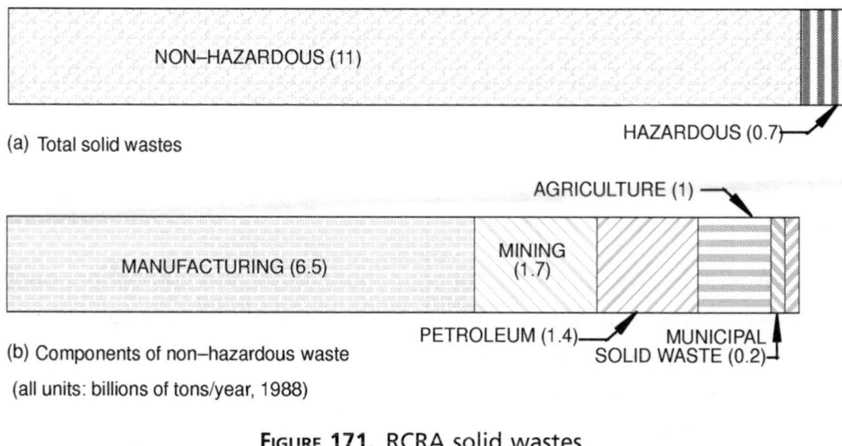

Figure 171. RCRA solid wastes. (OTA, 1992)

materials (Figure 172). Waste prevention is addressed by reducing the waste, toxicity, and energy use in the materials flow cycle and improving product life. Use of less material reduces not only product costs, as shown in Hundal (1993a), but also the wastes, emissions, and energy use at each stage. For example, improvements in lead-acid battery design have led not only to reduction in lead use by a third, but also to elimination of arsenic and antimony in the batteries. Optimal use of materials addresses

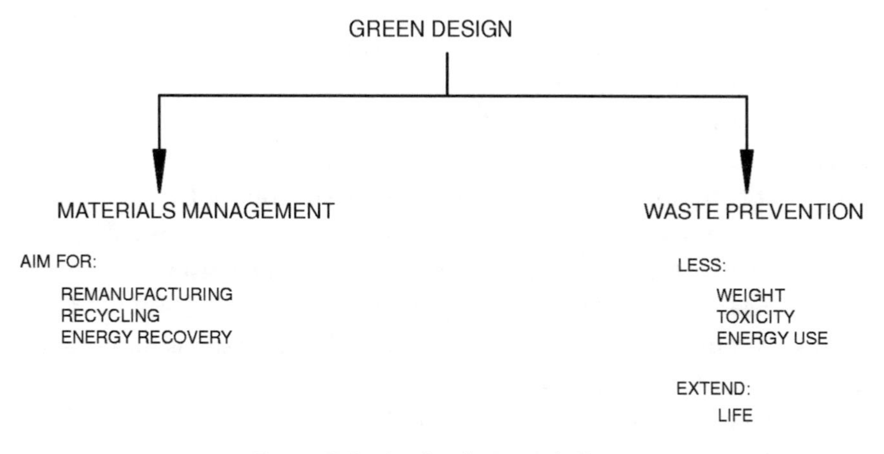

Figure 172. Goals of green design. (OTA, 1992)

issues such as remanufacturing and recycling and more efficient disposal possibilities which facilitate energy recovery and composting.

ENVIRONMENTAL LAWS

Environmental laws vary by country. These laws aim to address such problems as waste disposal, risks to human health and to ecosystems, global warming, and ozone depletion. In general, the public's environmental awareness in a given country, which eventually translates into laws, is proportional to the country's national wealth. Thus, one sees the greatest progress in this direction in North America, Western Europe, and Japan. It should also be noted that under present conditions of economies based upon consumption, the amount of waste produced in a country is dependent on the size of its economy. Nevertheless, it is incumbent on U.S. manufacturers to be cognizant of environmental laws of the countries to which they export.

In the United States, the major laws at the national level are the Clean Water Act, the Clean Air Act, and RCRA. In addition, there are a number of executive orders that can significantly affect the scope of the laws, e.g., the 1991 Federal Recycling and Procurement Policy, which mandates the use of recycled material in federal purchases. The RCRA regulates only hazardous industrial waste; the management of nonhazardous solid waste is left to the individual states. Table 49 lists the major federal environmental laws, with a brief description and the years of their enactment. U.S. policies in general focus on regulating industrial waste streams. In many areas, the United States is ahead of European countries, such as in the control of automobile emissions and the phasing out of CFCs.

In Europe environmental laws vary by country, but with the emergence of the European Community there is expected to be increasing commonality. European laws tend to focus on environmental attributes of products. European countries also lead in developing labeling systems for products (ecolabeling). There are trends toward uniform product standards, which tend, however, to set "minimum" standards. "Greener" (read "richer") countries are able to exceed these standards, while poorer countries have difficulty meeting them.

TABLE 49 Environmental Laws in the United States

Law	Year
Rivers and Harbors Act Control construction, allow navigation	1899
Atomic Energy Act Civilian use of atomic energy Mining, production, power, medical uses	1954
Motor Vehicle Control Program Control of CO, NO_X	1967
National Environmental Policy Act Created Environmental Protection Agency (EPA) Environmental impact of federal projects Environmental impact statement (EIS)	1970
Clean Air Act Control of TSP,[1] SO_2, CO, NO_2, O_3, Pb (airborne)	1970
Clean Water Act Chemical, physical, biological Restoration to "fishable and swimmable" condition	1972
Marine Protection, Research and Sanctuaries Act Ocean dumping of wastes Permit required for transportation	1972
Federal Insecticide, Fungicide and Rodenticide Act Storage and disposal of pesticides; protection of groundwater Banning of pesticides	1972
Solid Waste Disposal Act	1965
Resource Recovery Act	1970
Resource Conservation and Recovery Act (RCRA) Generation, transport, storage of hazardous wastes Emphasizes conservation and recycling	1976
Hazardous and Solid Waste Amendments (to RCRA) Protection of groundwater Landfills, underground tanks	1984
Pollution Prevention Act • Source reduction • Facilitate: government businesses • Methods of measurement • Training programs • Review regulations • Federal procurement practices	1990

[1]TSP: Total suspended particulates.

Germany enacted its Waste and Packaging Law in 1991. It puts the responsibility for recovery and recycling of packaging wastes on the manufacturers and retailers. The Germans had aimed for 80% recycling by 1995; however, the system has been overwhelmed by the amount of waste collected (Cattanch et al., 1994). In the future Germany expects to extend collecting and recycling to durable goods. This may be costly and difficult to implement. It should be mentioned that publications on DFR (denotes a union of DFRec and DFRem) in the German technical literature date back to the early 1980s. Rules for DFR have been compiled in a publication of the German engineering society, the VDI (VDI Guideline 2243, 1991).

Netherlands is in many ways a leader in the European environmental movement. Its National Environmental Policy Plan is the most detailed and comprehensive of any country's. It aims for a "process integrated environmental technology" to achieve "sustainable development." Netherlands favors investment in clean technologies and voluntary agreements with industry to achieve environmental goals. It has established waste reduction targets for 29 priority waste streams. The use of life cycle analysis to assess environmental impacts under various scenarios has been developed to a high degree.

The Nordic countries (Denmark, Sweden, Norway, and Finland) use a combination of laws and voluntary agreements to achieve their environmental goals. These include phasing out of toxic chemicals, banning nonrefillable beverage containers, and reducing consumption of nonrenewable materials, heavy metals, and toxic substances.

DFE AND THE PRODUCT REALIZATION PROCESS

We will look at how the product realization process described earlier may be affected by environmental concerns. Present-day technical products tend to be complex in order to fulfill the functional and other requirements imposed on them. As an example, consider the case of the simple snack bag. Figure 173 shows the cross section through the foil (OTA, 1992). Five different materials are used in it, each contributing to its desired properties, to fulfill the required functions.

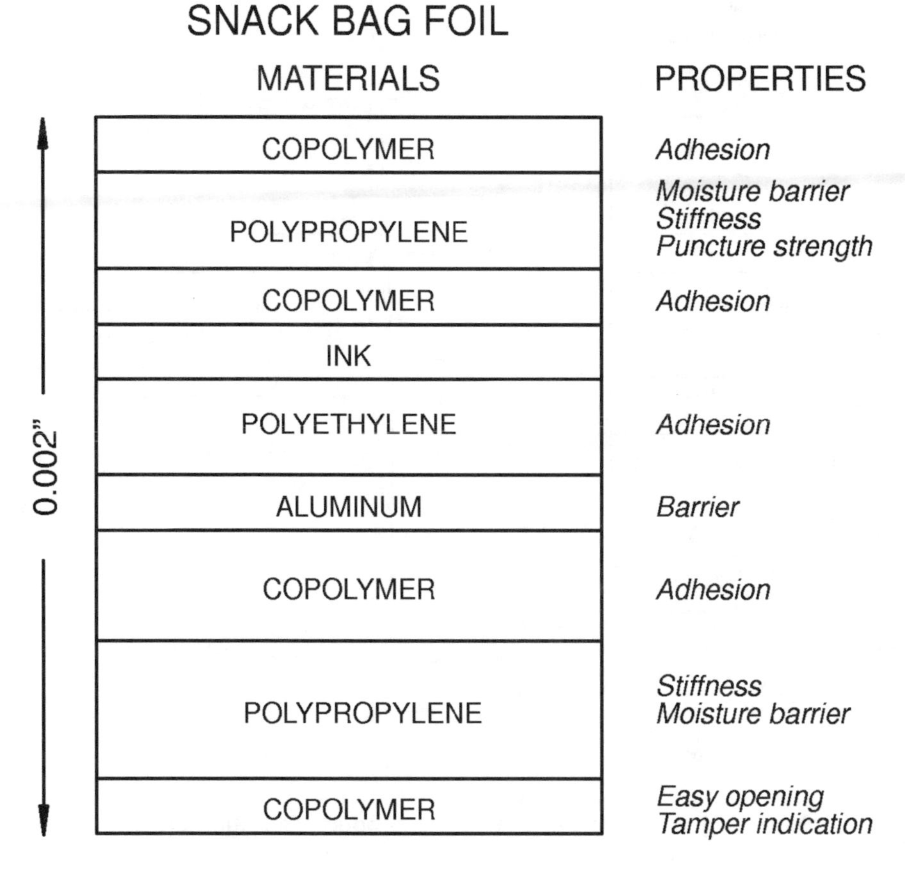

FIGURE 173. Complexity of a present-day product. (OTA, 1992)

The product realization process was shown in Figure 39 (Chapter IV), which is reproduced and enhanced here as Figure 174. DFE requires the involvement of more parties in the process, e.g., the material suppliers and waste handlers. In addition to the concerns of cost, manufacturability, and quality, environmental considerations require that we also be concerned with the following:

- Toxicity of materials used, during manufacturing, operation, and disposal
- Recyclability, reusability, and remanufacturability of the product

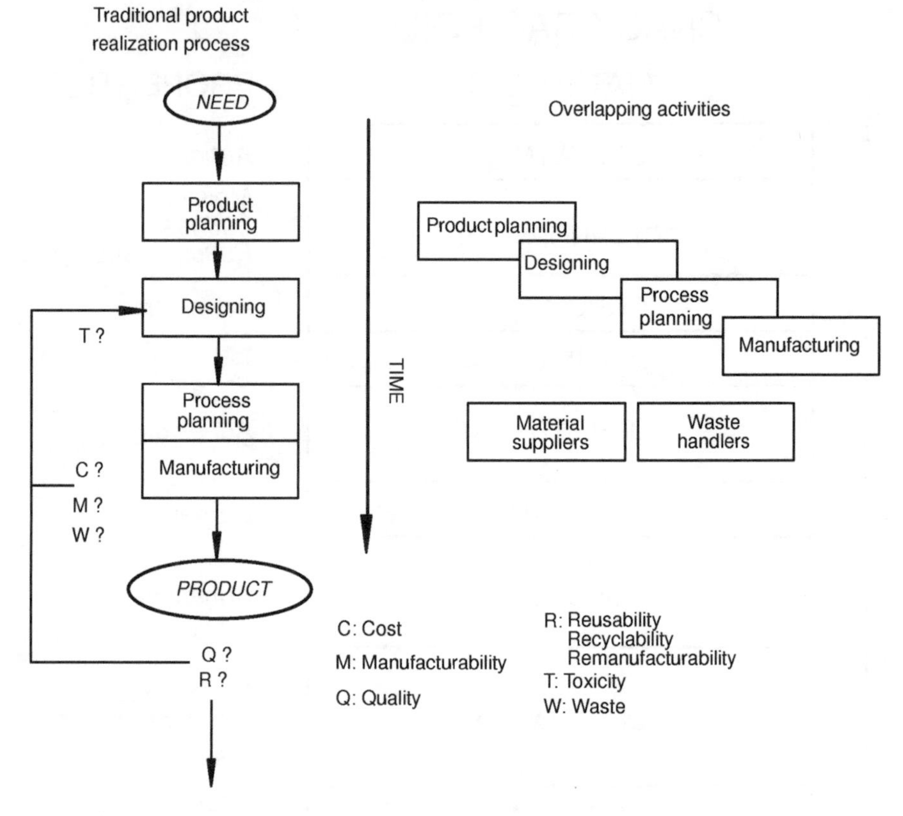

Figure 174. The effect of DFE on the product realization process.

- Waste produced during the various steps of material flow shown in Figure 170

The various steps of systematic design (from Chapter III) were indicated in Figure 170. How DFE affects each of these is described below.

Task Clarification

After the problem statement has been derived in generalized, solution-neutral terms, the next step is the generation of the requirements or specifications list. Requirements vary according to (1) the life phase of the product and (2) their type. Table C-1 in Appendix C shows these categories, which include the "environmental" type. In the past, the environmental requirements were concerned primarily with how the product itself could be protected during use or transportation. DFE imposes requirements on the impact of the product on the environment in its various life phases, chiefly, production, use, and disposal. The requirements list should integrate requirements dictated by materials use and recycling; it should list technologies and strategies available for reprocessing, and the possible recycling procedures.

Conceptual Design

The conceptual phase looks at the functional requirements of the product. The functions are listed, and complex functions are broken up into simpler subfunctions. These are arranged in the form of a function block diagram or function structure with inputs and outputs for the complete system as well as for the individual subfunctions. It is possible at this stage to form functional variants by rearranging, combining, and subdividing the functions. Up to this point little thought is given to the solutions, the shapes, or the hardware that might be involved. The next step in conceptual design is to look for solutions for each subfunction. By considering different physical processes it is possible to obtain a number of solutions for each function. The subsolutions thus found are then combined in a systematic and rational way to obtain a number of solutions or concept variants for the task.

Implications of environmental factors at this stage are as follows. Since the product exists only at an abstract level, we have the greatest freedom in choosing what it might look like to fulfill the given functions. At the function structure level we should strive for function separation rather than integration. The resulting solution structures lead to modular design in which individual modules can be easily replaced and recycled, as well as modernized with advancing technology. This is contrary to the rules for low-cost design (Hundal, 1993a), but conforms to the methods of time-driven product development (Hundal, 1993b). The chosen concept should use physical processes that call for fewer different materials, and foresee joinings that are easy to separate.

Embodiment and Final Design

It is at the embodiment stage that decisions regarding materials and shapes are made. Thus, this is the design stage that has the greatest influence on the materials flow cycle, the production processes that are used, and many of the details as well. Design of the product in a modular form allows for easy disassembly. The arrangement of the overall layout can make it easier to perform the necessary steps in remanufacturing, as described in the next major section. The principle of division of tasks (function separation) used in embodiment design (see Chapter VII) should concentrate wear on parts that can be easily reconditioned. Surface treatments should be designed for the total life of the product, including remanufacturing. Otherwise they should be easily renewable during remanufacturing. Design should aim for extending product life, i.e., for ease of maintenance. Providing easy lubrication and cooling will extend part life, and design for easy replacement of parts, preferably by the user, extends product life. Use of same and similar parts leads not only to lower-cost design (Hundal, 1993a), but also to ease of remanufacturability. The types of information that can be useful to the designer at this stage are, for example, the properties of joints, such as provided in VDI Guideline 2243 on the design of products for recycling, an excerpt from which is shown in Figure 175 (Hundal, 1994). This information can be used to complement the data shown in Figure 145 in Chapter IX.

MANUFACTURING

The most important points in manufacturing are the minimization of toxic by-products and also nontoxic waste. The designer determines the type and amount of scrap by the choice of shapes and materials. Manufacturing technologies such as near net shapes and high-precision forging and casting, sheet metal stamping, and plastics molding are examples of low waste generation. Optimum layout of stampings aims to minimize scrap. Fewer different materials used enable easier recycling.

RECYCLING AND REMANUFACTURING

Recycling and remanufacturing offer alternatives to waste generation and material consumption. Recycling of parts and materials reduces the need

TYPES OF JOINTS → / PROPERTIES ↓		Adhesive	Welded	Velcro	Bolted	Snap
STRENGTH	Axial	AVERAGE	GOOD	AVERAGE	GOOD	AVERAGE
	Transverse	AVERAGE	AVERAGE	POOR	AVERAGE	AVERAGE
	Fatigue	POOR	AVERAGE	POOR	GOOD	POOR
COST OF	Joining	AVERAGE	POOR	POOR	GOOD	POOR
	Separating	AVERAGE	POOR	POOR	GOOD	POOR
	Excision	GOOD	POOR	GOOD	GOOD	GOOD
RECYCLABILITY		AVERAGE	AVERAGE	GOOD	AVERAGE	GOOD

PROPERTIES: GOOD AVERAGE POOR

FIGURE 175. Properties of various types of joints. (Hundal, 1994)

for virgin material, thus reducing the waste associated with extraction. Recycling, typically after reconditioning of a part, may lead to a return to the original use, or to lower value-added use. An example, one of many, is the use of waste rubber from used automobile tires in road surfacing mix; see Figure 170. Materials usually degrade during recycling. It is important to note that recycling of a part may use more energy and cause more pollution than production from virgin material.

Remanufacturing does not necessarily mean reproduction of the original product. It affords the opportunity to upgrade its performance, e.g., by installing a faster CPU in a computer, or by substituting an electronic for an electromechanical coin changer in a vending machine. Design for recycling (DFRec) has received increasing attention from researchers. Rules have evolved for designing for recyclability, as they have for other attributes, such as cost. They conform to the general rules for systematic design. The steps in remanufacturing are disassembly, cleaning, sorting, checking, reconditioning, and reassembly. Design for remanufacturing involves rules to help in each of these stages.

Disassembly

Disassembly is not simply the opposite of assembly; rather it requires special considerations of its own. For example, the desirable properties of components for assembly, say, insertion, are different from those for the corresponding disassembly step, extraction. The disassembly process should be laid out during design. Just as design for assembly (DFA; Boothroyd and Dewhurst, 1987) techniques have led to reduction in the number of parts and assembly steps, design for disassembly (DFDA) should reduce the number and complexity of disassembly steps. At present, many products are designed not to be disassembled, in order to save on initial cost and discourage user ingress due to liability problems.

Design rules for disassembly are:

1. Arrange the subassemblies for easy disassembly.
2. Use joints that are easy to separate (Figure 175).
3. The joints should have the same life span as the whole product.

Cleaning

It has been estimated that in remanufacturing 90% of the parts have to undergo cleaning (Steinhilper, 1990). Design rules for the cleaning phase are:

1. Design of parts should provide for easily accessible reentrant corners and cavities or should avoid these altogether.
2. It should enable the rational design of the cleaning line.
3. Markings on parts should withstand cleaning.
4. Use of only environmentally friendly cleaning agents should be required.
5. Surfaces to be cleaned should be smooth and wear-resistant.
6. All deposits, impurities, and other materials should be removable without damage to parts.

Sorting

After cleaning, a part can be classified as being usable as is, usable after reworking, or unsuitable for further use. A sorting problem which occurs not only in manufacturing but in original manufacture as well is that due to parts that are similar but not exactly the same. Design rules for ease of sorting are:

1. Parts, particularly similar-looking ones, should be identified for easy sorting and classification.
2. Parts that fulfill the same function should either be identical or be clearly identifiable as different.

Checking

Before any reworking of a part can be done, it is necessary to check whether it is at all needed, and if so, to what extent. Thus the checking operations require that:

1. Wear and corrosion of parts should be easy to verify.
2. Data such as material properties, load limits, tolerances, and adjustments should be available.

Reconditioning

Shape of the parts should permit the use of jigs and fixtures for reconditioning, e.g., machining, material deposition, and insert replacement. Threaded bushings should be easily replaceable. In this respect, through bolts are better than blind studs.

Reassembly

In reassembly the same general rules as for DFA apply. The remanufacturing facility should have access to knowledge of the assembly procedures of the original product. Reassembly should be simple and unambiguous, and permit the use of mass production techniques. In general, DFA benefits both the original assembly operations and reassembly, as well.

• • • • •

EXAMPLE
COPYING MACHINES

Xerox Corporation remanufactures parts from its copiers—for example, motors and power supplies—resulting in savings of $200 million per year. The company has standardized designs leading to the use of fewer different parts in different models, thus applying at the same time one of the rules for low-cost design. The remanufacturing lines at Xerox have been set up in parallel with new-product lines to achieve the same level of high quality.

• • • • •

EXAMPLE
RECYCLING IN THE AUTOMOBILE INDUSTRY

Automobiles are one of the most highly recycled products. Figure 176 shows the average material makeup of present-day automobiles. About 75% by weight of materials is recovered and recycled, which includes most of the metal component (Holt, 1993). The remaining 25% is landfilled, amounting to 1% of the MSW in the United States. This material consists of plastics, rubbers, and glass. Design and material changes to increase the recycling and improve the recyclability of plastics are under way. A study has found resin identification on 37% of plastic parts and plastics from newer cars to be 42% less expensive to recover.

The profitability of material recovery depends on the efficiency of the recovery operations, the price of recovered materials, and the cost of disposal in landfills. The three primary operations in automobile recycling are:

- Dismantling of the automobile. This is the most expensive operation at the present time.
- Shredding and separating the iron component.
- Separating the nonferrous components.

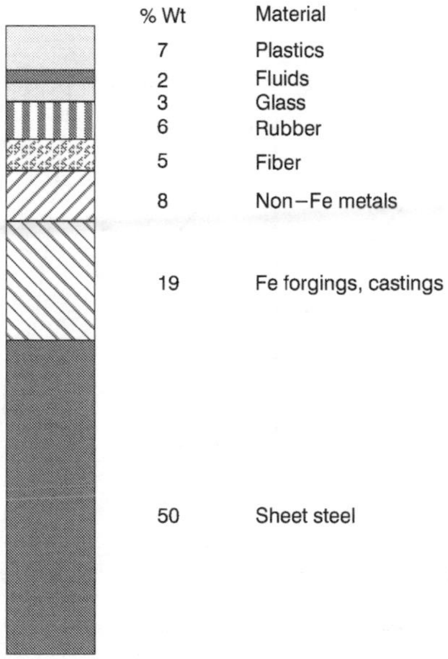

% Wt	Material
7	Plastics
2	Fluids
3	Glass
6	Rubber
5	Fiber
8	Non–Fe metals
19	Fe forgings, castings
50	Sheet steel

FIGURE 176. Materials content of average automobile.

Figure 177 shows the profitability of the three operations as a function of landfill fees. Figure 178 shows the components of the different costs and the net profit per vehicle for the three operations.

In Germany the landfill situation is far more acute than in the United States. The proposed legislation there to require car makers to take back their products at the end of the life cycle has spurred the manufacturers into taking a systems approach to design/manufacturing, materials supply, and disassembly. The goal of one manufacturer, BMW, is to produce a car out of 100% recycled parts by the year 2000. They intend not only to make cars that can be easily disassembled, but also to encourage materials suppliers to accept recycled materials from salvage yards. This idea of closing the materials loop is also being pursued by U.S. automobile manufacturers.

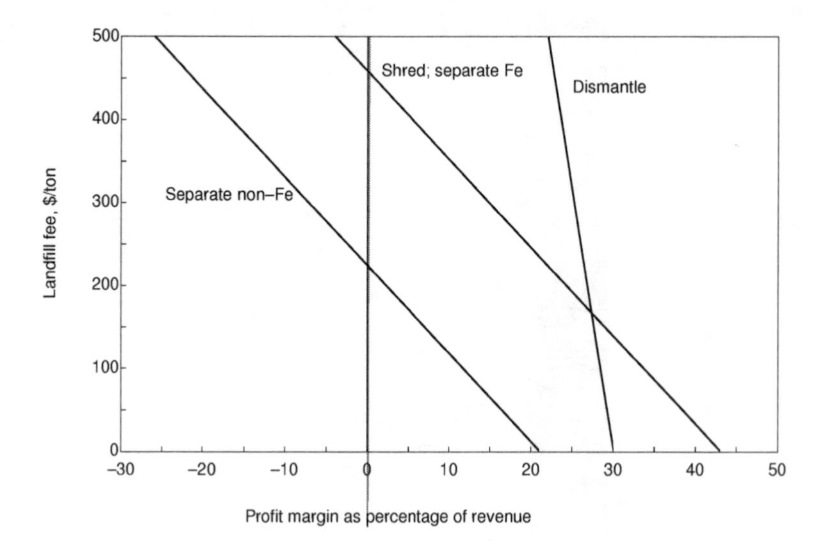

FIGURE 177. Profitability of recycling operations.

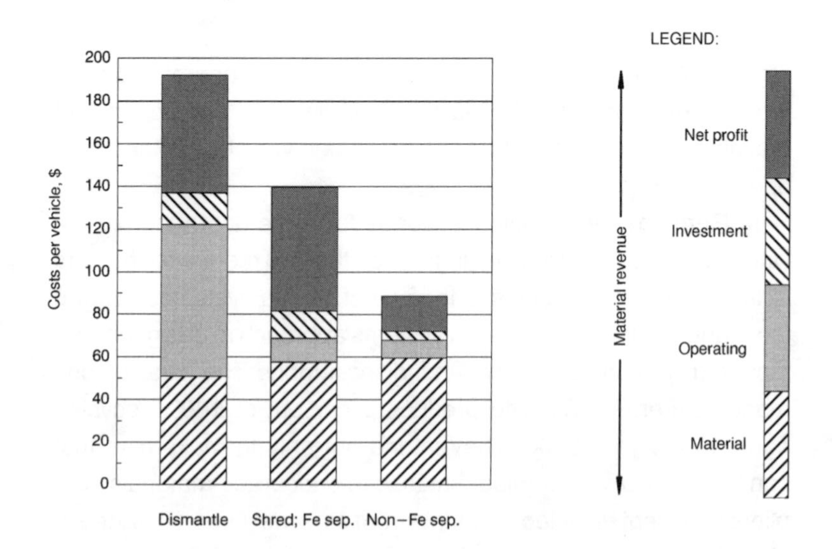

FIGURE 178. Economics of operations during recycling.

A life cycle analysis shows the comparison of the effects of using hazardous versus nonhazardous materials and processes on the cost of a product. The costs of disposal and storage of hazardous materials and environmental site audits are more than compensated for by the environmental overhead incurred by using nonhazardous materials (Kainz et al., 1995). Figure 179 shows a 9.34% increase in the total product cost when hazardous materials are used.

Using nonhazardous materials		Using hazardous materials	
Direct labor and materials	45.35%	Direct labor and materials	45.35%
Ordinary overhead	44.82%	Ordinary overhead	44.82%
Environmental overhead	9.83%	Hazardous material disposal	9.50%
		Hazardous material storage	3.56%
		Site audits	6.12%
Total product cost	100.00%	Total product cost	109.34%

FIGURE 179. Economics of using hazardous versus nonhazardous materials. (All figures as percentage of total product cost using nonhazardous materials)

• • • • •

SUMMARY

Environmental protection is a complex and controversial subject. Thus, DFE inherits the conflicts inherent therein. Decisions on environmental issues require value judgments and involve public policy. Trade-offs that depend on economic values are influenced by the country's tax policies. The goal of engineering design has been and still is to design and produce products and systems to fulfill the needs of society. DFE enlarges the scope of the problem to include the natural environment.

Engineers and designers need information to operate successfully. Governments need to determine priorities (e.g., recycling versus energy conservation) and relative risks of different materials to health and ecosystems. An example of a difficult issue is the snack bag (Figure 173), which is nearly impossible to recycle to its original use yet conserves materials and energy and lengthens the shelf life of food. For an automobile's lifetime, 10% of the energy is used in manufacturing, another 10% in disposal, and the remaining 80% during its use (Holt, 1993); thus, the greatest potential in energy saving lies in making cars more efficient. A list of products and processes posing high risks is needed. There is a need to find safer substitutes for hazardous substances.

DFE by itself does not make products cheaper, nor does it make them more expensive. On the other hand, the overall costs, when one considers the environmental impacts, can be lowered. Environmental attributes must be considered at all stages of the design, as are other attributes, including cost, manufacturability, and quality. The application of DFE requires knowledge more than any other resource. Products designed for recyclability are more easily serviceable. Remanufactured products can be of as high a quality as new products—and cheaper.

Products containing high-technology components that are likely to show consistent and frequent improvements in performance should be designed so that they may be easily disassembled. Such products cannot be designed for long life—they tend to quickly become obsolete. Thus they should be designed for easy renovation.

There are a number of problems in this field which require further study and research. Two of these areas follow.

Material Flow and Management

Reliable data are needed on material balances and flow of wastes. All parties in the materials flow cycle—material processors, designers, manufacturers, and waste handlers—require such information. Technologies for recycling and incineration need improvement. Typically, materials that are easy to recycle are the least toxic and the cheapest to begin with. More information is needed on the compatibility of materials for recycling, including the suitability of adhesives, paints, and coatings. Figure 180 (Hundal, 1994) shows an excerpt from VDI Guideline 2243 on the compatibility of plastics. Compatibility of aluminum sheet alloys is given in Holt (1993).

BASE MATERIAL	MATERIAL ADDED PE	PVC	PC	PP	ABS
Polyethylene	FULLY COMPATIBLE	INCOMPATIBLE	INCOMPATIBLE	FULLY COMPATIBLE	INCOMPATIBLE
PVC	INCOMPATIBLE	FULLY COMPATIBLE	INCOMPATIBLE	INCOMPATIBLE	COMPATIBLE IN LARGER QUANTITIES
Polycarbonate	INCOMPATIBLE	COMPATIBLE IN MINUTE QUANTITIES	FULLY COMPATIBLE	INCOMPATIBLE	FULLY COMPATIBLE
Polypropylene	COMPATIBLE IN MINUTE QUANTITIES	INCOMPATIBLE	INCOMPATIBLE	FULLY COMPATIBLE	INCOMPATIBLE
ABS	INCOMPATIBLE	COMPATIBLE IN LARGER QUANTITIES	FULLY COMPATIBLE	INCOMPATIBLE	FULLY COMPATIBLE

INCOMPATIBLE
COMPATIBLE IN MINUTE QUANTITIES
COMPATIBLE IN LARGER QUANTITIES
FULLY COMPATIBLE

FIGURE 180. Compatibility of plastics. (Hundal, 1994)

Better Design Methods

Improvements in methods and knowledge are needed in the area of design for recyclability, which includes design for disassembly. New fastening and joining techniques must be developed. The aims of the existing and well-advanced fastening and joining technology are, in general, to produce joints that stay intact. This technology must now be extended to develop joints that are strong yet can be easily separated. Information such as that in VDI Guideline 2243 needs to be more widely disseminated and further enhanced. Since, in general, products will have longer life, material response to long-term loading needs to be investigated. Products that experience rapid technological change would need to be designed for easy disassembly, since they will have a short useful life.

BIBLIOGRAPHY

Beitz, W. "Design for the Ease of Recycling." *Proc. ICED 90*, 3: 1551–1560. Zurich: Heurista, 1990.

Beitz, W. "Design for the Ease of Recycling—General Approach and Industrial Application." *Proc. ICED 93*, 2: 731–738. Zurich: Heurista, 1993. In this paper and the preceding, Beitz presents highlights from VDI Guideline 2243 on DFR, including general rules for DFR, materials recyclability, and joint design, with examples of design of a drill and a dishwasher.

Boothroyd, G., and P. Dewhurst. *Product Design for Assembly*. Wakefield, R.I.: Boothroyd Dewhurst, Inc., 1987.

Cattanch, R. E., J. M. Holdreith, D. P. Reinke, and L. K. Sibik. *Handbook of Environmentally Conscious Manufacturing*. Chicago: Irwin, 1994. Provides extensive coverage of environmental laws in the United States and elsewhere.

Henstock, M. E. *Design for Recyclability*. London: Institute of Metals, 1988. Henstock looks at recycling of steel scrap in automobiles and gives design rules on disassembly, material compatibility, and standardization. It is one of the few technical books on the subject.

Holt, D. J. "Recycling and the Automobile." *Automotive Engineering* (October): 42–73. Warrendale, Pa.: SAE International, 1993. A series of articles in this special issue of *Automotive Engineering* looks at the current status of recycling in the automotive industry, including the recovery of plastics and the costs of their recycling and design for aluminum recycling.

Hundal, M. S. "Rules and Models for Low-Cost Design." *Design for Manufacturability 1993*, DE-Vol. 52: 75–84. New York: ASME, 1993a.

Hundal, M. S. "Engineering and Management for Rapid Product Development." *Proc. ICED 93*, 1: 588–595. Zurich: Heurista, 1993b.

Hundal, M. S., "DFE: Current Status and Challenges for the Future." *Design for Manufacturability 1994*, DE-Vol. 67: 89–98. New York: ASME, 1994.

Jorden, W., and F. Gehrmann. "Produktgestaltung für das Funktionserhaltende Recycling" (Product Design for Function Preserving Recycling). *Proc. ICED 90*, 3: 1561–1568. Zurich: Heurista, 1990. The authors discuss the disassembly and modularity of products, flowcharts for assembly and disassembly, design rules for joints, and rules for remanufacturability.

Kainz, R. J., M. S. Simpson, and W. C. Moeser. "Life Cycle Management: A Solution for Decision Making." *Automotive Engineering* (February): 107–113. Warrendale, Pa.: SAE International, 1995. This paper on life cycle management compares the effects on the cost of a product of using hazardous and nonhazardous materials and processes. The environmental overhead incurred by using nonhazardous materials is more than compensated by the savings in costs of disposal and storage of hazardous materials and environmental site audits.

Keoleian, G. A., and D. Menerey. "Life Cycle Design Guidance Manual." EPA/600/R-92/226, 1993.

"Konstruieren Recyclinggerechter Technischer Produkte" (Design for Recycling of Technical Products). VDI Guideline 2243. Düsseldorf: VDI Verlag, 1991. See the papers by Beitz for a discussion and explanation in English.

Künne, B. "Konstruktion von Maschinen unter Verwendung von Recyclingaggregaten" (Design of Machines Using Recycled Units). *Proc. ICED 90*, 3: 1580–1587. Zurich: Heurista, 1990. Examples of design of machines using recycled parts and assemblies.

Kuuva, M., and M. Airila. "Design for Recycling." *Proc. ICED 93*, 2: 804–811. Zurich: Heurista, 1993. Presents a report on a project on DFR in Finland and shows case studies from the appliance, vending systems, and automobile industries, and on recycling firms.

Navinchandra, D. "Design for Environmentabitity." *Design Theory and Methodology 1991*, DE 31: 119–125. New York: ASME, 1991. The author suggests rating the "greenness" of a design on the basis of one of a number of indicators, e.g., degradability, separability, or life cycle cost. The effect on the pertinent indicator is then determined by evaluating one of six "changes": substitution, separability, shape redesign, enclosure, design for recovery, and material selection.

Navinchandra, D. "ReStar: A Design Tool for Environmental Recovery Analysis." *Proc. ICED 93*, 2: 780–787. Zurich: Heurista, 1993. In this paper Navinchandra discusses a program for product disassembly and environmental recovery.

Office of Technology (OTA), U.S. Congress. "Green Products by Design: Choices for a Cleaner Environment." OTA-E-541. Washington, D.C.: U.S. Government Printing Office, 1992. A "must read" for anyone interested in the subject of DFE. Many of the figures in this chapter are from this book. Discusses both the technical and the policy-making aspects of the problem.

Olesen, J., and T. Keldmann. "Design for Environment—A Framework," *Proc. ICED 93*, 2: 747–754. Zurich: Heurista, 1993. Design tools which can be integrated into the product development process. The authors stress the importance of knowledge on the part of the designer regarding the appropriate use of the design tools.

Steinhilper, R. "Produktrecycling—Praxis und Perspektiven (Product Recycling—Practice and Perspectives)." *Proc. ICED 90*, 3: 1598–1605. Zurich: Heurista, 1990. Discusses remanufacturing and reconditioning as requirements on design, and the industrial practice of remanufacturing, with examples based on machine tools, vending machines, robots, and automobiles.

Thurston, D. L., and A. Blair. "A Method for Integrating Environmental Impacts into Product Design." *Proc. ICED 93*, 2: 765–772. Zurich: Heurista, 1993. A methodology for integrating environmental considerations into the design process is shown and is applied to the problem of choosing materials for beverage containers on the basis of the materials' attributes—e.g., energy use, waste generated, durability.

Van der Horst, T., and A. Zweers. "Environmentally Oriented Product Development: Various Approaches to Success." *Proc. ICED 93*, 2: 739–746. Zurich:

Heurista, 1993. A variety of approaches for judging the environmental impact of products, along with their advantages and disadvantages, are presented in this paper. Approaches include: (1) closing the materials cycle, (2) energy indicators, (3) hazardous waste, (4) life cycle analysis, (5) environmental-economic, (6) environmental marketing, (7) environmental legislation, and (8) conceptual design approaches. The study shows the results of the "ecodesign" program at the TNO Product Center in the Netherlands. The authors show the importance of establishing criteria in evaluating a given approach, whether a reduction in materials use, energy conservation, or another approach. In the last section, they emphasize the importance of design.

Nomenclature

a	acceleration
A	area, cross-sectional area
b	y intercept
c_i	individual estimated cost of part, step, etc.
C	cost, general
C_{dev}	development cost
C_{fix}	machine, setup, or initial cost
C_{init}	initial cost
$C_{\text{lab}, d}$	direct labor cost
C_{LC}	total life cycle cost
C_{mach}	machining cost
C_{maint}	maintenance cost
C_{mat}	total materials cost
$C_{\text{mat}, d}$	materials direct cost
$C_{\text{mat}, o}$	materials overhead cost
$C_{\text{M.C.I.}}$	cost of manually cast gray C.I.
C_{mfg}	manufacturing cost
$C_{\text{mfg}, v}$	variable manufacturing costs
C_{MP}	cost of hand-made pattern
C_{op}	operating cost per unit weight and time
C_{post}	postprocessing cost
C_{pr}	total production cost
$C_{\text{pr}, o}$	production overhead cost
C_{prep}	preparation cost
$C_{\text{pt as}}$	cost of part assembly
$C_{\text{pt pr}}$	cost of part production
C_{rel}	relative (manufacturing) cost
C_T	cumulative cost after T hours of operation
C_{tool}	tool cost
C_v	cost per unit volume
C_W	welding cost
$C_{W, \text{tot}}$	total welding cost
d, D	diameter
CER	cost estimating relationship
$e_{i, \text{al}}$	accuracy (allowable error) for estimating individual costs

e_{tot}	total relative error
E	energy, electric field, error, modulus of elasticity
E_i	error in c_i
E_{in}, E_{out}	input energy, output energy
F	force, function
F_i	fixed costs for process i
g	gravitational acceleration
G	universal gravitational constant, modulus of rigidity
h	heat transfer coefficient
H	height
k	constant, thermal conductivity
L	length, stroke, loss, Lagrangian
m	mass, slope
$\dot{m}$	dm/dt
M_{in}, M_{out}	material input, material output
P	pressure, power, probability
P_i	probability for part i to be acceptable
q, Q	heat flow rate, electric charge, volume flow rate
r	radius
R	gas constant, resistance, radius
S	ultimate tensile strength (UTS)
S_{in}, S_{out}	input signal, output signal
t	time, thickness
T	operation life, target value, temperature
ΔT	tolerance limit
U	criterion function
v	velocity
v_{ij}	value of criterion i for item j
V	volume, velocity, voltage
V_i	variable costs per unit, for process i
V_r	radial velocity
w_i	weight of criterion i
W	weight
Y	yield
Δ	distance, displacement
ϵ	permittivity
ϵ_i	error at data point (x_i, y_i)

λ	Lagrange multiplier
μ	viscosity, permeability
ρ	density
σ	Stefan-Boltzmann constant
ϕ	ratio
ϕ_l	linear size ratio
$\phi_1, \phi_2, \ldots, \phi_p$	regional constraints
Φ	magnetic field
$\psi_1, \psi_2, \ldots, \psi_m$	functional constraints
ω	angular velocity

Design Rules

- **Cost-Driven Design**
- **Time-Driven Versus Cost-Driven Design**

We summarize here the various rules for cost-driven design. The commonalities and conflicts between cost- and time-driven design will be enumerated.

COST-DRIVEN DESIGN

The designer can call on several means to help in the decision making of cost-driven design. The most often used method is to follow traditional so-called design rules. These rules are based upon experience and generally not quantified or expressible in algorithmic form. Examples of design rules are: "use few part types," "avoid sudden cross section changes"—in general, the "dos and don'ts" of design. The rules can sometimes contradict each other, in which case further analysis is required for decision. Rules relate certain parameters to properties of interest. They form the basis of expert systems. Using the design rules helps in decision making, but it can create a mind-set and preclude innovative solutions—just the opposite of what one strives for in design. In designing to cost, we usually calculate actual or absolute costs. However, often a more definitive comparison of alternative designs can be made on the basis of relative costs. The simplest example is the relative cost of materials. However, objects can also be compared on the basis of function (e.g., fastenings, couplings, bearings) or production (e.g., machined entities, production processes). Relative costs allow quick and rough calculation and do not change much over time (as opposed to absolute costs).

An important step in designing is the selection of one solution from a number of alternatives. This occurs during both concept development and embodiment. The selection takes place under technical as well as economic constraints. The basic rules reducing product costs may be summarized as follows:

- At the problem definition stage, ask for fewer demands—only the minimum accuracy and tolerances and conformance to standards.

- At the concept stage, use concepts that lead to smaller sizes and lighter construction.
- Use higher speeds for power transmission, thus reducing torque and consequently the amount of material required.
- Use parallel paths for flow of energy (example: planetary gear trains).
- Use robust physical effects, e.g., mechanical and hydrostatic energy.
- Use concepts with simple construction and fewer parts by use of function integration, especially for small products and/or large quantities.
- Use smaller parts for one-of-a-kind products. For large quantities, smaller size always leads to lower costs.
- Use same and/or similar parts.
- Produce in large quantities. This results not only in an economy of scale, but also the opportunity to use more optimum manufacturing processes.
- Reduce complexity—use fewer parts and production operations.
- Reduce size, and thereby the material volume.
- Use safety devices (examples: overrunning clutch, relief valve) so that the product does not have to be designed for high loading levels that occur only occasionally.
- Use higher-strength materials and/or surface treatment to reduce the product size and generally the manufacturing costs.
- Reduce scrap generation.
- Use fewer machining operations, e.g., through integral design or near-net-shape forming. This also reduces setup costs.
- Do not design “left-” and “right-handed” parts. Make the two sides identical.
- Do not build to more stringent specifications than necessary.
- Look at life cycle cost to decide if longer life or more frequent replacement will be cheaper.

TIME-DRIVEN VERSUS COST-DRIVEN DESIGN

The procedures involved in time-driven development of products and rules for low-cost products can work at cross-purposes at various stages of implementation. As an example, modular products call for function division and can be developed faster, whereas one of the underlying requirements of low-cost design is function integration. This and similar issues need to be resolved.

We have seen the procedures that enable a product to be developed and brought to market faster, and in the previous section we saw the rules for lowering product costs. We look now at areas in which there is commonality between the two goals, and also where the two types of demands work at cross-purposes and what techniques may be incorporated for low-product cost design and manufacturing in a rapid-development environment.

Concurrent product and process development is the most important factor in time-driven product development. It reduces product costs for the following reasons:

- The development time is shortened, thus reducing design cost. The latter is a part, albeit small, of the total cost, as was shown in Figure 2 in Chapter I.
- There is less probability of mistakes and misunderstandings as the results of one activity are passed on to the next downstream group—e.g., from design to manufacturing—due to continuous feedback. Thus there are fewer mistakes to be corrected under pressure of time.
- Fewer design iterations are necessary.

On the other hand, one of the rules of time-driven development is to make products modular. Having more modules provides the following advantages:

- More flexibility is possible in design.
- Each module is simpler.
- Each module carries fewer functions.
- Modules are more independent and thus can be designed concurrently.
- Modules can be upgraded as technology improves.
- Repairs are simpler and cheaper.

Disadvantages of having more modules are that there are more parts and more interfaces and therefore lower reliability. Low-cost design, on the other hand, calls for function integration: fewer parts and fewer modules. This approach, however, provides less flexibility.

Design rules for low cost that apply as well in accelerated product development include:

- At the problem definition stage, ask for fewer demands.
- Reduce complexity.
- Use same and/or similar parts.
- Use existing, proven technology. This reduces development costs as well as risk and, therefore, the life cycle costs.
- Use parts and systems already designed for other products. This reduces design costs and risk.

The Requirements List

- System Interface
- Importance of the Requirements
- Types of Requirements
- Summary

After a design problem statement has been derived in generalized, solution-neutral terms, the next step is to generate a requirements or specifications list. The requirements and the time spent on developing them vary according to the extent of the design. They help to determine, essentially, what function(s) must be fulfilled by the product and what attributes it should or should not have. The collection of requirements constitutes a complete and abstract description of the design task.

Requirements need to be stated in the language of the designer. This is more precise than, say, the shop or colloquial language in which the problem may have been (and generally is) presented in the first place. The requirements are stated in solution-neutral terms; they should not restrict the product to a certain form unless it is specifically so desired. To develop a list of requirements, it is useful to first categorize them, which can be done in several ways. The chief categories and methods of classification are:

- Requirements dictated by the system interface
- Relative importance of the individual requirements
- Requirements at different life phases of the product
- Types of requirements—e.g., technical, economic, or aesthetic

These topics will be described next.

SYSTEM INTERFACE

The system or product being designed almost always interfaces with some existing system or is to be located within it. For example, the interface for a temperature-indicating device is the surrounding medium and the persons reading the instrument. As another example, a control system being designed for an engine has to interface with the engine. Thus, a part of the requirements list must enumerate the pertinent properties at the interface between the system being designed and the existing system.

IMPORTANCE OF THE REQUIREMENTS

The requirements on a system vary in weight and priority. Some are more important than others. An important classification of requirements is as *demands* versus *wishes*.

Demands

Demands are those requirements that *must* be satisfied. If these are not met, the product is unacceptable. Demands may be positive or negative, properties the system must have or must not have. Demands can be further classified according to the way "quantities" are described.

1. Fixed demands are those that allow for no variation or tolerance: "the bicycle shall have 10 speeds"; "the bottling machine shall handle 1-liter bottles"; etc.
2. Min/max demands are those calling for some minimum or maximum value: "engine output shall be at least 5 hp"; "maximum cost per unit shall be $20"; etc.

Wishes

Wishes are those properties which are desirable for the product to have, but are not absolutely necessary. These are requirements that might, for example, increase the performance or lower the cost. Examples of these are "machine may also handle 1/2- and 2-liter bottles" and "weight of the bicycle should be kept low."

While clarification of the task is the first step in the design process, it is really a continuous operation and proceeds hand in hand with the problem solution. As more information is gained, the requirements list can be refined and modified as needed. The requirements list plays a vital role in the eventual evaluation of various concepts and solutions and of the final design. Designs can be accepted or rejected depending on whether they do or do not fulfill the fixed demands. A choice can then be made among the remaining designs according to which best fulfill the min/max demands and the wishes, in that order.

TYPES OF REQUIREMENTS

Requirements on a product can be broadly classified by type as (1) engineering (technical), (2) economic, (3) ergonomic, (4) legal, (5) environmental, and (6) other (e.g., aesthetic).

Before dealing with each of these in detail, it should be pointed out that the requirements in the different categories apply in differing degrees in the different life phases of the product. The life phases and requirement types can be thought of as forming a matrix, as shown in Table C1. Thus, for example, there are engineering requirements that pertain to manufacturing and to use of the product. The same is true for economic and other categories. Environmental requirements for the production phase would call for the use of nonpolluting processes or for easily separable inserts in molds, for example.

The most important requirements to be considered during conceptual design are engineering requirements pertaining to product use that must be met—i.e., demands.

ENGINEERING (TECHNICAL) REQUIREMENTS

The most important engineering requirements are the functional and the operational, and pertain to product use. Functional requirements specify what the product must do. Operational requirements point to how it must fulfill its purpose. The considerations in asking questions which lead to establishing the engineering requirements for the different life phases are

TABLE C1 Matrix of Life Phases and Types of Requirements

Life Phases	Types of Requirements					
	Engineering	Economic	Ergonomic	Legal	Environmental	Other
Planning						
Design						
Production						
Marketing						
Product use						
Disposal						

given in Table C2. More details on technical requirements encountered for defining the product are given in a later section.

Preparing the Engineering Requirements List

The list in Table C3 may be used to help prepare the engineering requirements in the various categories.

Economic Requirements

Economic requirements pertain to the costs incurred in the various life phases of the product. These are also included in Table C2.

Ergonomic Requirements

Ergonomic requirements deal with the human/product interface. Starting with manufacture, and proceeding through disposal of the product, we need to consider the interaction of humans with it. The requirements in each of these phases include:

Manufacture

- Ease of handling parts
- Ease of assembly
- Unambiguous recognition of parts and features

Marketing

- Ease of transportation
- Ease of handling
- Size of doorways and other openings

Product use

- Safety
- Unambiguous controls and displays
- Ease of handling
- Does not tire user
- No unwanted noise and vibration

Maintenance

- Ease of disassembly and reassembly
- Unambiguous recognition of parts
- Ease of handling parts

TABLE C2 Development of Engineering and Economic Requirements

Life Phase	Engineering Requirements	Economic Requirements
Planning Design	State of technology Company's state of knowledge Possibility of modular design	Planning, design, licensing costs
Manufacturing	Production processes required Quantities to be manufactured Assembly requirements Need for special equipment Testing equipment availability	Production, assembly, testing costs
Marketing	Storage requirements (space, climate, etc.) Packaging Transportation requirements (size, shock, vibration)	Storage, packaging, testing costs
Product Use	On-site assembly facilities Startup requirements Functions of the product: conversion of energy, signals, material Geometric requirements Operating expertise required Degree of automation Maintenance requirements Repairability	Installation, operation, maintenance, repair costs
Disposal	Recycling possibilities Waste disposal Reuse of product	Costs of recycling, scrapping

Disposal and salvage

- Ease of disassembly
- Unambiguous recognition of parts
- Ease of handling parts
- Identification of materials

TABLE C3 Checklist for Preparing the Engineering Requirements List

Category	Items
Geometry	Length, width, height, diameter, area, volume, space requirement, number, arrangement, connection
Energy	
Mechanical kinematics	Displacement, velocity, acceleration Type: steady, oscillation Direction
Kinetics	Force/moment: direction, amplitude Weight, load, friction Stability
Electrical	Voltage, current, power DC/AC, frequency
Thermal/fluid	Temperature, pressure Heating, cooling
Miscellaneous	Storage, efficiency, conversion
Material	Flow rate Properties: physical, chemical, etc. State: solid, liquid, gas
Solids	Ductile, plastic, brittle Uniform, composite Size (lumps, powder)
Signal	
Type	Mechanical, electrical, optical, etc. Analog or digital
Function	Storage, indication, alarm

Low cost and ergonomic requirements may not always be compatible. Figure C1 shows two designs of the electrical plug for an appliance. The design in Figure C1a is simpler and cheaper but is not as easy to grasp, because of its smooth surfaces. The design in Figure C1b is more ergonomical but costlier. Here (as indeed in any design) we need to look at the overall cost, the life cycle cost, and customer satisfaction.

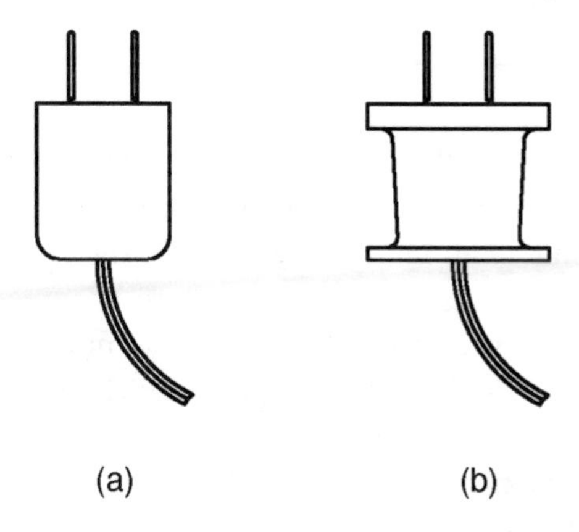

FIGURE C1. Low-cost versus ergonomic design.

LEGAL REQUIREMENTS

Additional requirements result from laws, regulations, warranties, standards that might apply to the product manufacture, and liabilities pertaining to marketing and use.

ENVIRONMENTAL REQUIREMENTS

Environmental requirements were discussed in detail in Chapter XI.

OTHER REQUIREMENTS

Aesthetic appeal is an important consideration, especially in the case of consumer products. A product's various impacts on society should also be taken into consideration. Certain organizational requirements, and constraints of various types, need to be considered. These include personnel considerations during design, manufacturing, servicing, and customer contact, and times required and deadlines to be met in the various activities. Ethical, cultural, and political issues also need to be addressed.

SUMMARY

The requirements list is the most important document prepared at the beginning of a design project. It should be prepared with the involvement of a cross-functional team (marketing, sales, design, manufacturing, etc., and perhaps even the customer) so that there is agreement among the various parties. A requirements list ensures that important aspects of the problem are not overlooked. A systematic development of the requirements list involves considering various types of requirements that span the different life phases of the product, prioritizing the requirements, and looking at both the system and its interfaces.

Functional Analysis and Variants

The systematic design method calls for the preparation of a requirements list for the design problem, followed by development of a function structure. The functional description of a product is a description at an abstract level, in solution-neutral terms. At the functional stage, different design possibilities can be explored by developing functional variants. Two major avenues for this are subdividing functions (divergence), and combining and eliminating functions (convergence). The latter can lead to simpler and cheaper designs through function integration and reduction. We shall discuss functional variants with the help of an industrial design example.

FUNCTIONS

Functions are the most important of the technical requirements for product use. The functional description of a product is a description at an abstract level, in solution-neutral terms. In contrast to this, detailed drawings describe a product in a concrete fashion. Functional analysis and the function structure, which are the subject of this appendix, not only help directly in designing, but also enable us to classify systems and study them in a systematic manner. This is particularly true for complex systems in which complex functional relationships occur. By breaking up functions into simpler subfunctions, we can subdivide a problem into more manageable parts.

A function in its simplest form can be expressed as a verb, e.g., "sense," "amplify," "convey." All functions involve the processing or transformation of one, two, or all three of the fundamental quantities *energy, material*, and *signal*. A signal is the physical embodiment of information, and can be considered to be energy at a low level. Figure D1 shows the general function block with its inputs and outputs.

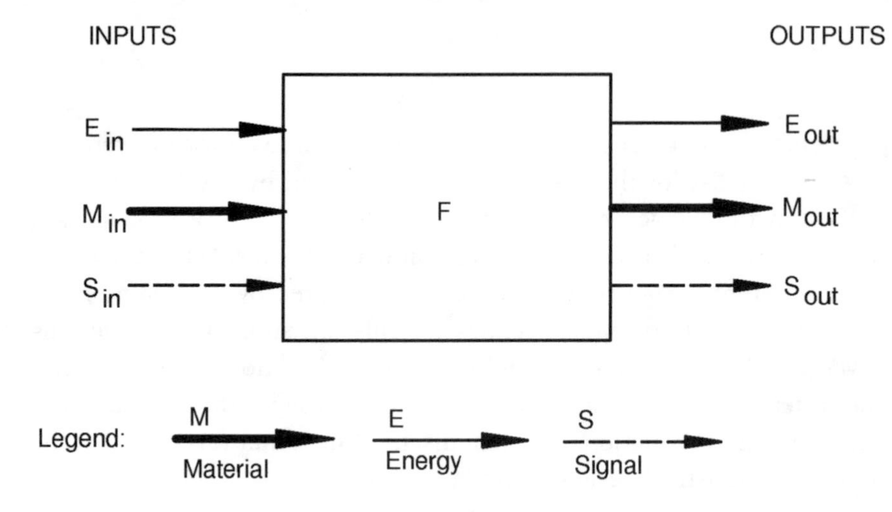

FIGURE D1. A function block.

• • • • •

EXAMPLE FUNCTION BLOCK DIAGRAM FOR A COFFEEMAKER

As an example, we consider a coffeemaker. The primary function of a coffeemaker is to "brew" coffee. The inputs to this device are:

1. Material: coffee, water
2. Energy: electrical energy
3. Signal: switching the device on

The outputs of the device are:

1. Material: brewed coffee, coffee grounds
2. Energy: thermal energy dissipated
3. Signal: a light indicating that the coffee is ready

These quantities are shown in the overall block diagram in Figure D2.

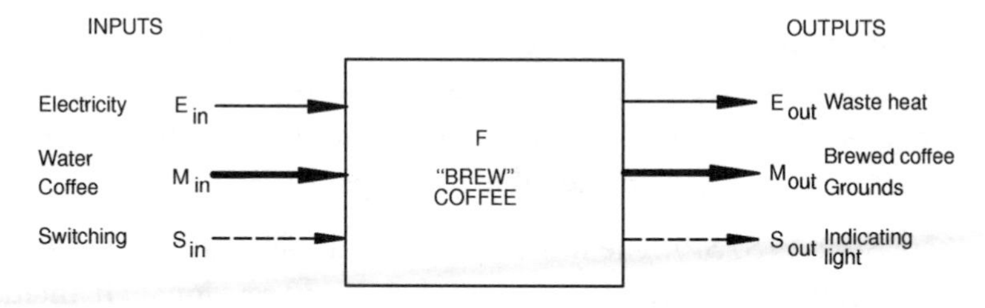

FIGURE D2. Overall function block diagram for a coffeemaker.

• • • • •

Since numerous functions occur in physical systems, it is helpful to classify these into "basic" or "elementary" functions. The number of such functions varies, depending on the degree of abstraction, or, conversely, on how close to physical functions they are.

The setting up of a function structure is of the greatest importance in the case of new designs. The basic concept is not known in these cases, and the time devoted to looking for possible subfunctions and their positioning in the system is well spent. In the case of adaptive designs, an initial structure can be produced by analyzing the existing product, and variations of this can be obtained by using methods described in the section "Functional Variants."

CLASSIFICATION OF FUNCTIONS

Basic functions are classified here under the following categories:

- store/supply
- connect
- branch
- channel
- change magnitude
- convert

These are close to the basic functions of Roth (1982). All physical functions can be shown to belong to one of these categories or to be formable as a combination of them. Brief descriptions of the basic functions follow. The abbreviations given for each type will be used in the figures and in subsequent discussion. Solutions—actual design concepts—based on these functions are given in the tables in Appendix F.

- *Store/Supply (STSP)*. The function "store/supply" implies the storage, and eventual supplying, of material, energy, or signal. It includes the functions "store," "empty," "supply," and "receive" for energy, material, or signal. The functions "hold," "stop," and "release" are also included under this basic function. Any of these can imply a lapse of time in execution of the "store/supply" function.
- *Connect (CONN)*. The function "connect" applies whenever two or more quantities are brought together. It includes physical functions such as "mix," "switch," or "compare" and arithmetic operations. The function has more than one input and one output.
- *Branch (BRCH)*. "Branch" is the opposite of the basic function "connect." It has one input and more than one output. It includes functions such as "separate," "cut," or "count."
- *Channel (CHNL)*. The basic function "channel" includes physical functions such as "transmit," "transport," and "convey." It applies to energy, material, or signal.
- *Change Magnitude (CHMG)*. The basic function "change magnitude" implies a change in magnitude while the form remains the same. It is used for energy or signal, and for material properties.
- *Convert (CVRT)*. "Convert" applies whenever the form of the output is different from that of the input, as in "convert" pressure to displacement. It can imply a change of state of material or a change in the form of energy or signal.

FUNCTIONAL VARIANTS

In methodically developing function structures for the purpose of simplification of devices, there are two basic ways to proceed:

- Start with the simplest function structure. Look for its simplest physical realization. Check whether it fulfills the requirements; if not,

enhance the function structure and repeat the process until a satisfactory solution is obtained. This is the traditional approach.
- Develop a comprehensive function structure that has functions meeting all of the individual requirements. Then simplify the structure before proceeding to solutions.

After a function structure has been developed, the next step is to develop functional variants, a process which can lead to discovery of new solutions not only for new tasks but also for adaptive designs of existing products, once a preliminary function structure has been obtained through analysis; see Hundal and Byrne (1990). The methods for producing functional variants are the following:

- Relocating functions in an existing block diagram and moving the system boundary
- Subdividing complex functions (divergence)
- Combining or eliminating functions (convergence)

These procedures will be elaborated upon next.

Relocating Functions

After a tentative block diagram has been developed, functional variants can be generated by rearranging subfunctions within the system, perhaps introducing feedbacks while doing so. An extension of this idea is the relocation of functions from inside the system to outside it, or vice versa. For example, an internal power supply may be replaced by an external power source. This amounts to moving the system boundary.

Subdividing Functions

A complex function can be re-formed in terms of its basic functions, arranged serially and/or in parallel. As an example, consider the function "amplify force," which has the variable force as both input and output. It can be broken into two-function chains such as "convert force to moment and moment to force," or "convert force to pressure and pressure to force." It can also be expressed as a longer chain, e.g., force to motion to current to force. See Figure D3 for such an example. Subdivision offers a larger number of subsolutions, and thus a greater flexibility in design. However, it can also lead to more components and thus higher cost.

Figure D4 shows the general case of function subdivision. An overall function F (Figure D4a) is subdivided into three functions F_1, F_2, and F_3 (Figure D4b). The subfunction F_2 is shown further subdivided in part c. The latter subdivision involves a feedback loop.

COMBINING OR ELIMINATING FUNCTIONS

Let us assume that a comprehensive function structure has been derived, either from the analysis of an existing system or during conceptual design. It can then be simplified in two ways:

- By eliminating functions, i.e., performing a *reduction*
- By eliminating connections, i.e., combining functions

In the latter case we are performing an *integration*. Certainly, some caution must be exercised that some important function is not eliminated, or that the resulting function is realizable. A structure thus reduced or integrated calls for fewer subsolutions and components, leading to a lower-cost design, as has been amply shown in this book. The process of function structure simplification also underlies value analysis, as shown by Fowler (1990). The processes of reduction and integration will now be illustrated by an example.

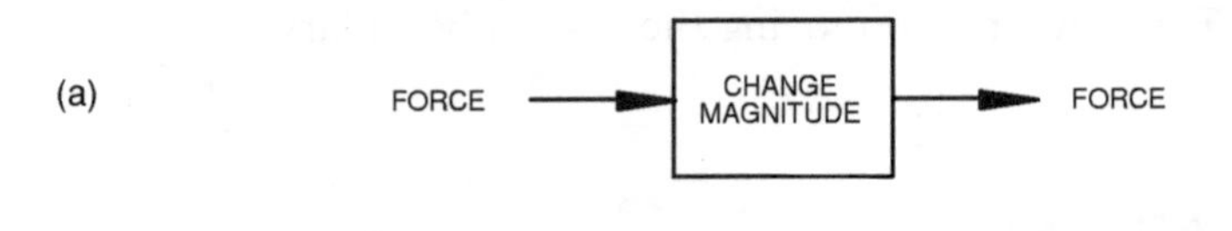

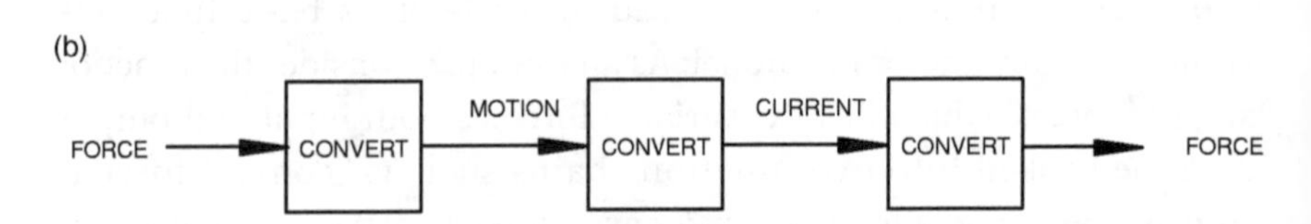

FIGURE D3. Subdivision of a function.

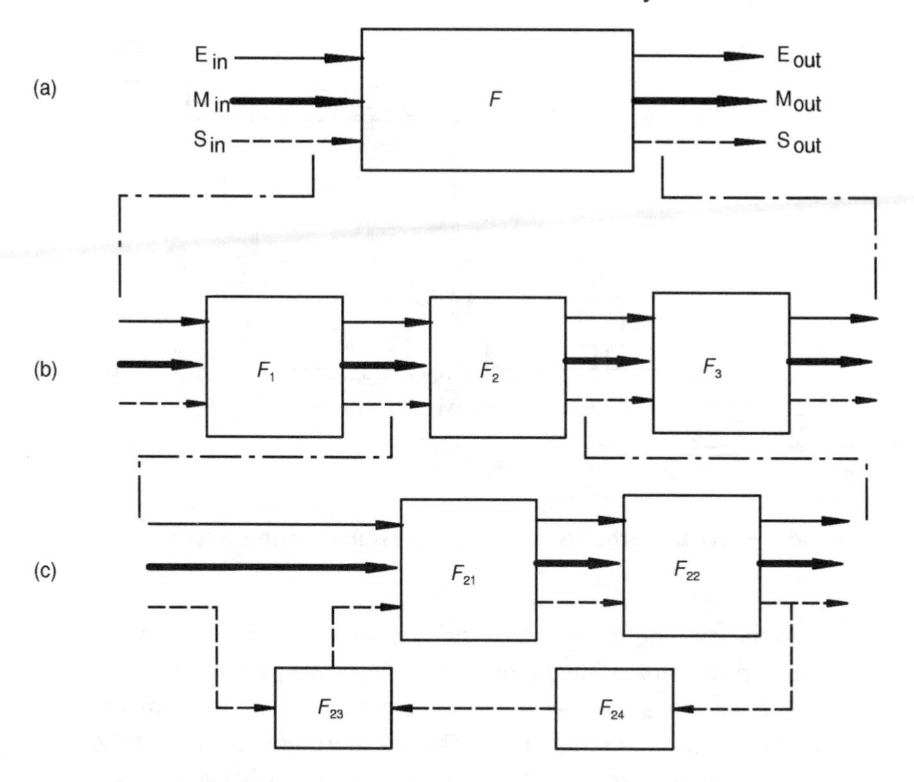

FIGURE D4. Subdividing functions—the general case.

• • • • •

EXAMPLE LIQUID PRESSURE VARIATOR

We consider an industrial system which is to supply liquid at varying pressure. The device is meant for testing filters that will be subject to varying pressures during operation. The system takes the liquid at atmospheric pressure and produces an output at pressures varying in the range of 50 to 200 psi. The waveshapes can be triangular, sinusoidal, or of any other practical form. Figure D5 shows the schematic of such a system. For the sake of clarity, mechanical or hydraulic components are used for control, and not all components are shown.

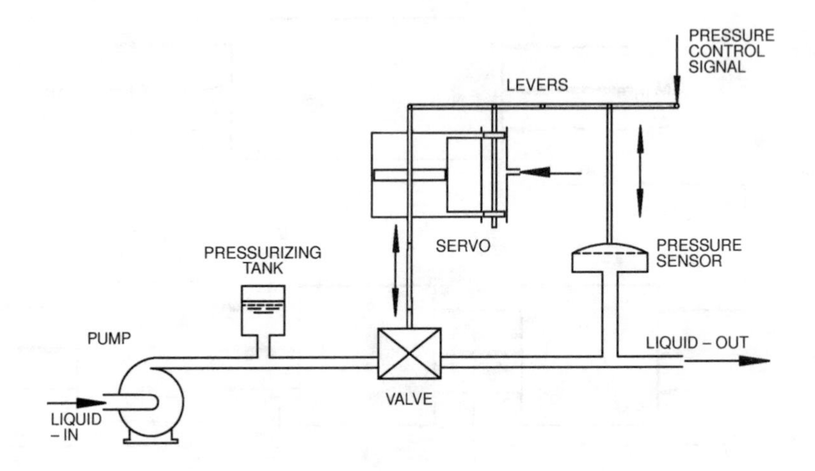

FIGURE D5. Schematic of liquid pressure–varying system.

In the existing design, a centrifugal pump feeds a valve via a pressurizing tank. Both the pump and the valve perform the function "change magnitude" on the liquid pressure. The tank embodies the function "store/supply." The pressure at the outlet of the valve is sensed, implying the function "convert" from liquid pressure to pressure signal. A signal generator supplies the desired waveshape signal, which is compared with the sensed pressure signal. This is the embodiment of the function "connect." The resulting error signal drives a hydraulic servo, which in turn actuates the valve. The servo also embodies the function "convert," from signal to mechanical energy. The system requires electrical energy, which is converted to mechanical energy and supplied to the pump. Electrical energy is also needed to drive a hydraulic pump, which powers the servo.

Figure D6 shows a detailed function structure of this system. The most important functions are described in Table D1. The control signal represents the desired input pressure P_i. Output is the fluid flow at varying pressure P_c. We wish to look for possibilities for simplifying this system design, as well as for other design variants.

• • • • •

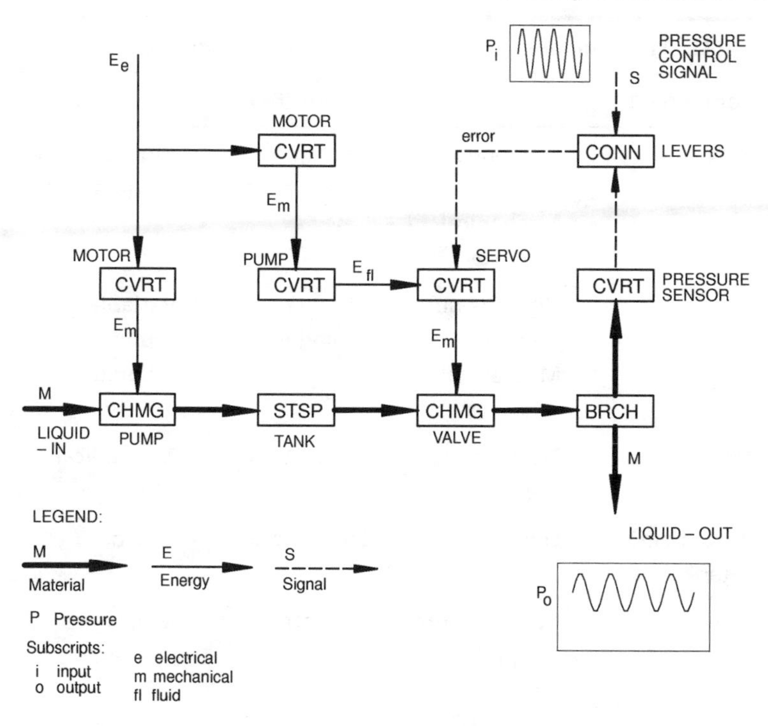

FIGURE D6. Detailed function structure.

OBTAINING FUNCTIONAL VARIANTS THROUGH SIMPLIFICATION

The simplification of the function structure in Figure D5 will now be performed by carrying out reduction and integration operations.

FUNCTION REDUCTION

Reduction implies eliminating functions. The reduction steps in function structure modification are shown in Figure D7.

- The first step in reduction is on the servo, where the conversions of electrical energy to mechanical and fluid energy are eliminated. This can be realized by using an electrical servo. The result is shown in Figure D7a.

TABLE D1 Components and Their Functions in Figure D6

Component	Input(s)	Function	Output
Pump	Low-pressure liquid Mechanical energy	Change magnitude	High-pressure liquid
Valve	High-pressure liquid Mechanical energy	Change magnitude	Variable-pressure liquid
Motor	Electrical energy	Convert	Mechanical energy
Pressurizing tank	Liquid	Store/supply	Liquid
Servo	High-pressure liquid Error signal	Convert	Mechanical energy
Pressure sensor	Liquid	Convert	Pressure signal

- As the next step, we can eliminate the pressurizing tank, as in part b of the figure. The implications of this on the operation of the system should be investigated.
- A final step in reduction is shown in part c, where all but "change magnitude" and "convert" functions have been eliminated. A physical device that can change low-pressure liquid to varying-pressure liquid is a reciprocating pump. The frequency of pressure waves can be controlled by varying motor speed. However, the waveshape is fixed. The amplitude can be varied by a simple valve.

Function Integration

Next we look at the possibilities of integration, some of which are illustrated in Figure D8. Some of the structures shown here combine integration and reduction.

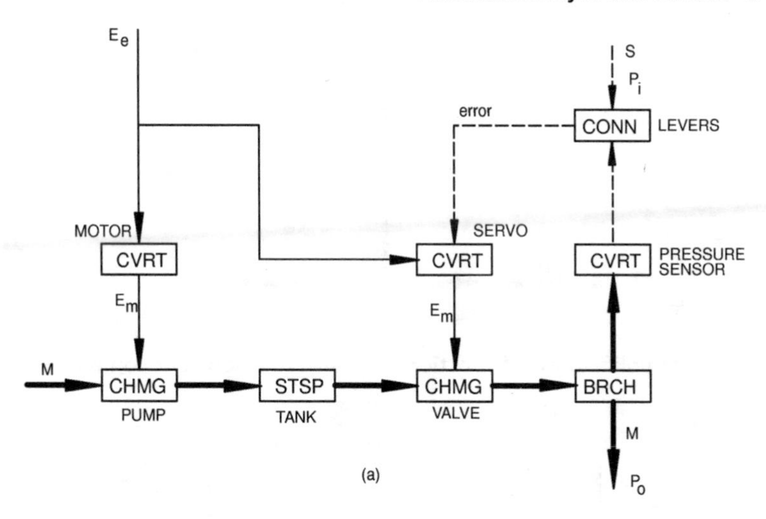

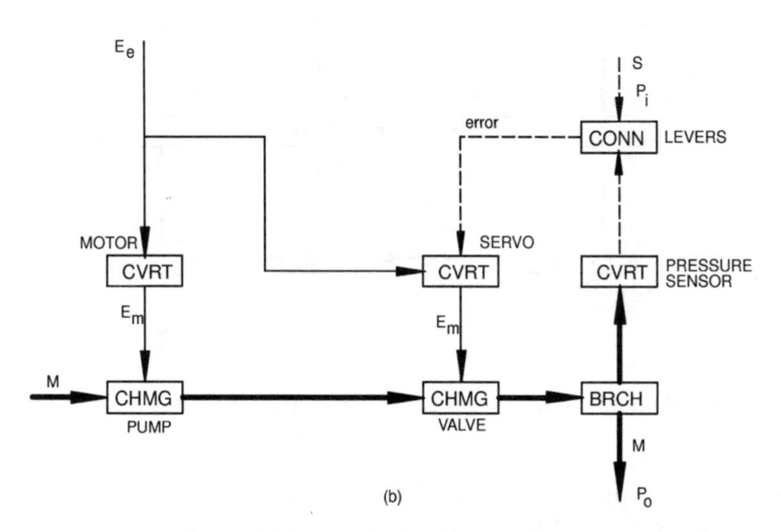

FIGURE D7. Simplification of function structure through function elimination.

- We can integrate the pressurizing tank with the pump (Figure D8a).
- As a next step, the comparator action may be integrated into the servo (Figure D8b).
- If the servo is eliminated and the energy from the pressure sensor is utilized in actuating the valve, a reduction to the system, shown in part c, is obtained.

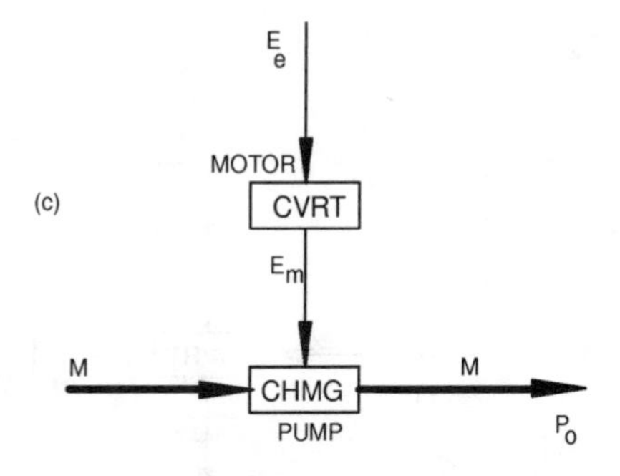

FIGURE D7. Simplification of function structure through function elimination. (cont.)

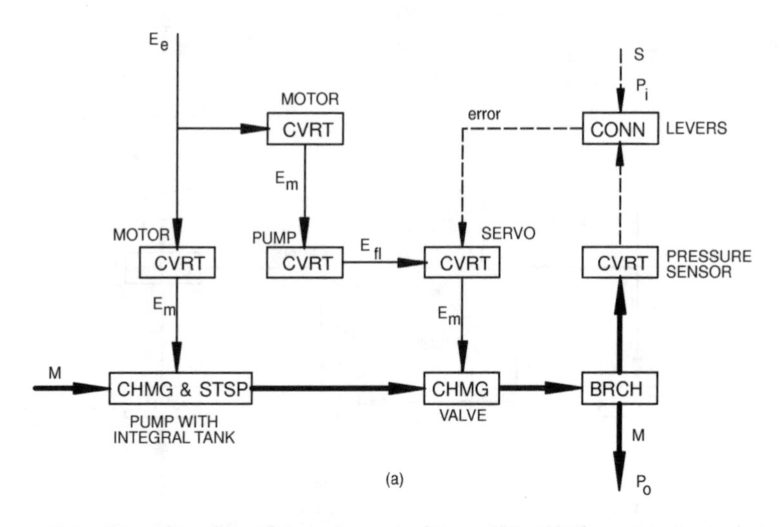

FIGURE D8. Simplification of function structure through function integration.

- Finally, an open-loop system is shown in part d, which implies that the control signal P_i has sufficient energy to actuate the valve.

SUMMARY

The development of a function structure is a key step in systematic design. At the functional stage, different design possibilities can be explored by developing functional variants. This process is useful both for new products and when applied in adaptation of existing designs for which the

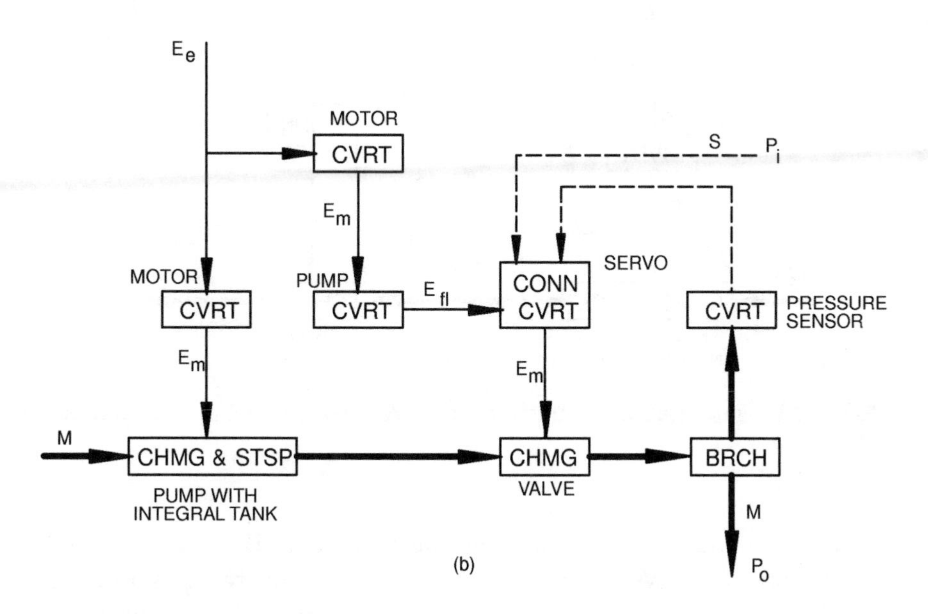

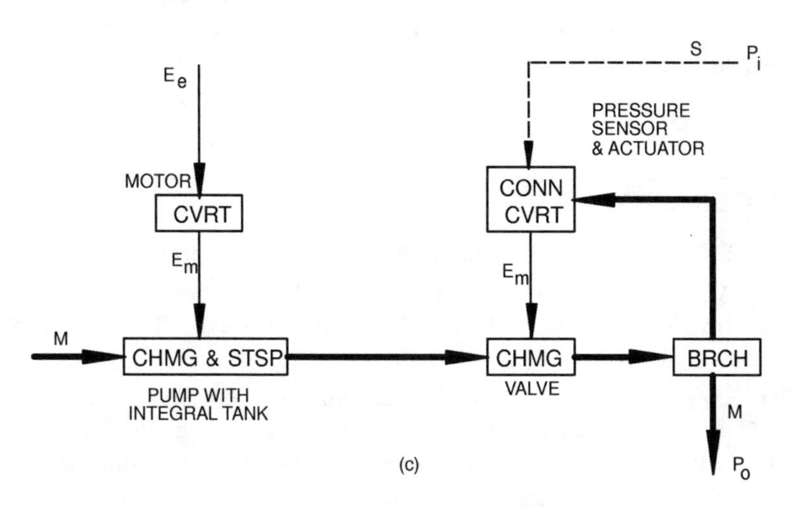

FIGURE D8. Simplification of function structure through function integration. (cont.)

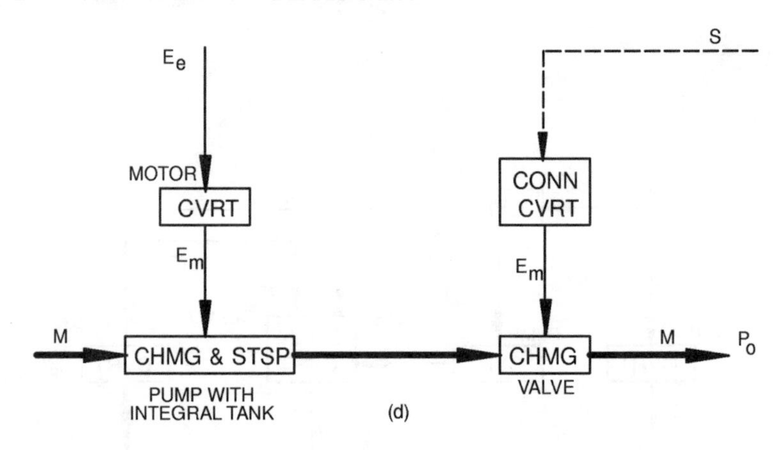

Figure D8. Simplification of function structure through function integration. (cont.)

function structure has been obtained through analysis. Since functional description takes a system to a more abstract stage, the variants produced and subsequent solutions found are independent of the hardware and arrangement in the original system. The structure can be simplified in two ways: by eliminating functions, i.e., performing a reduction; and by eliminating connections, i.e., combining functions. A simpler structure calls for fewer components, thus leading to a lower-cost and more robust design.

BIBLIOGRAPHY

Fowler, T. C. *Value Analysis in Design*. New York: Van Nostrand Reinhold, 1990.

Hundal, M. S., and J. F. Byrne. "Computer-Aided Generation of Function Block Diagrams in a Methodical Design Procedure." DE 27: 251–258. New York: ASME, 1990.

Koller, R. *Konstruktionslehre für den Maschinenbau (Mechanical Engineering Design)*. 2nd ed. Berlin and New York: Springer Verlag, 1985. Koller goes directly to 12 basic physical functions and their inversions—thus, 24 in all.

Rodenacker, W. G. *Methodisches Konstruieren (Methodical Designing)*. 3rd ed. Berlin and New York: Springer Verlag, 1984. Rodenacker envisions the function structure as having purely logic elements. He uses three—"connect," "separate," and "channel," the first two being based on two-valued logic. He then defines a physical structure containing elements incorporating physical effects, as well as a shape structure which shows a significant degree of embodiment.

Rodenacker, W. G., and R. Baumgarth. "Die Vereinfachung der Geräte Beginnt mit der Funktionsstruktur" (The Simplification of Devices Begins with the Function Structure). *Konstruktion* 28: 479–482 (1976). An example of function structure manipulation for the simplification of a design.

Roth, K. *Konstruieren mit Konstruktionskatalogen (Designing with Design Catalogs)*. Berlin and New York: Springer Verlag, 1982. Also 2nd ed. (2 vols.) 1994. Roth defines four basic functions: "change," "connect" (including branch), "channel" (including convert), and "store."

Physical Effects

- **Tables of Physical Effects**
- **Physical Effects in Sketch and Equation Form**

TABLES OF PHYSICAL EFFECTS

Table E1 shows the various physical effects, laws, and phenomena that relate each of 12 types of physical quantities with itself and other physical quantities. In some cases, solutions (devices) are mentioned in order to suggest a context in which the effect may be applied.

PHYSICAL EFFECTS IN SKETCH AND EQUATION FORM

Tables E2 and E3 illustrate some of the selected physical effects by use of sketches and show the corresponding equations which describe the laws. Some of the physical effects have already been illustrated in the design example of the temperature measurement device; see Figures 24 and 25 in Chapter III.

PHYSICAL EFFECTS FOR GENERATING A FORCE

Table E2 lists some of the solutions to the very commonly occurring function "generate force."

PHYSICAL EFFECTS FOR OTHER FUNCTIONS

Table E3 shows a few other physical effects in equation and sketch form.

INPUT AND OUTPUT

An equation that describes a physical law relates the various parameters and variables. It does not dictate which is the input and which is the output. As an example, we look at the physical law

$$F = k\Delta \tag{E-1}$$

applied to a spring, where F is the force, Δ the deflection, and k the stiffness of the spring. In words, this equation reads, "The deflection of a spring is proportional to the applied force." The system is shown in

TABLE E1 Table of Physical Laws and Effects Relating Physical Quantities

Input (or Change in)	Output (or Change in)					
	Length, Area, Volume	Velocity	Acceleration	Force, Pressure, Mechanical Energy	Mass, Moment of Inertia, Density	Time, Frequency
Length, area, volume	Lever, wedge, capillary effect, cohesion, adhesion	Continuity, eccentricity, Torricelli's law, toughness, differentiation*	Centrifugal acceleration	Hooke's law, Poisson's law, shear, torsion, Coulomb's first and second laws, buoyancy, Boyle's law, gravity, capillary pressure	Eccentricity	Elasticity (clamped length), gravity (pendulum length), transmission time
Velocity	Weissenberg effect, integration*	Lever, wedge, impact	Coriolis acceleration, charge in magnetic field, differentiation*	Kinetic energy, impulse, angular momentum, Bernoulli's law, toughness, Coriolis acceleration, turbulence, fluid resistance		
Acceleration	Mass/spring	Integration*	Lever, wedge	Newton's law		

TABLE E1 Table of Physical Laws and Effects Relating Physical Quantities

Input (or Change in)	Output (or Change in)					
	Length, Area, Volume	Velocity	Acceleration	Force, Pressure, Mechanical Energy	Mass, Moment of Inertia, Density	Time, Frequency
Force, pressure, mechanical energy	Hooke's law, Poisson's law, shear, torsion, Coulomb's first and second laws, buoyancy, Boyle's law	Kinetic energy, impulse, angular momentum, Bernoulli's law, toughness, Magnus effect	Newton's law	Lever, wedge, friction, hysteresis, cohesion, adhesion	Boyle's law	Natural frequency
Mass, moment of inertia, density		Sound wave	Newton's law	Gravity, Newton's law, Coriolis force, centrifugal force		Natural frequency
Time, frequency	Resonance, standing waves	Dispersion		Resonance, absorption		Beat frequency
Sound	Acoustic excitation			Sound pressure		
Heat, temperature	Thermal expansion, water anomaly	Molecular motion, sound wave		Thermal expansion, surface tension, osmotic pressure	Gas law	Natural frequency (quartz)

(continued)

TABLE E1 Table of Physical Laws and Effects Relating Physical Quantities (*continued*)

Input (or Change in)	Output (or Change in)					
	Length, Area, Volume	Velocity	Acceleration	Force, Pressure, Mechanical Energy	Mass, Moment of Inertia, Density	Time, Frequency
Electric resistance						
Electric field, voltage, current	Electrostriction	Electrokinetic effect	Charge in electric field	Biot-Savart law, electrokinetic effect, hysteresis, Coulomb's first law		Josephson effect
Magnetic field, inductance	Magnetostriction	Induction, eddies		Biot-Savart law, hysteresis, Coulomb's second law, ferromagnetism		
Light, EM waves				Light wave pressure		

Input (or Change in)	Output (or Change in)					
	Sound	Heat, Temperature	Electric Resistance	Electric Field, Voltage, Current	Magnetic Field, Inductance	Light, EM Waves
Length, area, volume	Length change	Plastic deformation, convection	Conductor length and area, electrolyte	Piezo effect, electrode separation	Magnetic shielding, air gap	Interference, scattering, absorption

TABLE E1 Table of Physical Laws and Effects Relating Physical Quantities (*continued*)

Input (or Change in)	Output (or Change in)					
	Sound	Heat, Temperature	Electric Resistance	Electric Field, Voltage, Current	Magnetic Field, Inductance	Light, EM Waves
Velocity	Doppler effect, stick-slip	Convection		Induction law, electrokinetic effect	Charge velocity, Barnett effect	Doppler effect, flow-induced double refraction
Acceleration				Electrodynamic effect, Talmann effect		Charge, electromagnetic wave
Force, pressure, mechanical energy	Stick-slip, pressure wave	Friction, plastic deformation, convection, hysteresis, Joule-Thompson effect	Force-sensitive materials	Piezo effect, static electricity, ionization	Magnetoelasticity, permeability, magnetic anistropy	Refractive index, stress-induced double refraction, friction (flint)
Mass, moment of inertia, density					Permeability	Refraction (slip lines)
Time, frequency	Dispersion	Dielectric loss, eddy current	Skin effect	Josephson effect		Scattering
Sound	Reflection, absorption, interference	Friction (ultrasonic welding)				Sears effect (bending of light waves), absorption (shadow)

(*continued*)

TABLE E1 Table of Physical Laws and Effects Relating Physical Quantities (*continued*)

Input (or Change in)	Output (or Change in)					
	Sound	Heat, Temperature	Electric Resistance	Electric Field, Voltage, Current	Magnetic Field, Inductance	Light, EM Waves
Heat, temperature	Thermophone	Melting, vaporization, condensation, solidification	Conductors, semiconductors, super-conductors, thermal ionization	Thermionic emission, thermal effect (bimetal)	Permeability, Curie point	Stefan-Boltzmann law, liquid crystals, refraction
Electric resistance				Ohm's law		
Electric field, voltage, current	Thermophone	Electric arc, Joule heating, eddy current loss, Ohm's law	Varistor, FET, tunnel effect	Transformer, amplifier, transduction	Magnetization	Electro-luminescence, liquid crystals, X rays, lasers, scintillation
Magnetic field, inductance		Demagnetization	Lorentz effect	Hall effect, induction law, MHD, magnistor	Hysteresis, saturation effect	Faraday effect, Cotton-Mouton effect
Light, EM waves		Radiant heating, microwaves	Photoresistance, photodiode, ionization	Photoelectricity		Refraction, double refraction, interference, polarization, conduction, absorption, dispersion

*The mathematical operations can be fulfilled by different physical effects.

TABLE E2 Design Catalog for the Function "Generate Force"

Physical Law	Equation	Sketch
Gravitation (earth)	$F = mg$	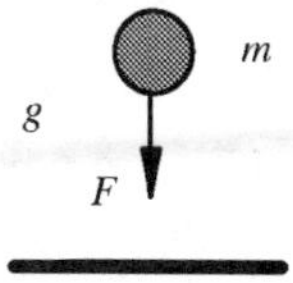
Gravitation (general)	$F = G\dfrac{m_1 m_2}{r^2}$	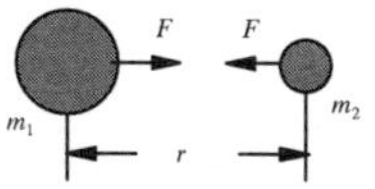
Gravitation (buoyancy)	$F = \rho g V$	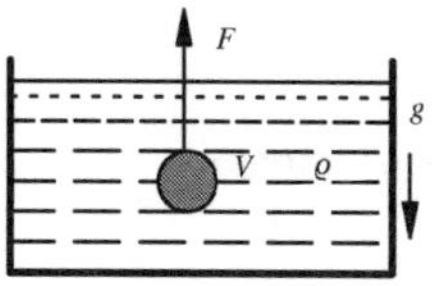
Newton's law	$F = ma$	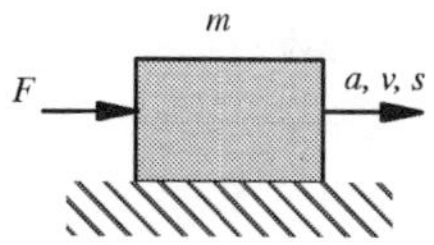
Coriolis effect	$F = 2m\omega V_r$	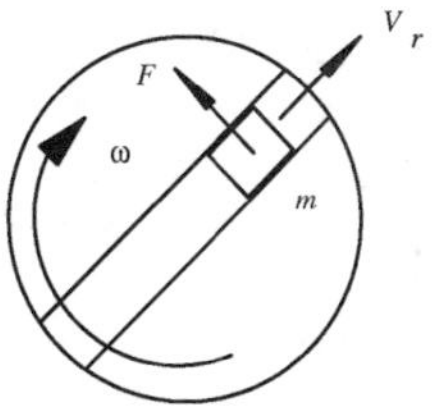
Jet reaction	$F = \dot{m}V_r$	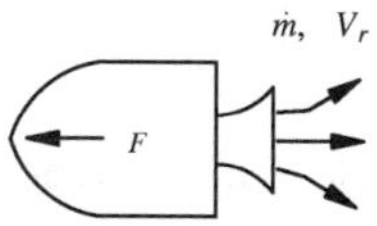
Viscous friction	$F = A\mu \dfrac{dV}{dy}$	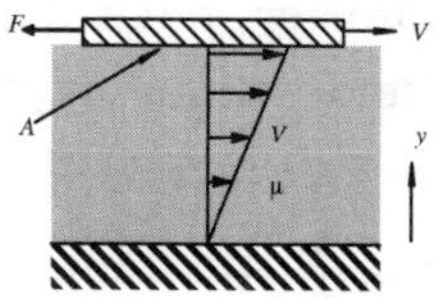

(continued next page)

TABLE E2 Design Catalog for the Function "Generate Force" (*continued*)

Physical Law	Equation	Sketch
Magnus effect	$F = 2\pi R^2 \omega v L \rho$	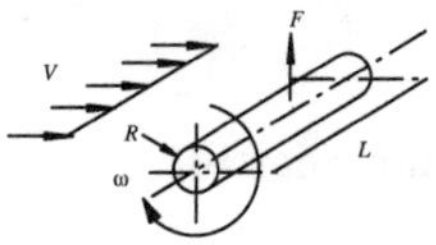
Osmosis	$F = \dfrac{AnRT}{V}$	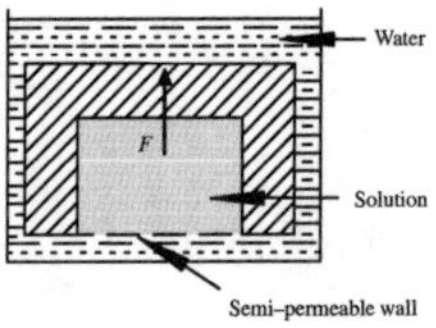
Thermal expansion	$F = E\alpha A\,\Delta T$	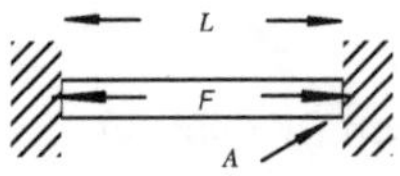
Coulomb's law (1)	$F = QE$	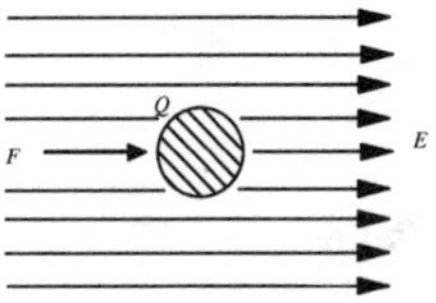
Coulomb's law (2)	$F = \dfrac{qQ}{4\pi\epsilon_o\epsilon_r\Delta^2}$	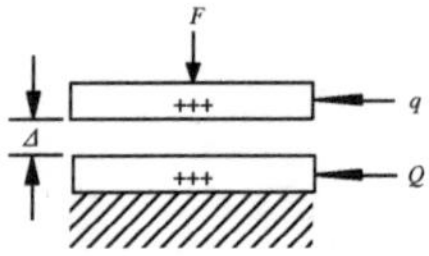
Magnetic force	$F = \dfrac{\Phi_1\Phi_2}{4\pi\mu_o u_r\Delta^2}$	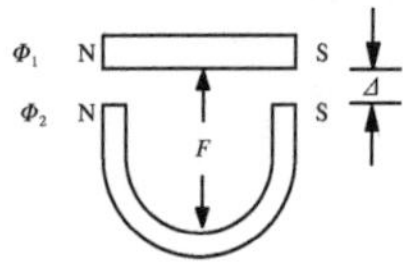

Figure E1. The parameter k contains further parameters of physical processes for a spring. For a helical spring these are the modulus of rigidity, wire diameter, pitch diameter, and the number of coils. As far as the "input" and "output" of this system are concerned, the following possibilities exist:

- Force F is the input; distance Δ is the output. Then the spring has the purpose of a "connection member" between force and distance, for example, when used for purposes of measurement.

TABLE E3 Design Catalog for Other Functions

Physical Law	Equation	Sketch
Continuity law	$V_2 = \frac{A_1}{A_2} V_1$	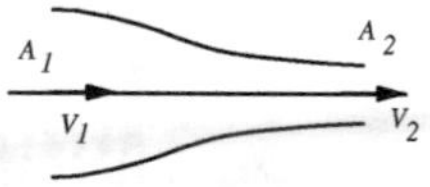
Heat conduction	$Q = \frac{kA}{L}(T_o - T_i)$	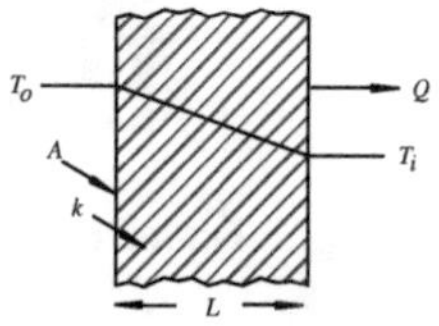
Heat convection	$Q = hA(T_o - T_i)$	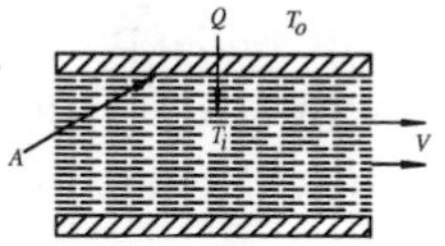
Heat radiation	$Q = \sigma A(T_o^4 - T_i^4)$	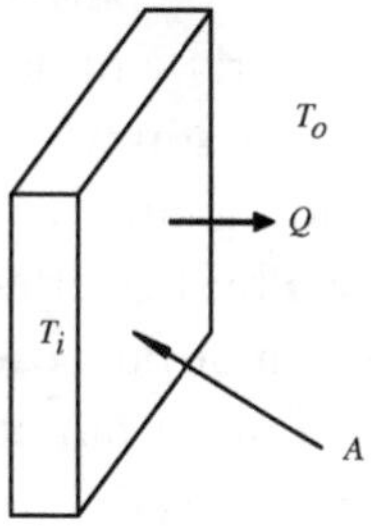
Force/resistance	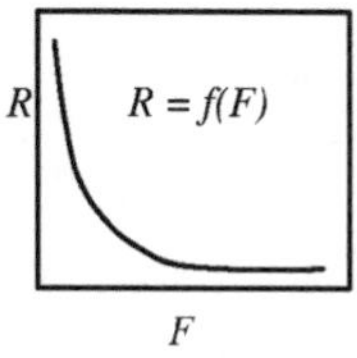 $R = f(F)$	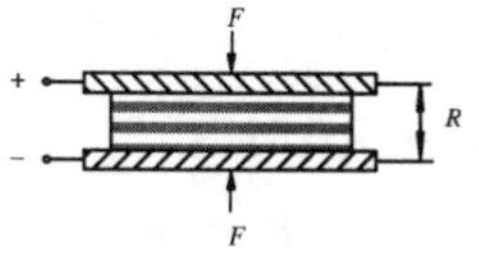

- One force is the input, another is output. Then the spring works as a "constraining member."
- Motion of one end is the input, motion of the other is output. Then the function of the system is "transmission," or "conveying."

Thus the physical system "spring" can be thought of as a connection, separation, or conveying member; that is, it can fulfill any of these functions, for which other systems also exist.

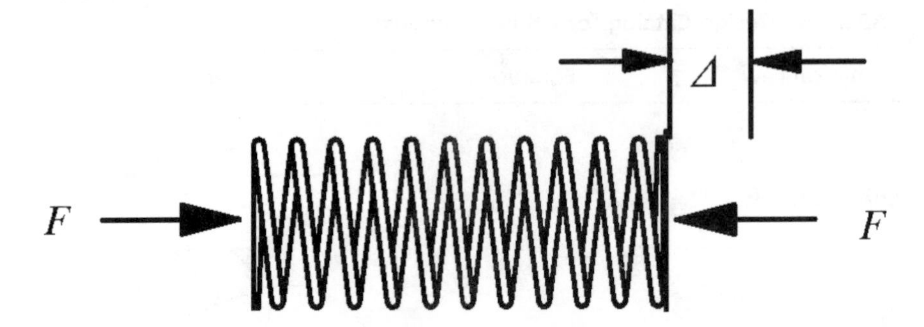

FIGURE E1. Spring as a multifunctional element.

In conclusion, a physical consideration alone is not sufficient for making a generalized formulation of the purpose or function of a device or its elements. Perhaps it is just because of the simplicity of the purpose of a machine or part—often reflected in its name—that its function is not particularly emphasized by the supplier or manufacturer. That may well be the reason why engineers and designers have difficulty accepting, understanding, and using the logical functions of machines and parts as necessary. Physical effects and systems as such are purpose-independent. The function or purpose to be realized is determined by the choice of the input and output quantities of a system.

BIBLIOGRAPHY

Hix, C. F., and R. P. Alley. *Physical Laws and Effects*. New York: Wiley (no publication date). Describes over 300 physical laws and effects.

Koller, R. *Konstruktionslehre für den Maschinenbau (Mechanical Engineering Design)*. 2nd ed. Berlin and New York: Springer Verlag, 1985. Provides extensive tables of solutions which realize the various physical phenomena.

Roth, K. *Konstruieren mit Konstruktionskatalogen (Designing with Design Catalogs)*. Berlin and New York: Springer Verlag, 1982; also 2nd ed. (2 vols.), 1994. A comprehensive book on systematic design. Includes extensive tables of physical effects and solutions. The book shows this information in the form of design catalogs. Roth's tables also show information on items such as order of magnitude, external energy need, important design parameters, geometric requirements, and typical applications.

Solutions for Basic Functions

- **Tables of Solutions**

A design solution for a function reflects the physical effect—friction, Bernoulli's principle, piezoelectricity, or any such physical phenomenon (see Appendix E)—and the elementary form in which it is used. In looking for solutions for a given function, we might first consider pertinent physical effects and then, using shape design principles, arrive at possible solutions. The alternative is to look for solutions directly, since most people are used to thinking in terms of physical shapes. The first approach is at a more basic level and can lead to a greater variety of solutions, including unusual ones, since less prejudice is introduced. Solutions may be stored for the designer's perusal in tables as shown in this appendix, or in computer memory.

The degree of abstraction with which solutions are viewed depends on:

- The design step at which they are used—e.g., at the function structure, physical effect, working principle, or solution principle level
- The number of solutions that are stored in the tables

This number depends on the degrees of freedom the designer wants in generating variants, and how convenient it is to access the solutions list. A large number of solutions, each using only one solution principle, allows us to generate a larger number of variants, but may be unwieldy.

We should bear in mind that many solutions recur, thus limiting to some extent the number of solutions that must be stored. We proceed from the function level where one of the basic categories—convert, change magnitude, channel, connect, branch, and store/supply—has been identified. The quantities handled (material, energy, signals), and the input and/or output and their physical forms, are likewise known. On the basis of this information, the tables given in this appendix present solutions to satisfy the requirements. Although not given in these tables, each solution should include a sketch, if appropriate; the underlying physical effect (see Tables E2 and E3); equations, if any; and pertinent remarks. The choice of solutions for a function is narrowed by the use of descriptors included in the solution tables. Many functions, such as "amplify (force)," or "fasten (materials)," to cite only two, have a large number of solutions, of course.

TABLES OF SOLUTIONS

The number of solutions given in the following tables is, of course, only a sampling. The designer can add solutions specific to his or her areas of interest.

TABLE F1 Solutions to the Basic Function "Store-Supply"

Function	Input and Descriptors	Solution	Output
Store/supply	Energy		Energy
	Mechanical		
	Torque/rotation	Flywheel	
	Mechanical		
	Torque/rotation	Spring	
	Mechanical		
	Force/translation	Lift-weight	
	Mechanical		
	Force/translation	Spring	
	Mechanical		
	Fluid pressure	Accumulator	
	Mechanical		
	Fluid pressure	Accumulator	
	Thermal	Heat-mass	
	Thermal	Liquify-solid	
	Thermal	Vaporize-liquid	
	Electrical		
	DC	Battery	
	Electrical		
	DC	Capacitor	
Stop	Energy		None
	Mechanical	Frame	
	Electrical	Insulation	
	Thermal	Insulation	
Store/supply	Material		Material
	Solid	Surface	
	Liquid	Tank	
	Gas	Tank	
Store/supply	Signal		Signal
	Digital	Register	
Hold/release	Material		Material
	Solid	Clamp	

TABLE F2 Solutions to the Basic Function "Branch"

Function	Input and Description	Solutions	Outputs
Cut	Material	Shears	Material
		Tool	
		Thermal device	
Branch	Material		Material
	Liquid	Tee fitting	
Count	Material		Material + signal
Separate	Material		Material
	Solid		
	Granular	Sieve	
	Solid + liquid	Filter	
Branch	Signal		Signals
	Mechanical	Gearbox	Mechanical
	Torque/rotation		
	Mechanical	Linkage	Mechanical
	Torque/rotation		
	Mechanical	Walking beam	Mechanical
	Force/translation		
	Electrical	Junction	Electrical
Branch	Energy		Energy
	Mechanical	Gearbox	Mechanical
	Torque/rotation		
	Mechanical	Linkage	Mechanical
	Torque/rotation		
	Mechanical	Walking beam	Mechanical
	Force/translation		
	Electrical	Junction	Electrical

TABLE F3 Solutions to the Basic Function "Connect"

Function	Inputs and Descriptors	Solutions	Output
Connect	Energy		Energy
	Mechanical Torque/rotation	Gearbox	Mechanical
	Mechanical Torque/rotation	Linkage	Mechanical
	Mechanical Force/translation	Walking beam	Mechanical
	Electrical	Junction	Electrical
Compare	Signals		Signal
	Electrical	Comparator	
Mark	Material + signal		Material
	Material—solid		
	Signal—mechanical	die	
Control-flow	Material + signal		Material
	Material—liquid		
	Signal—mechanical	valve	
Switch	Energy + signal		Energy
Pack	Materials		Material
Mix	Materials		Material
Add	Signals		Signal
Subtract	Signals		Signal
Multiply	Signals		Signal
Divide	Signals		Signal
AND	Signals		Signal
OR	Signals		Signal

TABLE F4 Solutions to the Basic Function "Channel"

Function	Input and Descriptors	Solutions	Output
Transmit	Energy		Energy
	Electrical	Conductor	Electrical
	Mechanical		
	Torque/rotation	Shaft	Mechanical
	Mechanical		
	Force/translation	Bar	Mechanical
	Mechanical		
	Force/translation	Cable	Mechanical
	Fluid	Pipe	Fluid
Transmit	Signal		Signal
	Electrical	Conductor	Electrical
Transport	Material		Material
	Fluid	Pipe	Fluid
	Solid	Conveyor belt	Solid
	Solid	Pneumatic tube	Solid

TABLE F5 Solutions to the Basic Function "Change Magnitude"

Function	Input and Descriptors	Solutions	Output*
Crush	Material Solid Brittle	Crusher, hammer	Material
Form	Material Solid Ductile	Press	Material
Coalesce	Material		Material
Change magnitude	Material Fluid		Material Fluid
	Pressure Low	Reciprocating pump	Pressure High Sinusoidal
	Pressure Low	Centrifugal pump	Pressure High Constant
	Pressure High	Valve	Pressure Low
	Density Low	Pump	Density High
Change magnitude	Energy		Energy
	Mechanical Torque/rotation	Gearbox	Mechanical
	Mechanical Torque/rotation	Linkage	Mechanical
	Mechanical Force/translation	Lever	Mechanical
	Mechanical Force/translation	Linkage	Mechanical
	Mechanical Force/translation	Wedge	Mechanical
	Mechanical Force/translation	Differential cylinder	Mechanical
	Mechanical Fluid—pressure	Differential cylinder	Mechanical
	Electrical	Amplifier	Electrical
	Electrical—AC	Transformer	Electrical

*With the function "change magnitude," outputs and inputs will always be of the same nature.

TABLE F6 Solutions to the Basic Function "Convert"

Function	Input	Function	Output
Convert	Energy		Energy
	Mechanical		
	Displacement		Force
		Elastic element	
	Displacement		Electrical
		LVDT	
		Piezoelectric	
	Displacement		Heat
		Friction	
	Force		Displacement
		Elastic element	
		Cam/follower	
	Force/displacement		Torque/rotation
		Belt/pulley	
		Screw/nut	
		Rack/pinion	
		4-bar linkage	
		Brake	
	Force/displacement		Pressure
		Piston/cylinder	
		Diaphragm	
	Force/displacement		Stress
		Elastic element	
	Force/displacement		Electrical
		Piezoelectric	
	Torque/rotation		Force/displacement
		Belt/pulley	
		Screw/hut	
		Rack/pinion	
		Cam/follower	
		4-bar linkage	
	Torque/rotation		Electrical
		Generator	
	Torque/rotation		Pressure
		Fluid motor	

TABLE F6 Solutions to the Basic Function "Convert" *(continued)*

Function	Input	Function	Output
Convert	Pressure/vacuum	Piston/cylinder Diaphragm	Force/displacement
	Pressure/vacuum	Diaphragm	Stress
	Stress	Strain gauge	Electrical
	Density	Buoyancy element	Force/displacement
	Electrical	Solenoid Piezoelement Linear motor	Force/displacement
	Electrical	Rotary motor	Torque/rotation
	Electrical	Induction coil	Magnetic
	Electrical	Heat coil	Heat
	Electrical	Lamp LED	Light
	Electrical	Speaker	Sound
	Magnetic	Ferromagnetic	Force/displacement
	Magnetic	Ferromagnetic	Torque/rotation
	Thermal	Solid—expansion	Force/displacement
	Thermal	Thermocouple Thermo-resistance	Electrical Voltage Resistance

(continued next page)

TABLE F6 Solutions to the Basic Function "Convert" (*continued*)

Function	Input	Function	Output
Convert	Thermal		Pressure
		Fluid—expansion	
	Light		Electrical
		Photocell	
	Sound		Electrical
		Microphone	
Convert	Energy		Energy
	Chemical		Electrical
		Fuel cell	
Sense	Energy		Signal
	Mechanical		Electrical
	Torque	Strain gauge	
	Mechanical		
	Rotation	Encoder	
Sense	Material		Signal
	Fluid		
	Pressure	Strain-gauge	Electrical
	Velocity	Venturi	Pressure
Change state	Material		Material
	Solid	Heater	Liquid
	Liquid	Cooler	Solid
Integrate	Signal		Signal
	Electrical	Circuit	
Differentiate	Signal		Signal
	Electrical	Circuit	
Process	Signal		Signal
	Electrical	Circuit	

APPENDIX G

Computer Use in Design

Computers, due to their capacity to perform numerical calculations at high speeds, store large amounts of data, and control graphic devices have found a great deal of use in design activities. Computers are also capable of making decisions when logical relationships can be expressed in algorithmic form. Activities are mostly creative in nature at the beginning of the design process and become more routine as it proceeds toward the detail design stage. Computers have therefore traditionally found the least use in conceptual design, more in embodiment design, and the most use in performing the routine activities in the detail design phase.

The use of digital computers in design may be broadly classified into the following categories: explicit numerical computations, computer-aided design (CAD), and artificial intelligence (AI). Although all computer use for design may be put under the title "computer-aided design," we shall restrict this term to imply special-purpose software for graphics, drafting, analysis, and the modeling of object geometry. Such software, of course, requires numerical computation in its operation. However, the latter is transparent to the user. Modern CAD software uses each of these three categories as its building blocks.

The operation of a digital computer can be understood from the block diagram shown in Figure G1. The computer consists of a memory unit and a central processing unit (CPU). The latter contains the control unit, which has a clock and controls all the operations in the computer, and an arithmetic and logic unit (A&LU), which does the "number crunching" and performs the logic operations during the running of a program. The input/output unit (I/O) transfers the program from an external source into the memory, reads the data into the memory as dictated by the program, and transfers the output generated by the program to an external entity, such as a file or a printer.

NUMERICAL COMPUTATIONS

The earliest digital computers were developed for the specific purpose of doing numerical calculations. Thus, the historical use of computers in

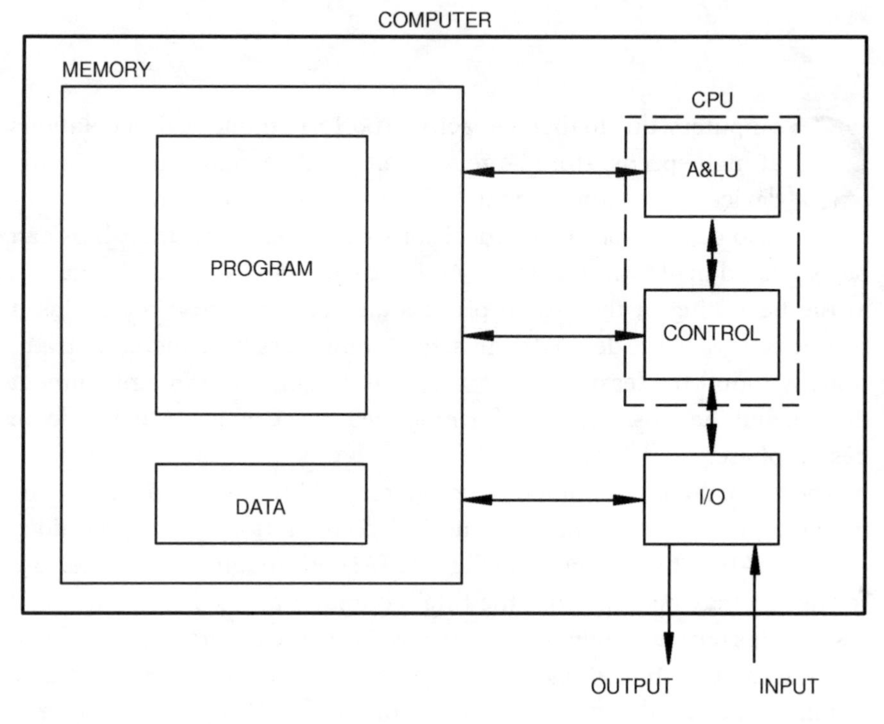

FIGURE G1. Block diagram of a digital computer.

engineering has been for analysis, and analysis as an aid for design. The software available for numerical computation includes:

- High-level programming languages such as Basic, Fortran, Pascal, or C
- Spreadsheets
- Equation-solving software

In numerical or conventional (as opposed to AI) computing, the program consists of one or more algorithms; see Figure G2. An algorithm is a sequence of steps which performs a specific task, such as finding the solution to an equation. The computer uses binary words to represent numbers, letters, and other symbols. While individuals may prefer to use a high-level programming language for specific applications, most routine computation can be carried out with spreadsheets and equation solvers.

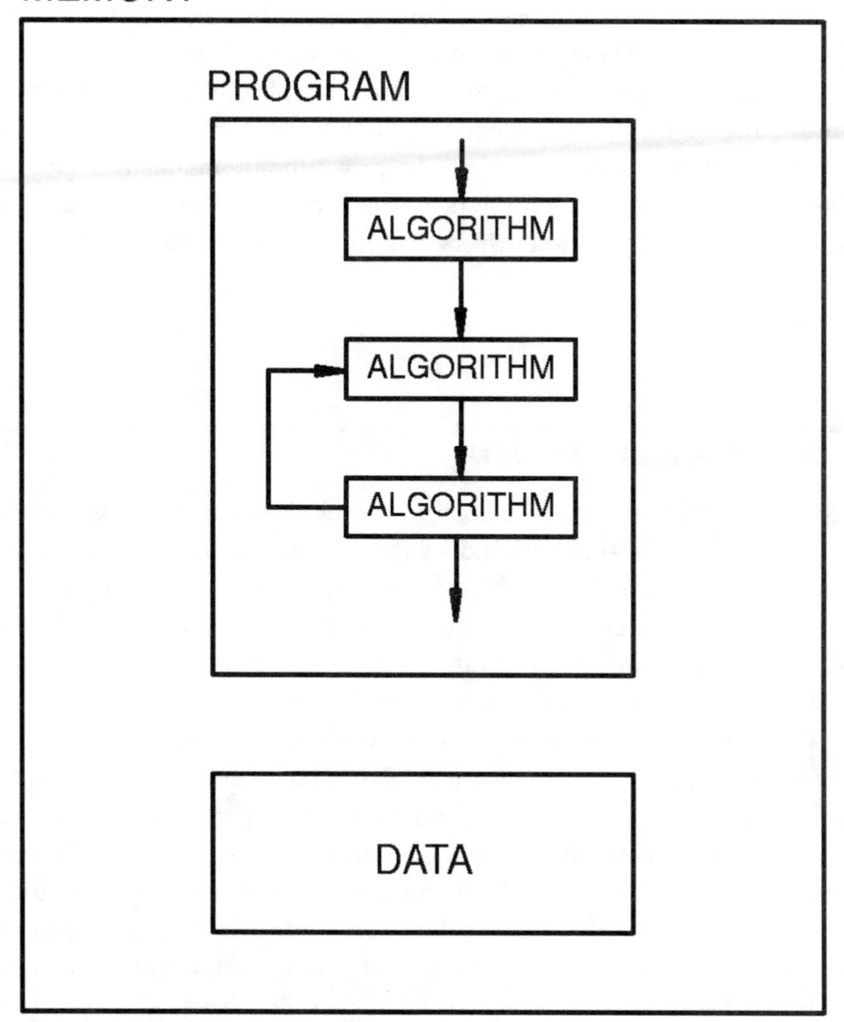

FIGURE G2. The conventional computing process.

Spreadsheets are two-dimensional grids of cells, similar to accounting ledgers. Each cell may contain text, a number, or a formula operating on numbers in other cells. Spreadsheets are useful for simple calculations. Many of the tables in this book—e.g., Tables 16 and 17—were prepared by using a spreadsheet program.

Equation solvers are of two types—numeric and symbolic. Numeric equation solvers are programmed by typing in the data, equations, and logic expressions. Their programming for procedures such as matrix operations and numerical integration is much simpler than using programming languages. Symbolic equation solvers require equations to be entered in purely symbolic form. The program automatically solves for the variable(s), for which numerical values have not been defined. Both spreadsheets and equation solvers can produce output in graphic form.

COMPUTER-AIDED DESIGN

Under the term *computer-aided design* we include topics such as geometric modeling (of lines, surfaces, and solids), transformations, and visualization; tolerancing; and assembly—as well as analysis techniques such as finite element and boundary element analysis, fluid dynamic simulation, and kinematic-chain analysis (and synthesis). Geometric modeling and finite element analysis are discussed in the pages that follow.

Figure G3 shows the evolution of CAD techniques from the time such software became commercially available. The earliest versions of CAD programs were simple drafting tools. Soon the importance of 3-D representation of solid objects was realized, leading to wire frame models. True solid modeling was achieved only after the techniques of mathematical representation of surfaces and solids were perfected. The developments up to this point come under the title of geometric modeling. The next step was feature-based design, in which a solid object is made out of features—such as slots, holes, or protrusions—instead of lines and points. The latest developments in this field involve the melding of CAD with AI (see the next major section), thus creating a knowledge-based CAD system. Recent years have seen increasing integration of solid modeling with manufacturing [CAD/CAM (computer-aided modeling), rapid prototyping] as well as with analysis tools such as finite element software. In the future we can expect this trend to continue and more flexible and user-friendly systems to be developed.

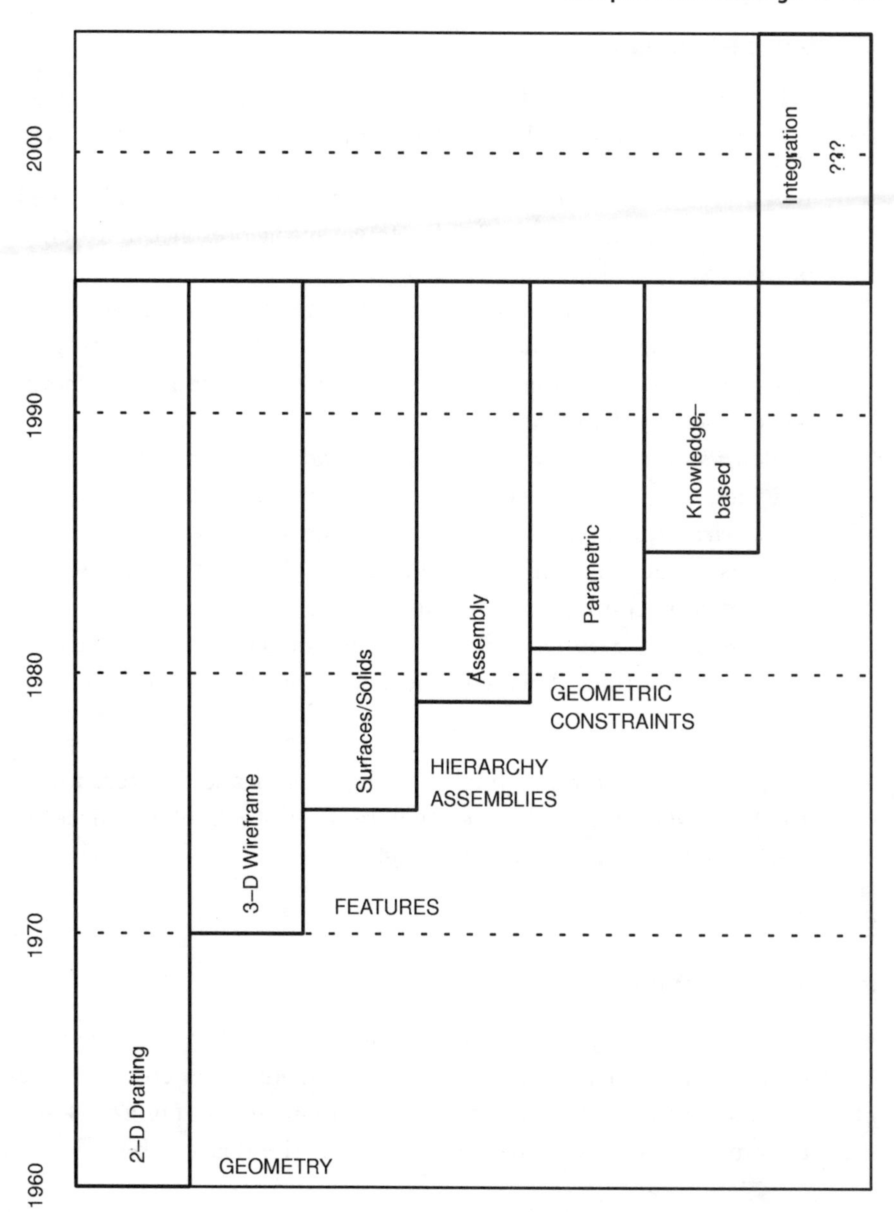

Figure G3. Developments in computer-aided drafting and design.

Geometric Modeling

Geometric modeling includes both 2-D modeling, in which only lines exist (i.e., drawings), and 3-D modeling. Solid objects can be represented by:

1. *Wire frame models.* In wire frame models, all corners (and edges) are shown, and all lines are visible. However, the presence or absence of material is ambiguous.
2. *Solid modeling.* In solid modeling, objects are described by volumes. Material presence is defined unambiguously and the object properties—mass, center of mass, inertia, etc.—can be found. Chief methods of solid modeling are:
   a. *Boundary representation (B-Rep).* Boundary surfaces are created which, when combined, define the volume (object).
   b. *Constructive solid geometry (CSG).* Uses basic elements (primitives)—rectangular solid, cylinder, sphere, etc. Objects are formed by combining elements using Boolean algebra.
   c. *Sweeping (of a surface).* Volume is generated by rotation or translation of a surface. This is the most elementary method of creating a solid.

Many of the 3-D figures in this book, e.g., Figures 81 and 82, were made using the Pro/ENGINEER software (by Parametric Technology Corporation, Waltham, Mass.) for solid modeling.

Parametric Modeling

A parametric modeling system allows the user to enter relationships between different dimensions. Thus, if one dimension is changed, the program changes the related dimensions automatically. In Figure G4, for the four dimensions shown, the user can define the relationship

$$a = b + c + d \tag{G-1}$$

in the program. Thus, only three of these dimensions need to be defined; the fourth is determined automatically.

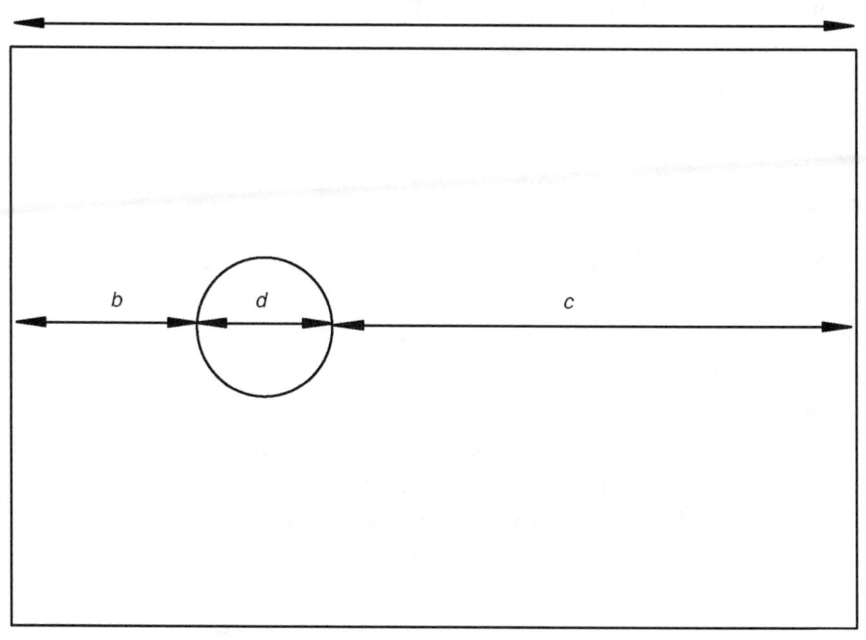

FIGURE G4. A parametric relationship.

KNOWLEDGE-BASED CAD

While parametric modeling produces a feature-based geometric model in which dimensions may be altered, knowledge-based (KB) CAD software allows one to create a true engineering model of the device. Such a model is based on rules which mimic human designers' actions. A human designer starts with the requirements and then uses his or her knowledge, along with information from and about manufacturers' catalogs, previous designs, handbooks, and analyses. The end results of the designer's efforts are drawings and documentation about the design, manufacturing procedures, installation, and operation and maintenance instructions. Knowledge-based CAD software, on the other hand, has in its knowledge base (see under "Artificial Intelligence") rules for creating the geometry and part arrangement, planning the manufacturing, and performing the required analyses. It is able to interrogate databases containing the same type of information that a human designer would use and produce all the needed reports, CAD models, manufacturing procedures, and installation, operation, and maintenance instructions. The initial cost, training, and commitment required for KB-CAD software are significantly greater than for conventional CAD.

Finite Element Analysis

Finite element (FE) analysis is the single most widely used design analysis method in use today, for which a number of different commercial and public-domain software packages are available. FE analysis is used, among other applications, for static and dynamic, linear and nonlinear deflection and stress, buckling, thermal, and electromagnetic problems.

Finite element theory is based upon the minimization of potential energy in a solid continuum. The volume is divided into elements. The apexes of the elements are called *nodes*. The elements can be one-, two-, or three-dimensional. For analyzing realistic mechanical parts, only two- and three-dimensional elements are used.

The degrees of freedom (number of coordinates required to describe the displacement) at a node depends on the number of dimensions. In a 2-D problem, each node has three degrees of freedom—x and y displacements and rotation about the z axis. In a 3-D problem, each node has six degrees of freedom—x, y, and z displacements and rotation about the three axes. The equation for each element relates the geometry, the material properties, and the displacements and forces at each node. The equations for all the elements are then assembled into a master set of equations. Each row in these equations corresponds to a degree of freedom. In the next step the constraints are applied by removing the rows (and columns) corresponding to those degrees of freedom that are fixed. The final set of equations for static analysis is as follows:

$$[\mathbf{K}][\mathbf{x}] = [\mathbf{F}] \tag{G-2}$$

where

$[\mathbf{K}]$ = square stiffness matrix
$[\mathbf{x}]$ = (unknown) displacement vector
$[\mathbf{F}]$ = (known) vector of applied forces and moments

This set of equations is solved to determine the displacements [**x**]. Once these are known, the reactions at the fixed degrees of freedom can be found by using the equations that were removed in the previous step. From the displacements, the program calculates strains and stresses in each element. These can be outputted in both text and graphical format.

• • • • •

EXAMPLE
SOLID MODELING AND FINITE ELEMENT ANALYSIS

Figures G5 and G6 show an example of a model created by using the Pro/ENGINEER software and subsequent stress analysis obtained by using the finite element software ANSYS (by ANSYS, Inc., Houston, Pa.). The example is that of a rectangular plate with a hole at its center. It is a classical elasticity problem for which a closed-form solution is available. It is essential that such problems be used when a new FE program is tried. In this way a check on the numerical answers is possible. Due to the symmetry of the part, only one-quarter of it needs to be simulated. Such a model is shown in Figure G5, which was developed by using Pro/ENGINEER software.

The Pro/ENGINEER program is capable of generating a finite element "mesh" for a solid model, that is, dividing it into elements and assigning element and node numbers. Constraints and applied forces can also be indicated. Figure G6a shows the model of the plate generated in the ANSYS program. The complete plate is subjected to distributed forces at its two vertical edges. This is a plane stress problem; thus, 2-D triangular and quadrilateral elements are used. The two inner edges of the plate are constrained as shown in the figure. The node and element numbers are shown.

Figure G6b shows the displacements after the loads are applied, i.e., after Equation (G-2) is solved. Figure G6c is the plot of stress distribution, and Figure G6d is a zoomed-in picture of

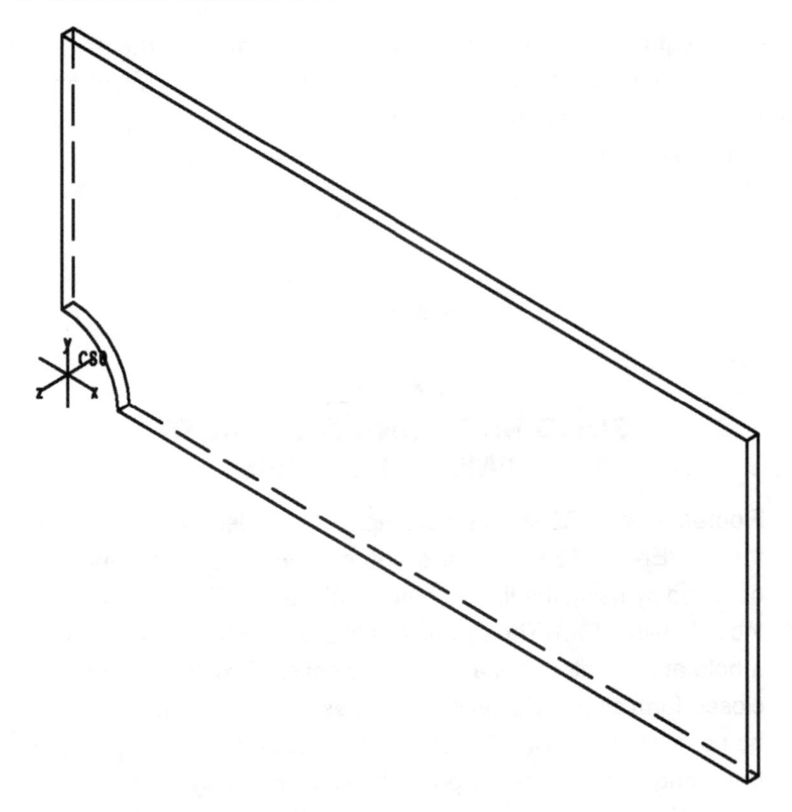

Figure G5. Solid model of a plate with hole—quarter section.

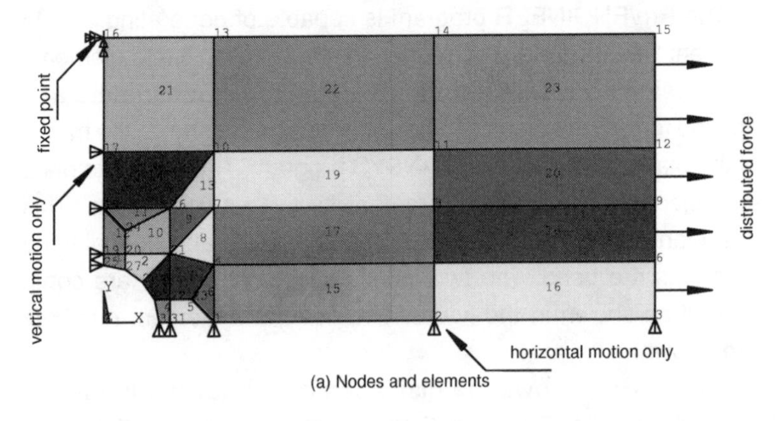

Figure G6. Graphic output from finite element analysis of plate with hole.

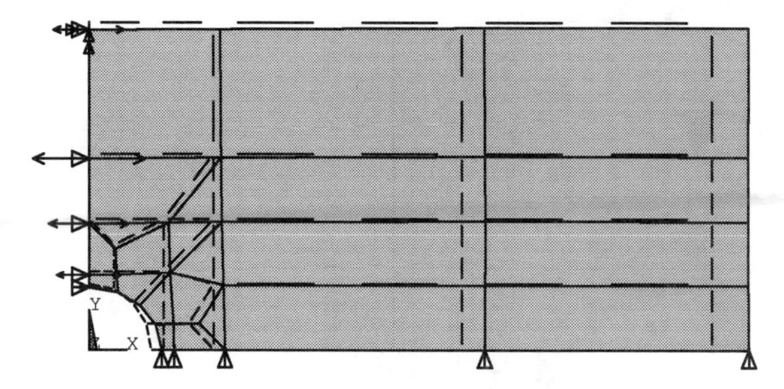

(b) Deformed shape

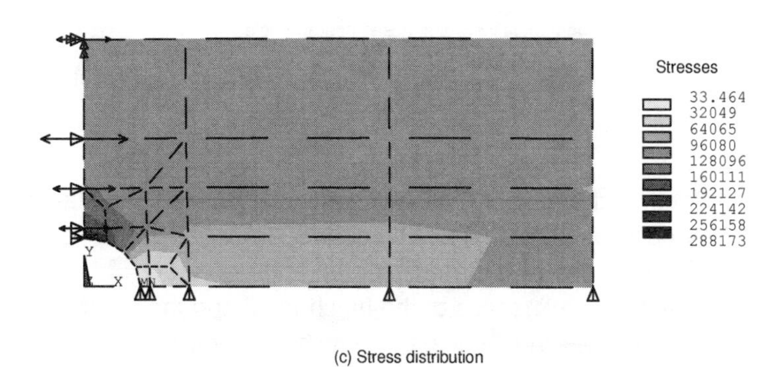

(c) Stress distribution

Figure G6. Graphic output from finite element analysis of plate with hole. (cont.)

the area around the hole. Note that the points of maximum and minimum stress occur around the hole.

The use of finite element software calls for a certain degree of judgment and caution on the part of the novice user. The following points should be considered:

- When starting with a new software program, try a simple problem for which answers are easily found by conventional methods and thus can be checked.
- Before deciding to use a finite element program on a problem, consider solving it by conventional methods—using a closed-form equation from a mechanics textbook or handbook (e.g.,

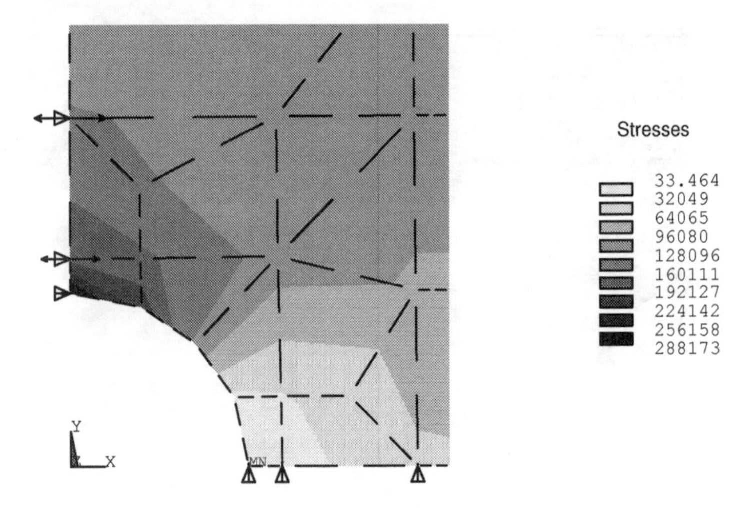

(d) Zoomed–in view of stresses near the hole

FIGURE G6. Graphic output from finite element analysis of plate with hole. (cont.)

Roark and Young, 1975). An equation shows the part played by each parameter. Sometimes the model may need to be simplified before a formula can be applied.

- Selecting the fineness of the mesh used and the number of constraints—and, hence, the number of elements and nodes and the number of degrees of freedom, respectively—requires the use of engineering judgment. A fine mesh is needed where a rapid change in stress is expected. An unnecessarily large number of degrees of freedom requires longer computation time and large memory and may not lead to more accurate results.
- The numerical results obtained from a finite element solution must always be checked by an independent procedure. By using a simplified model and a closed-form equation, one can at least ascertain that the results are in the ballpark.
- Use the finite element program only for the critical portions of the structure, e.g., where stress concentrations are expected.

● ● ● ● ●

ARTIFICIAL INTELLIGENCE

Artificial intelligence is a field of computer science which aims to make computers more intelligent—to make them more closely emulate the human mind with the purpose of making them more useful. A unique characteristic of the human mind is intelligence. Intelligence is the ability to acquire and apply knowledge, to think and reason. Knowledge is what we perceive, discover, understand. Knowledge is more than information. Information is raw data—facts and figures. Knowledge, on the other hand, is made up of information that has been analyzed, codified, or otherwise reformatted into useful concepts and ideas. Intelligence implies the ability to use knowledge. We examine a situation and conceive ways knowledge can be applied to it. Now, in regard to the first word of the title of this section: *artificial* implies imitation or simulation, i.e., human-made, not naturally occurring. In short, AI is the intelligence given to a computer. The study of artificial intelligence has two goals: (1) make computers smarter, so that they will help humans in thought, reasoning, and problem solution; and (2) understand the principles of intelligence—how the human mind works.

Conventional computing, as described earlier, is algorithmic. Algorithms implement virtually all programs. They give computers a form of intelligence. However, the intelligence is fixed for a particular problem. Conventional computing does not have reasoning capability.

AI computing, on the other hand, is symbolic: it uses symbolic programming. Binary words are used to represent symbols (instead of numbers, as in the case of conventional computing). A symbol is anything that represents something else. In AI computing, symbols are used to represent nonnumerical and nontext information: objects, concepts, ideas, facts, judgments, relationships, rules, and other forms of human knowledge. Symbols can be manipulated in various ways. Symbolic information is used to form a knowledge base (KB). Manipulation of the symbolic KB is analogous to thinking and reasoning.

Strategies for using knowledge in symbolic form are search techniques and pattern recognition techniques. Programming languages for using algorithmic techniques—Basic, Fortran, C, Pascal—are called *procedural languages*. In these languages, syntax and structure describe exactly what is to be done: step-by-step details must be given of *what* happens and *when*. Languages for AI computing are LISP and Prolog. These are *declarative languages*. Programs in these languages tell the computer *what* to do, not *how* to do it. It should be noted, however, that AI programs can be written in one of the procedural languages.

Figure G7 shows the essential difference between conventional and AI computing (compare it with Figure G2). The AI program is called the *inference engine*; it uses search and pattern recognition techniques. The data consist of the knowledge base, which has rules, goals, questions, etc., in symbolic form.

Artificial intelligence comprises the following specialties:

1. *General problem solving*, including solving puzzles, playing games, proving mathematical theorems
2. *Expert systems*, used for obtaining and codifying expert knowledge; enabling novices to use this knowledge
3. *Natural language processing*, used for speech or voice recognition, answering questions posed in conversational language, language translation, and acquiring knowledge by reading text
4. *Machine vision*, used for interpreting scenes (maps, work areas, etc.)
5. *Robotics*, for the development of intelligent robots
6. *Education*, for applications such as interactive learning and testing

It should be noted that many applications use more than one of these fields. For example, intelligent robots use machine vision, and computer-aided learning and testing use natural language processing.

Expert Systems

Expert systems (ES) is the largest and most active specialty in the artificial intelligence field. ES software allows the knowledge of experts in a given field to be used by other, less knowledgeable individuals. Experts working in a field acquire knowledge which is empirical, often called "rules of thumb," "tricks of the trade," etc. Such heuristic knowledge is difficult for them to explain and is not generally available in literature. Proven expert systems exist now which address problems such as diagnosing machine defects, locating oil drilling sites, diagnosing infections, designing power plants, preparing insurance policies, choosing and configuring

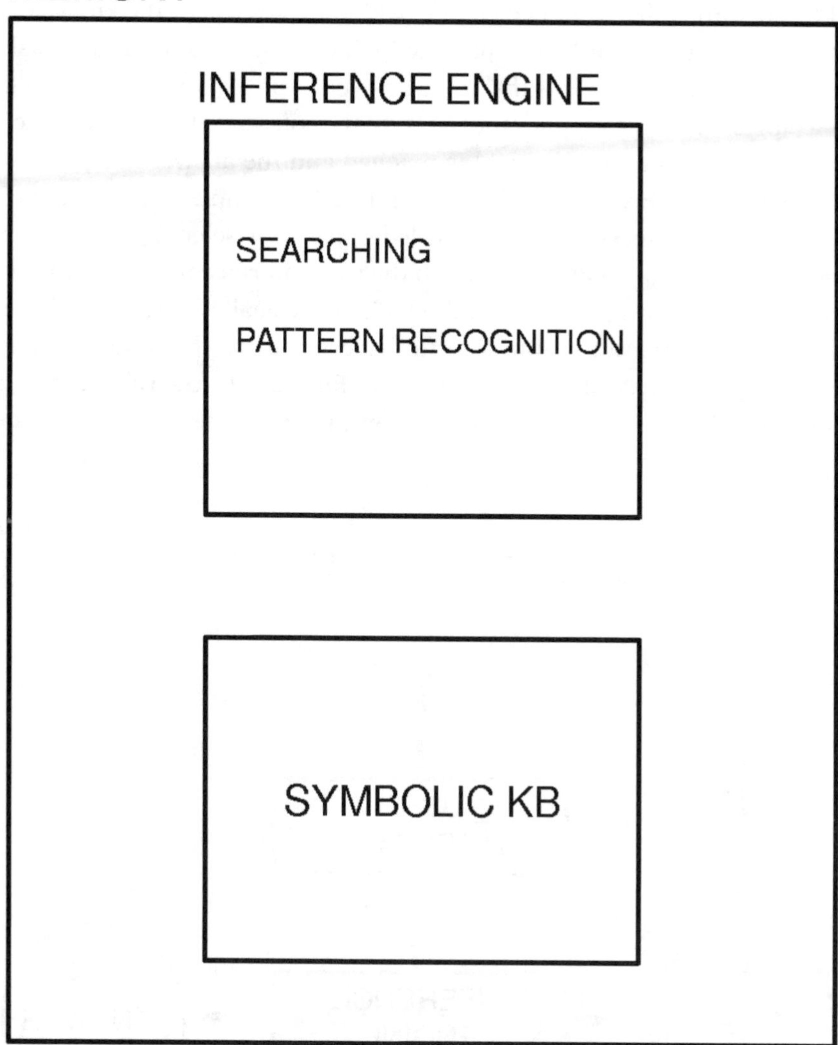

FIGURE G7. The AI computing process.

equipment for a computer room, or identifying chemical compounds from their molecular structure. New expert systems for design and manufacturing applications will be useful aids for product development in the future.

The working of an ES is shown by the block diagram in Figure G8. The program is the inference engine, which makes use of the knowledge in the KB. A user interface is provided which, among other things, asks the user questions, accepts the user's answers, and explains the workings of its logic. The user supplies the "goal" to the software. The user interacts with the program until the goal is reached and the answer is displayed.

The KB contains rules in IF-THEN form. As a simple example, Figure G9a shows an excerpt from a knowledge base for selecting the type of coupling for an application. Through the user interface the program asks the questions contained in the KB, which are typically answered by a yes or a no. The program's aim is to find the value for the `GOAL` variable, in this case `TYPE_COUPLING`. As shown in Figure G9a, the value is found to be equal to `OLDHAM_COUPLING`. The program then prints out the

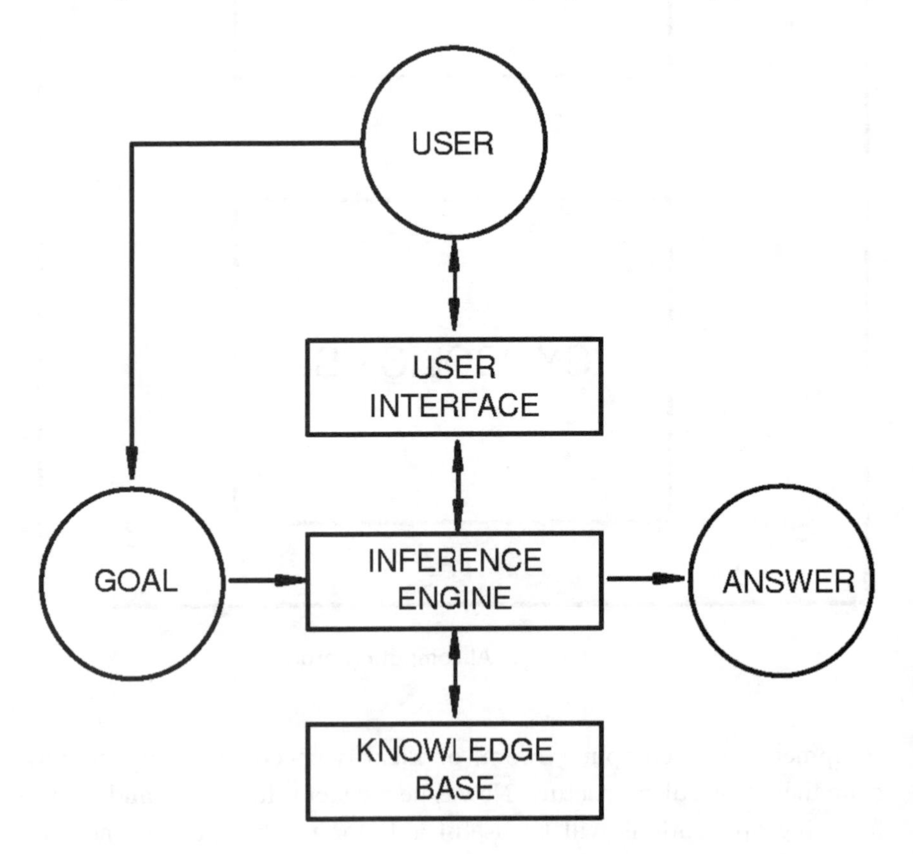

FIGURE G8. Expert system block diagram.

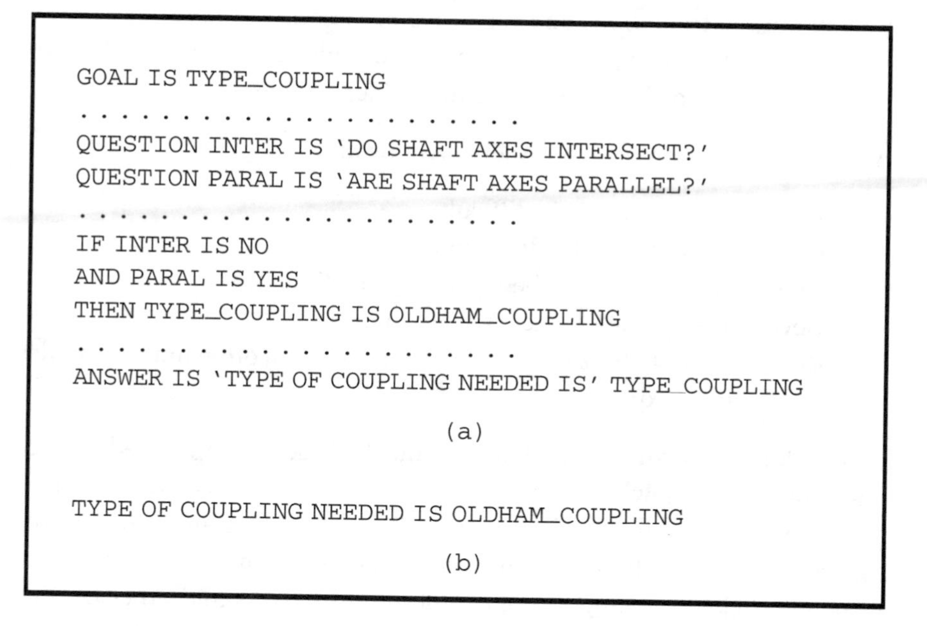

```
GOAL IS TYPE_COUPLING
. . . . . . . . . . . . . . . . . . . . . . .
QUESTION INTER IS 'DO SHAFT AXES INTERSECT?'
QUESTION PARAL IS 'ARE SHAFT AXES PARALLEL?'
. . . . . . . . . . . . . . . . . . . . . . .
IF INTER IS NO
AND PARAL IS YES
THEN TYPE_COUPLING IS OLDHAM_COUPLING
. . . . . . . . . . . . . . . . . . . . . . .
ANSWER IS 'TYPE OF COUPLING NEEDED IS' TYPE_COUPLING
                        (a)

TYPE OF COUPLING NEEDED IS OLDHAM_COUPLING
                        (b)
```

Figure G9. Portion of knowledge base for selection of type of coupling.

answer shown in part b. A typical design knowledge base will have thousands of rules in it.

The most significant challenge in ES development is the preparation of the knowledge base for a given problem. A new specialty called knowledge engineering has arisen for this purpose. After the human expert has been identified, the expert is interviewed by a knowledge engineer. Experts are generally reluctant to share their knowledge. They also find it difficult to explain precisely why and how they reach decisions. Thus, the knowledge engineer must ask questions to elicit knowledge in a systematic way. An advantage of such a question-and-answer session can be that it makes the experts think about why they do what they do. The results of such interrogation are coded into the knowledge base.

Among the benefits of expert systems are:

1. Knowledge and expertise of experts can be captured, which might otherwise be lost upon their retirement and death.
2. Nonexperts can make decisions nearly as good as the experts' own.
3. ES improves productivity and saves time.

4. It automates tedious and repetitive processes.
5. It permits new types of problems to be solved.
6. It makes knowledge available to a wider audience.

Among the disadvantages of expert systems are:

1. Development of a KB is difficult.
2. Good experts are difficult to find.
3. Extracting and coding the knowledge is difficult.
4. Development of a KB incurs high costs.
5. Results are not 100% reliable. The final decision often must be made by a human expert.

A melding of artificial intelligence and CAD software has led to the development of knowledge-based CAD software mentioned earlier. Such software allows capture and retention of the knowledge and expertise of experts. By using software of this type, an engineering department can respond quickly to changes in customer requirements and other conditions. This can be particularly helpful in responding to requests for quotations. The output of knowledge-based CAD software is always based on the latest and best engineering practices.

COMPUTER-AIDED COST ESTIMATION

A large number of computer programs are available for estimating and calculating costs. These vary in scope from calculating the manufacturing cost of a single part, to estimating the costs for large and complex projects. Many of these programs were developed by or for the Department of Defense, NASA, and other U.S. government agencies. The programs, often spreadsheet-based, contain or can access databases of items such as labor and material costs, and manufacturing processes and costs. Some are general-purpose models, whereas others are meant for specific tasks such as gas turbine engines, building projects, or HVAC systems.

It is not possible in this space to discuss or even name all the available software. For cost estimating and calculating during the various life phases of a product, the following are worthy of mention:

- The PRICE (Parametric Review of Information for Costing and Evaluation) software system (from Lockheed Martin PRICE Systems Company) uses company-specific data and expertise for life cycle costing and as an aid for design to cost. It can be used from the product planning stage through design and manufacturing up to operation and maintenance. The software has modules for electronic and mechanical systems, operation and maintenance costs, microcircuits, and software development and maintenance costs.
- HKB, a PC-based program (from Mirakon in St. Gallen, Switzerland), calculates the cost of machine parts (Ferreirinha et al., 1990, 1993). The interactive program requires as input the product description and manufacturing process description. The program determines the necessary processes, costs, and times for the different operations and prepares the cost structures and other documents. It is applicable at the various stages of the product realization process.
- ACEIT (Automated Cost Estimating Integrated Tools) is a comprehensive cost estimating software package developed for the Department of Defense (from Tecolote Research, Santa Barbara, California).
- BBEST, "black-box estimator" (from Tecolote Research, Santa Barbara, California), is an automated ACEIT model for estimating development and production costs of electronic systems. The user inputs as much data as available—the program assumes the rest, thus making it useful at the concept stage.

• • • • •

EXAMPLE
NASA AIRFRAME COST MODEL

Table G1 shows the results from an interactive on-line parametric model from NASA, available on the World Wide Web. It is suitable for use during the conceptual stage when detailed information is not available. As cost estimating relationships (CERs), the model uses nonrecurring engineering and tooling, development support, flight test, recurring engineering, tooling, manufacturing labor,

TABLE G1 Airframe Cost Calculation

Input Data		
Aircraft empty weight (lb)		100,000
Maximum speed (knots)		400
Number of flight test aircraft		2
Production quantity		200
Results		
	Hours	Cost ($M 96)
Nonrecurring costs:		
Engineering	11,014	1133
Tooling	6700	602
Development support		244
Flight test		26
Subtotal nonrecurring	17,714	2005
Recurring costs:		
Engineering	8831	908
Tooling	4925	443
Manufacturing	45,146	3799
Material		1600
Quality assurance	8404	702
Subtotal recurring	67,306	7452
Total	85,020	9457

manufacturing material, and quality assurance activities. The airframe cost includes wings, fuselage, empennage, nacelles, the air induction system, starters, exhausts, fuel control system, inlet control system, landing gear, secondary power, furnishings, environmental control, racks, mounts, intersystem cables and distribution boxes, and the integration and installation of the propulsion, avionics, and armament subsystems into the airframe.

This parametric model was developed from the data for 13 military aircraft launched in the 1960s and 1970s. These aircraft range in weight from 10,000 to 300,000 lb and in speed from 400 to 1300 knots.

● ● ● ● ●

SUMMARY

Computers have found extensive use in design activities. Thus far, however, the software for supporting the design process is limited in its application area to individual steps, chiefly for embodiment and detail design. Computers find the least use in conceptual design, and the most use in detail design. Exceptions to this are narrowly focused areas such as the conceptual design of kinematic chains, for which commercial software is available. A necessary prerequisite for computer-aided conceptual design is a new model of the design process, since the present models are suited only in a limited way for utilization in the computer. This circumstance is due to two important points. For one thing, the models that are used in the different design phases are not compatible with each other because of their origins from different scientific areas. A model therefore cannot be carried over directly from one design phase to the next. The second point is that the methods used during designing can be represented on the computer largely only under severe limitations, since the essential parts of activity are done by the creative designer. The job of such a model is to represent the relevant properties of a product as accurately as required. All developmental steps during designing relate to the same object—the product being designed—only in differing measures of its properties.

The phases of the design process are connected with each other by means of certain operational principles. For example, in the modeling phase of ascertaining functions and their structures, the functional relationship between the main inputs and outputs of a product are represented. In the next phase, the search for solution principles and their structure, the principal methods to find solution principles for subfunctions and realization of main function are described. For the computer-aided design task, there is a problem because of the different qualities of the information to be stored. On the one hand, in conceptual design, we are dealing with data and information on nonmaterial objects, such as the requirements list, the function structure, and the solution structure. On the other hand, in embodiment and detail design, we have to store real, material objects such as assemblies and parts. The development of comprehensive software for creating original concepts will have to await solution of these problems.

BIBLIOGRAPHY

Akin, J. E. *Computer-Assisted Mechanical Design*. Englewood Cliffs, N.J.: Prentice Hall, 1990. A comprehensive text on mechanical CAD techniques.

Chapra, S. C., and R. P. Canale. *Numerical Methods for Engineers*. New York: McGraw-Hill, 1985. A discussion of numerical methods—e.g., equation solving, integration, regression—with examples from different engineering fields.

Dym, C. L., and R. E. Levitt. *Knowledge-Based Systems in Engineering*. New York: McGraw-Hill, 1991.

Etter, D. M. *Engineering Problem Solving with MATLAB*. Englewood Cliffs, N.J.: Prentice Hall, 1993. Describes the use of an equation solver.

Ferreirinha, P. "Rechnerunterstützte Vorkalkulation im Maschinenbau für Konstrukteure und Arbeitsvorbereiter mit 'HKB' " (Computer-Aided Pre-Calculation in Machine Industry for Designers and Process Planners with "HKB"). *Proc. ICED 90*, 3: 1346–1353. Zurich: Heurista, 1990.

Ferreirinha, P., V. Hubka, and W. E. Eder. "Early Cost Calculation—Reliable Calculation, Not Just Estimation." *Design for Manufacturability 1993*, DE 52: 97–104. New York: ASME, 1993. This and the preceding paper discuss the HKB manufacturing cost estimating program.

Knight, C. E. *The Finite-Element Method in Mechanical Design*. Boston: P.W.S.-Kent, 1993. Good introductory book, with a program diskette.

Medland, A. J. *The Computer-Based Design Process*. London and New York: Chapman & Hall, 1992.

NASA Parametric Cost Estimating Reference Manual: Airframe Cost Model. World Wide Web
http://www.jsc.nasa.gov/bu2/airframe.html
List of a number of cost models and software.

National Research Council. *Computer-Aided Materials Selection during Structural Design*. Washington, D.C.: National Academy Press, 1995.

Pham, D. T., ed. *Expert Systems in Engineering*. Berlin and New York: Springer Verlag, 1988.

Roark, R. J., and W. C. Young. *Formulas for Stress and Strain*. 5th ed. New York: McGraw-Hill, 1975.

Rychener, M. D., ed. *Expert Systems for Engineering Design*. Boston: Academic Press, 1988.

Stewart, R. D., R. M. Wyskida, and J. D. Johannes, eds. *Cost Estimator's Reference Manual*. 2nd ed. New York: Wiley, 1995. This book has 19 chapters by various authors on cost topics—e.g., ABC, learning curves, parametric costing, design to cost, computer-aided cost estimating.

Tront, J. *Problem Solving with Borland's EUREKA*. New York: Wiley, 1987. Describes the use of an equation solver.

Waldron, M. B., and K. J. Waldron. *Mechanical Design: Theory and Methodology*. New York: Springer, 1996.

Wierda, L. S. "Linking Design, Process Planning and Cost Information by Feature-Based Modelling." *J. Engineering Design* 2(1): 3–20 (1991). Discusses feature-based design and the relationship of features to process planning, manufacturability, and costs.

Zeid, I. *CAD/CAM Theory and Practice*. New York: McGraw-Hill, 1991. A thorough treatment of CAD.

Regression Analysis

When data are obtained by observation, they are discrete in nature: the value of a dependent variable is known for a certain finite number of values of the independent variable. In order to use such data in mathematical models, curve fitting is necessary. Curve fitting yields a mathematical relationship from which the dependent variable may be calculated for any given value of the independent variable. Thus, we are able to reduce data from several observations—even hundreds—to an equation fully defined by one, two, or a few coefficients or exponents. The following discussion is of an introductory nature. Computer software (e.g., SAS) is available for solution of equations and general statistical analyses.

Let us consider the data given in Table H1. These are plotted in Figure H1 and are the relative manufacturing costs C_{rel} of a certain type of turbine, as a function of its power output in kW, normalized to that of a 50-kW turbine.

These points appear to lie approximately along a straight line. We could, by eye, draw a straight line which passes through most of the points, or one which has an equal number of points lying on either side of it. The equation of such a straight line would be of the form

$$C_{rel} = b + mP \tag{H-1}$$

TABLE H1 Relative Manufacturing Cost of a Power Turbine

Power, kW	Relative Manufacturing Cost
50	1.0
60	1.15
70	1.45
80	1.8
90	2.2
100	3.0

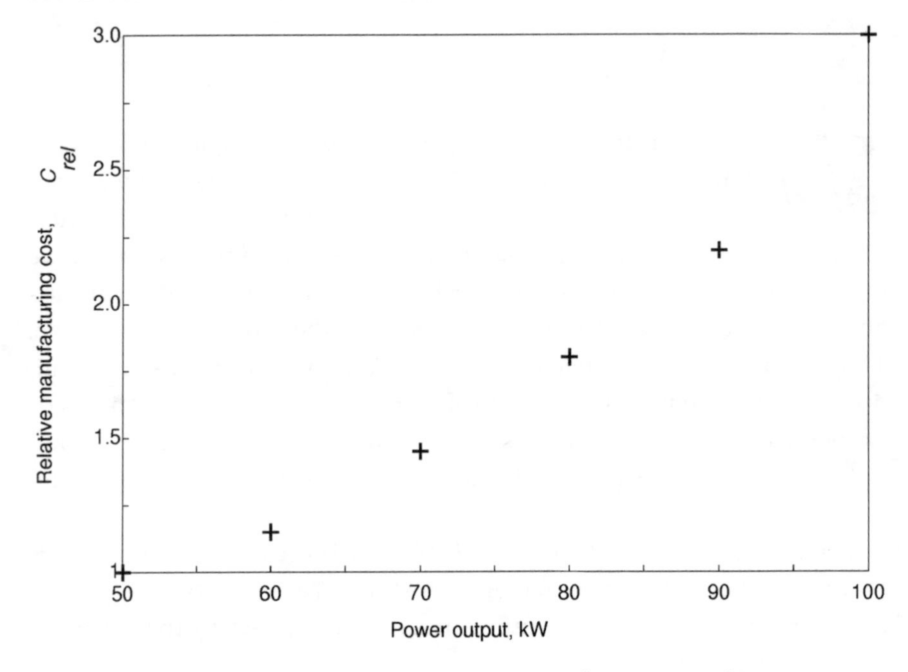

FIGURE H1. Relative manufacturing cost of a power turbine.

where C_{rel} is the relative manufacturing cost and P the power in kW. The constant b is the intercept of the line on the abscissa, and m is its slope.

LINEAR REGRESSION

The problem of finding the constants b and m for a straight line that provides the "best fit" to the data is called *linear regression*. The most common method of finding the best fit straight line is the least squares technique, which yields the straight line for which the sum of the squares of the errors between it and the data points is the minimum. The procedure will be illustrated with the aid of Figure H2.

Let there be n data points (x_i, y_i), $i = 1, 2, \ldots, n$. The equation for the line is

$$y = b + mx \tag{H-2}$$

for which we need to find b and m. The error ϵ_i at data point (x_i, y_i) is the vertical difference between the data point and the line; i.e.,

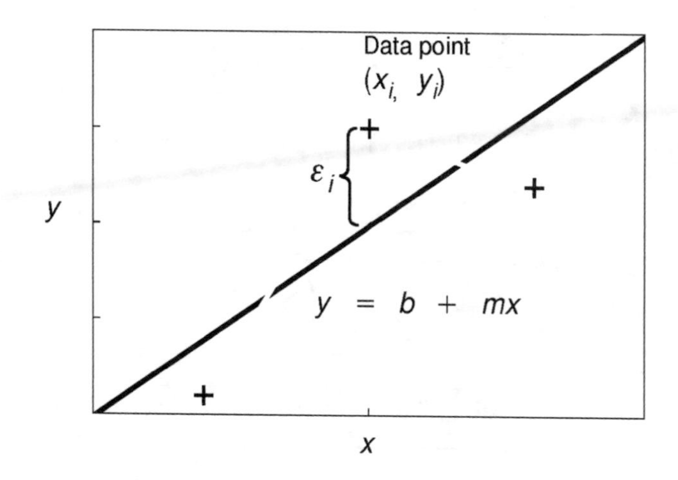

FIGURE H2. Fitting a straight line to data points.

$$\epsilon_i = y_i - (b + mx_i) \tag{H-3}$$

The quantity in parentheses in Equation (H-3) is the value of y at $x = x_i$. The sum of the squared errors is

$$E = \sum_i \epsilon_i^2 = \sum_i [y_i - (b + mx_i)]^2 \tag{H-4}$$

The summations are for $i = 1, 2, \ldots, n$. The minimum value of E is found by setting its derivatives with respect to b and m equal to zero:

$$\frac{dE}{db} = 0 \qquad \frac{dE}{dm} = 0 \tag{H-5}$$

After carrying through this mathematics on the expression in Equation (H-4), we obtain the following equations:

$$bn + m \sum_i x_i = \sum_i y_i \tag{H-6}$$

$$b \sum_i x_i + m \sum_i x_i^2 = \sum_i y_i x_i \tag{H-7}$$

Solving Equations (H-6) and (H-7) for b and m, we get the results:

$$b = \frac{\sum_i x_i^2 \sum_i y_i - \sum_i x_i \sum_i y_i x_i}{n \sum_i x_i^2 - \left(\sum_i x_i\right)^2} \tag{H-8}$$

$$m = \frac{n \sum_i x_i y_i - \sum_i x_i \sum_i y_i}{n \sum_i x_i^2 - \left(\sum_i x_i\right)^2} \tag{H-9}$$

After performing this arithmetic on the data in Table H1, we find the values of b and m in Equation (H-1) to be $b = -1.126$ and $m = 0.0386$; i.e.,

$$C_{\text{rel}} = -1.126 + 0.0386P \tag{H-10}$$

The line given by Equation (H-10) is shown in Figure H3, along with the data points. It is the "best" straight line fit to the data points, but as we can see it underestimates the data at the lower and upper levels and overestimates it in the middle.

There are remedies to improve the fit. One possibility is to fit two straight lines to the data—one, say, from 50 to 80 kW and the other from 80 to 100 kW. Another way to obtain an improved fit to the data points is to use a function other than a straight line, as described in the next sections.

GENERAL CURVE FITTING

The method of least squares fit is not limited to fitting straight lines to data points. We can fit a general curve of the form

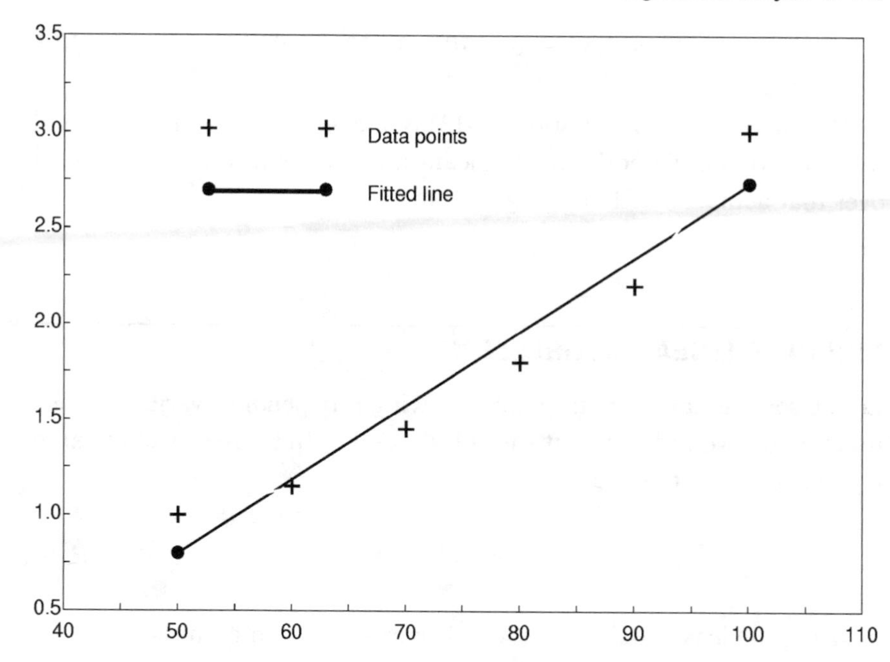

FIGURE H3. Linear regression line for turbine cost data points.

$$y = c_1 g_1(x) + c_2 g_2(x) + c_3 g_3(x) + \cdots \tag{H-11}$$

where the $g_i(x)$ are of one kind among many types of functions that may be used—polynomial, exponential, sinusoidal, etc. The coefficients c_i are determined in exactly the same manner as for linear regression. Equation (H-2) is a special form of Equation (H-11) where $g_1(x) = 1$ and $g_2(x) = x$.

The data points for the turbine cost in Figure H1 actually appear to lie more along a parabolic type of curve than a straight line. Let us use the following curve:

$$C_{\text{rel}} = c_1 + c_2 P + c_3 P^2 \tag{H-12}$$

Upon performing the least squares curve fit, where we must now solve the three equations in the three unknowns c_1, c_2, and c_3, we obtain the following:

$$C_{\text{rel}} = 2.35 - 0.059P + (6.52 \times 10^{-4})P^2 \qquad \text{(H-13)}$$

The curve given by Equation (H-13) is shown in Figure H4 along with the data points. We notice a significant improvement in the degree of fit over that achieved by linear regression.

MULTIPLE LINEAR REGRESSION

Let us say we have n data points showing a dependent variable y_i as a function of two independent variables x_i and z_i. In such a case we can fit a plane given by the equation

$$y = c + lx + mz \qquad \text{(H-14)}$$

by using the least squares method. The error at each point is

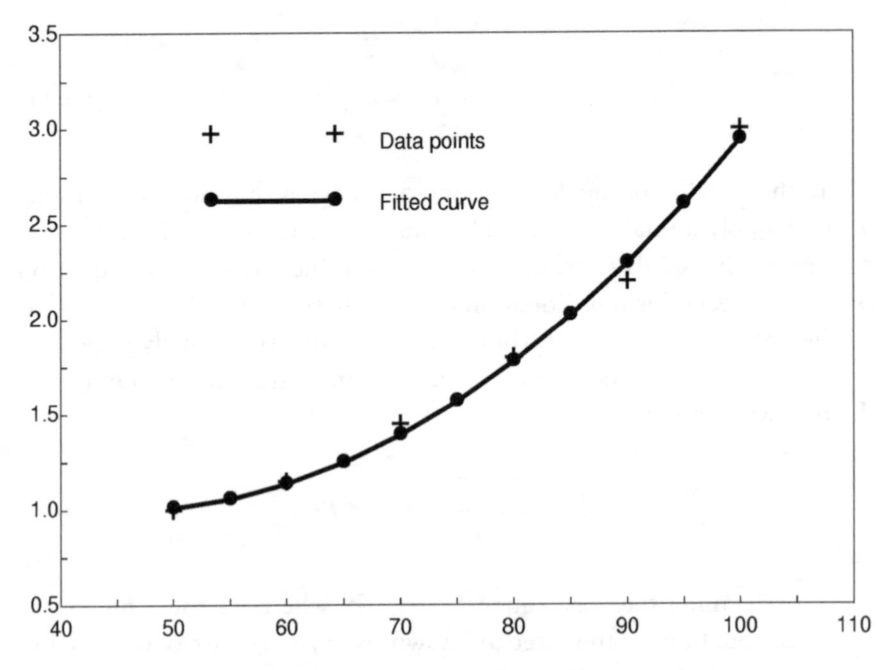

FIGURE H4. Parabolic curve fit for turbine cost data.

$$\epsilon_i = y_i - (c + lx_i + mz_i) \tag{H-15}$$

The sum of the squared errors is

$$E = \sum_i \epsilon_i^2 = \sum_i [y_i - (c + lx_i + mz_i)]^2 \tag{H-16}$$

The minimum value of E is found by setting its derivatives with respect to c, l, and m equal to zero:

$$\frac{dE}{dc} = 0 \qquad \frac{dE}{dl} = 0 \qquad \frac{dE}{dm} = 0 \tag{H-17}$$

which yields the following equations in c, l, and m:

$$cn + l \sum_i x_i + m \sum_i z_i = \sum_i y_i \tag{H-18}$$

$$c \sum_i x_i + l \sum_i x_i^2 + m \sum_i x_i z_i = \sum_i y_i x_i \tag{H-19}$$

$$c \sum_i z_i + l \sum_i x_i z_i + m \sum_i z_i^2 = \sum_i y_i z_i \tag{H-20}$$

After substituting the values of the data points (x_i, y_i, z_i), these three equations can be solved for c, l, and m.

GENERAL PRODUCT FUNCTION

A function of the form

$$y = bx^l z^m w^n \cdots \tag{H-21}$$

can be used for fitting to data points where values of y_i are given for the corresponding values of (x_i, z_i, w_i). Curve fitting by using this type of function is called *nonlinear regression*.

By taking the logarithms of both sides, Equation (H-21) becomes:

$$\log y = \log b + l \log x + m \log z + n \log w \quad \text{(H-22)}$$

Upon defining

$$Y = \log y \qquad B = \log b \qquad X = \log x \qquad Z = \log z \qquad W = \log w \quad \text{(H-23)}$$

Equation (H-22) can be written as

$$Y = B + lX + mZ + nW + \cdots \quad \text{(H-24)}$$

This is a linear equation for which we can determine the values of B, l, m, and w by applying the least squares method.

SUMMARY

We have seen how regression analysis can be used for developing parametric cost models. We looked at linear regression, general curve fitting, multiple linear regression, and general product functions. Costs are functions of product variables such as output, geometric size, and weight. We can show these relationships with the aid of graphs, or express them in equation form. Regression equations can be used with confidence as long as we stay within the limits of the data used to find the coefficients and exponents. Extrapolating outside the domain of observations is fraught with uncertainty.

BIBLIOGRAPHY

Chapra, S. C., and R. P. Canale. *Numerical Methods for Engineers*. New York: McGraw-Hill, 1985. A discussion of numerical methods—e.g., equation solving, integration, regression—with examples from different engineering fields.

Ertas, A., and J. C. Jones. *The Engineering Design Process*. New York: Wiley, 1993. Has chapters on optimization methods and statistical methods, including regression analysis.

Pacyna, H., A. Hillebrand, and A. Rutz. "Kostenfrüherkennung für Gussteile" (Early Cost Estimation for Cast Parts). VDI Berichte *Konstrukteure Senken Herstellkosten—Methoden und Hilfen* No. 457. Düsseldorf: VDI Verlag, 1982. Shows application of the general product function for estimating casting costs, e.g., Equation (52) in Chapter IX, "Manufacturing Processes and Economics."

Mathematical Methods for Optimization

The ultimate aim of all engineering design is to achieve an optimum design of the systems that are created. The optimization may be with respect to:

- Functions of the system
- A parameter of the system—e.g., weight or cost

Many engineering problems can be formulated to the point where the behavior of a system can be expressed in terms of a mathematical equation. It is at this point that the methods discussed in this section apply. We begin with an example, to show how optimization problems arise—and at what stage of the design process we might need to solve such problems.

• • • • •

EXAMPLE
DESIGN OF A LIQUID CONTAINER

Johnson (1978) has discussed the design of a container for liquids (e.g., beverages). This is one of several useful and practical optimization problems in his book. Johnson's message can be summarized as follows: before applying the mathematical schemes, study the system behavior—how do the different parameters affect the quantity being optimized?

We begin the design process by developing the requirements list. Using the categories suggested in Chapter III, this list will include items such as:

Category	Items to Be Considered or Specified
Geometry	Linear size, capacity
Material	See Chapter VIII
Safety	User safety
Ergonomics	Ease of handling, opening

Production	Quantities required; see Chapter IX
Storage	Ease and efficiency
Operation	Ease of opening, closing
Costs	Consider the market competition
Recycling	Recyclability

In a problem such as this the solution principle is already established; thus, there is little to do in the way of concept development.

The first major decision we need to make in the process is during embodiment, for selecting the shape. The container can have one of a number of shapes—cylindrical, rectangular, etc.; some shapes are shown in Figure I1.

Let us say that the ergonomic requirements are dominant and that the cylindrical shape is chosen for its ease of handling. The capacity of the container is determined by market studies. It is now desired to find the radius and height of the container such that it has the minimum amount of surface area. The surface area of the can is related to the amount of material used and the rate of heat transfer—both quantities of concern to the designer.

Let:

V = given volume of the container

H = its height

R = its radius

A = its surface area

The optimization problem is: Find H and R such that A is a minimum, for a given V. Thus, we need to minimize

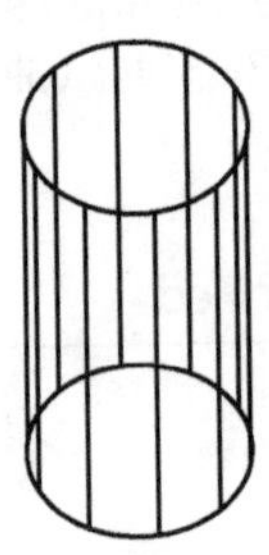
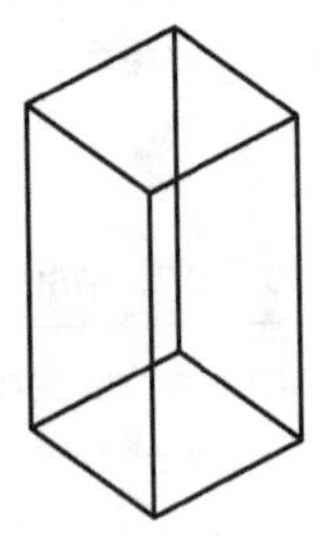
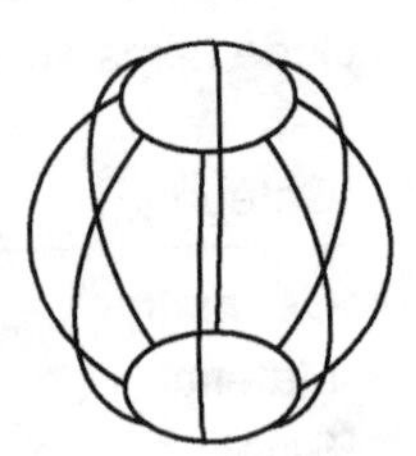

FIGURE I1. Possible shapes of the container.

$$A = 2\pi R^2 + 2\pi RH \tag{I-1}$$

subject to

$$V = \pi R^2 H = \text{constant} \tag{I-2}$$

There are other constraints on the size of the can:

$$0 \leq R \leq R_{max} \tag{I-3}$$

from an ergonomic viewpoint, and

$$0 \leq H \leq H_{max} \tag{I-4}$$

due to storage space limitation.

Since there are two variables, H and R, in Equation (I-1), we can eliminate one of these with the help of Equation (I-2). Let us assume for the moment that H is unconstrained, and eliminate it. We obtain

$$A = 2\pi R^2 + \frac{2V}{R} \tag{I-5}$$

In order to minimize A, we take its derivative with respect to R and set it equal to zero:

$$\frac{dA}{dR} = 4\pi R + \frac{2V}{R^2} = 0 \tag{I-6}$$

Solving Equation (I-6) for R gives

$$R = R_{opt} = \left(\frac{V}{2\pi}\right)^{1/3} \quad \text{for } A = A_{min} \tag{I-7}$$

As a side note, in order to *optimize*, we need to *minimize* or *maximize* some parameter(s). These two verbs can be combined into one word: *extremize*. We also note that the *optimum* value of a component or a measurement is not necessarily its *minimum* or *maximum* value. Consider the pipe optimization example in Figure 111 (Chapter VIII). In that example we wished to minimize the cost. In order to do so, we determined the optimum value of the pipe diameter and found it to be neither the largest nor the smallest possible value. In the present example, in order to minimize A, we found the optimum value of R.

Let us look at some plots. It always provides insight if we see the dependence of one variable upon another. Figure I2 shows the plot of the surface area of the can as a function of its radius, a plot of Equation (I-5).

The plot in Figure I2 provides the R_{opt} for A_{min} as long as R is unconstrained. If we apply the constraint given by Equation (I-3), we may have one of the possibilities shown in Figure I3. If R_{max} is greater than the value of R for which A is a minimum, the optimum value of R is as before, as if R were unconstrained, as shown in the graph in Figure I3a. As indicated in the figure, the domain of feasible solutions includes the unconstrained minimum.

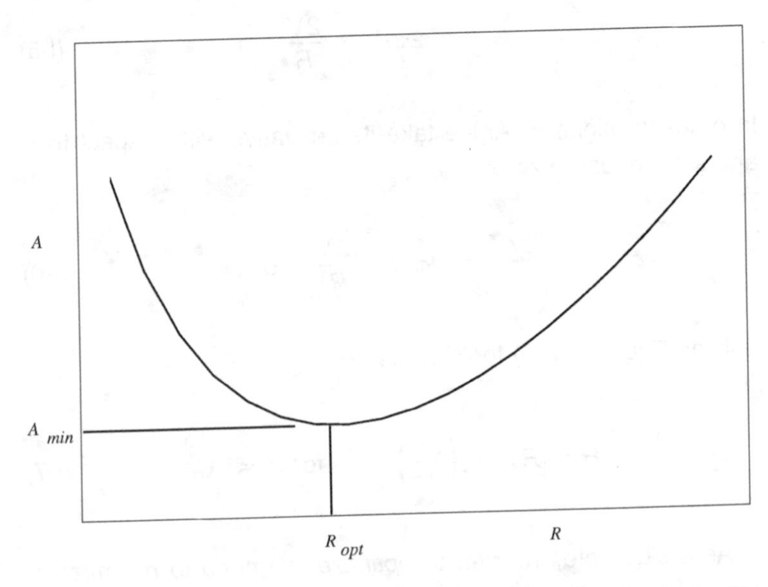

FIGURE I2. Can surface area versus radius.

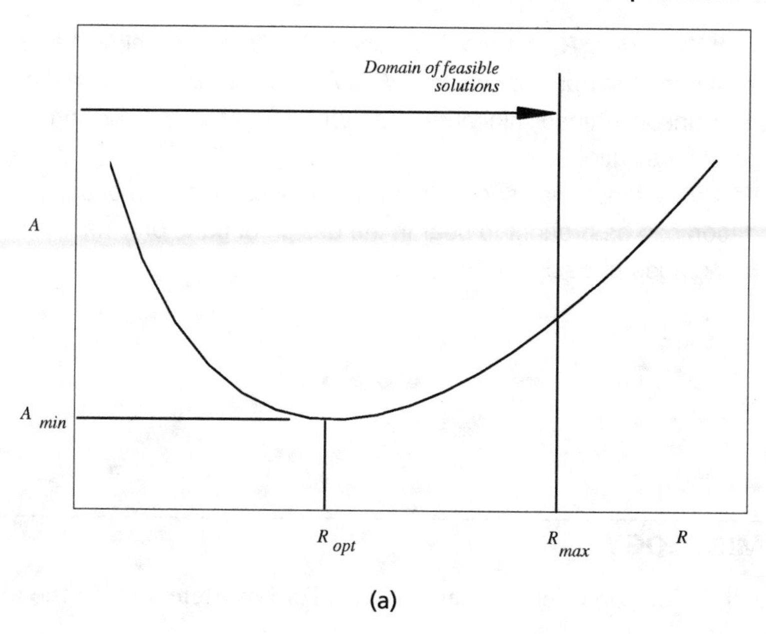

(a)

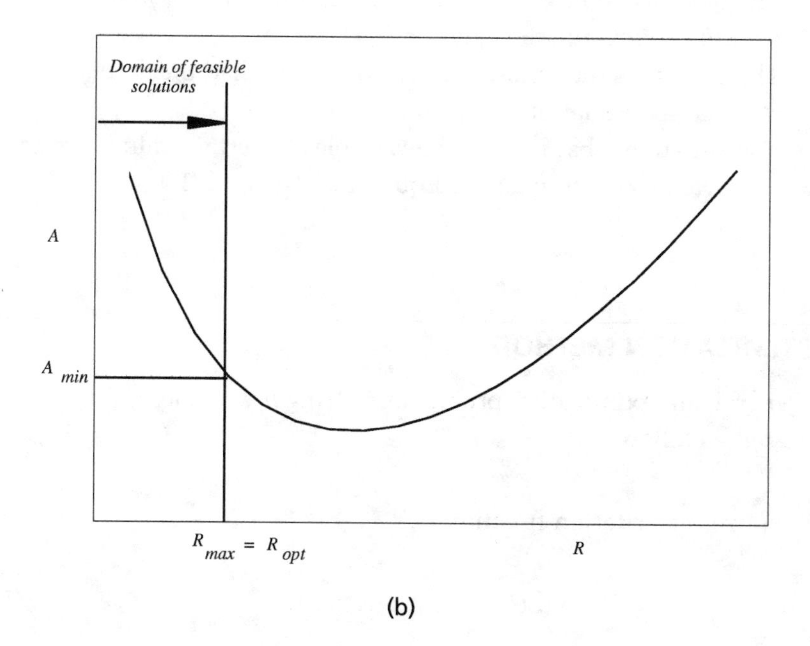

(b)

FIGURE 13. Minimum can surface area, with constraint on radius.

If, however, R_{max} is less than the value of R for which A is a minimum, the optimum value of R is R_{max}, as shown in Figure I3b. The unconstrained minimum is now no longer in the domain of feasible solutions.

This example has shown that mathematical methods of optimization can be brought to bear in the design of parts only near the last stages of design.

● ● ● ● ●

TERMINOLOGY

We will define some terms used in optimization. Referring to the above example:

- The quantity to be extremized is called the *criterion function* or the *objective function*; see Equation (I-1).
- The equations that relate the variables are called the *functional constraints*; see Equation (I-2).
- The equations that restrict the variables to certain values are called the *regional constraints*; see Equations (I-3) and (I-4).

OPTIMIZATION METHODS

In general, an optimization problem involving n variables $(x_1, x_2, \ldots, x_n)$ is stated as follows:

Extremize the criterion function U:

$$U = U(x_1, x_2, \ldots, x_n) \tag{I-8}$$

subject to the functional constraints

$$\begin{aligned}
\psi_1 &= \psi_1(x_1, x_2, \ldots, x_n) = 0 \\
\psi_2 &= \psi_2(x_1, x_2, \ldots, x_n) = 0 \\
&\cdots\cdots\cdots\cdots\cdots\cdots \\
\psi_m &= \psi_m(x_1, x_2, \ldots, x_n) = 0
\end{aligned} \tag{I-9}$$

and subject to the regional constraints

$$\begin{aligned}
\phi_1 &= \phi_1(x_1, x_2, \ldots, x_n) \leq L_1 \\
\phi_2 &= \phi_2(x_1, x_2, \ldots, x_n) \leq L_2 \\
&\cdots\cdots\cdots\cdots\cdots\cdots \\
\phi_p &= \phi_p(x_1, x_2, \ldots, x_n) \leq L_p
\end{aligned} \tag{I-10}$$

where the L_i are constants. The problem is then as follows: Find the values of the variables $(x_1, x_2, \ldots, x_n)$ such that U in Equation (I-8) is extremized, subject to the conditions in Equations (I-9) and (I-10).

FUNCTIONS OF ONE VARIABLE

We wish to find the extremum of $U = U(x)$. This case arises if there was one variable to begin with or the number of variables has been reduced to 1 by using the functional constraint equations. If there are no regional constraints, we set $\Delta U = 0$, i.e., its derivative with respect to x equal to zero:

$$\frac{dU}{dx} = 0 \tag{I-11}$$

which yields an equation in x. Its solution is a possible extremum of U. In order to check whether it is a minimum or a maximum we check the sign of the second derivative of U with respect to x.

Solution of an equation like Equation (I-11) is not difficult, since it contains only one variable. If a closed-form solution is not available, as was the case with Equation (I-6), it can be solved by an iterative technique such as the half-interval search, the secant method, or Newton's method. Textbooks on numerical methods, such as Chapra and Canale (1985), describe many such methods, and also provide computer programs for their implementation.

Functions of More Than One Variable

Let U be a function of two variables x and y; that is, we wish to extremize $U = U(x, y)$. For this also we set $\Delta U = 0$, i.e.,

$$\frac{\partial U}{\partial x}\Delta x + \frac{\partial U}{\partial y}\Delta y = 0 \qquad \text{(I-12)}$$

Now, if x and y are considered to be independent, we can set the coefficients of Δx and Δy separately equal to zero, which gives us two equations corresponding to (I-11):

$$\frac{\partial U}{\partial x} = 0 \qquad \text{and} \qquad \frac{\partial U}{\partial y} = 0 \qquad \text{(I-13)}$$

Carrying out these differentiations yields two equations in two unknowns, x and y. Thus, in general, to extremize $U = U(x_1, x_2, \ldots, x_n)$, if there are no functional or regional constraints, we set $\Delta U = 0$, which yields n equations:

$$\frac{\partial U}{\partial x_i} = 0 \qquad i = 1, 2, \ldots, n \qquad \text{(I-14)}$$

In mechanical design problems, equations such as (I-13), or, in general, equation set (I-14), tend to be nonlinear. Eliminating variables in order to reduce the number of equations is not always feasible, especially if the equations are complex. They can ordinarily be solved only by an iterative numerical scheme (Chapra and Canale, 1985). Commercial software packages are available for solving such equations, and indeed for solving extremum problems as defined by Equations (I-8), (I-9), and (I-10).

Method of Lagrangian Multipliers

The use of Lagrange multipliers is a means of avoiding the substitution of functional constraint equations in the objective function in order to reduce the number of variables. For the purpose of illustrating this method, let us consider a function of three variables:

Extremize the criterion function U:

$$U = U(x_1, x_2, x_3) \tag{I-15}$$

with two functional constraints

$$\psi_1 = \psi_1(x_1, x_2, x_3) = 0 \tag{I-16}$$

$$\psi_2 = \psi_2(x_1, x_2, x_3) = 0 \tag{I-17}$$

Theoretically it is possible to use Equations (I-16) and (I-17) and eliminate two of the variables from the expression for U, Equation (I-15). Instead, we form the Lagrangian L as follows:

$$L = U(x_1, x_2, x_3) + \lambda_1\psi_1(x_1, x_2, x_3) + \lambda_2\psi_2(x_1, x_2, x_3) \tag{I-18}$$

where λ_1 and λ_2 are called the Lagrange multipliers. We now extremize the Lagrangian L, instead of the criterion function U. We can justify this by saying that since ψ_1 and ψ_2 are each zero, as defined by Equations (I-16) and (I-17), extremizing L is the same as extremizing U.

Thus, we proceed as before and set $\Delta L = 0$; i.e.,

$$\Delta L = \Delta U + \lambda_1\,\Delta\psi_1 + \lambda_2\,\Delta\psi_2 = 0 \tag{I-19}$$

or

$$\left(\frac{\partial U}{\partial x_1} + \lambda_1\frac{\partial \psi_1}{\partial x_1} + \lambda_2\frac{\partial \psi_2}{\partial x_1}\right)\Delta x_1 + \left(\frac{\partial U}{\partial x_2} + \lambda_1\frac{\partial \psi_1}{\partial x_2} + \lambda_2\frac{\partial \psi_2}{\partial x_2}\right)\Delta x_2$$
$$+ \left(\frac{\partial U}{\partial x_3} + \lambda_1\frac{\partial \psi_1}{\partial x_3} + \lambda_2\frac{\partial \psi_2}{\partial x_3}\right)\Delta x_3 = 0 \tag{I-20}$$

Now, setting the coefficients of Δx_1, Δx_2, and Δx_3 in Equation (I-20) separately equal to zero gives us three equations:

$$\frac{\partial U}{\partial x_1} + \lambda_1 \frac{\partial \psi_1}{\partial x_1} + \lambda_2 \frac{\partial \psi_2}{\partial x_1} = 0$$

$$\frac{\partial U}{\partial x_2} + \lambda_1 \frac{\partial \psi_1}{\partial x_2} + \lambda_2 \frac{\partial \psi_2}{\partial x_2} = 0$$

$$\frac{\partial U}{\partial x_3} + \lambda_1 \frac{\partial \psi_1}{\partial x_3} + \lambda_2 \frac{\partial \psi_2}{\partial x_3} = 0 \qquad \text{(I-21)}$$

These three equations, together with Equations (I-16) and (I-17) for ψ_1 and ψ_2, give us the five equations to solve for the five unknowns x_1, x_2, x_3, λ_1, and λ_2. Thus, the method of Lagrange multipliers results in more equations than there were variables initially. These equations must ordinarily be solved by an iterative numerical scheme.

Consider the beverage can example again. We need to minimize

$$A = 2\pi R^2 + 2\pi RH \qquad \text{(I-22)}$$

with the functional constraint

$$\psi = \pi R^2 H - V = 0 \qquad \text{(I-23)}$$

Thus, the Lagrangian L is

$$L = A(R, H) + \lambda\psi(R, H) \qquad \text{(I-24)}$$

The equations corresponding to (I-21) are

$$\frac{\partial A}{\partial H} + \lambda \frac{\partial \psi}{\partial H} = 0$$

$$\frac{\partial A}{\partial R} + \lambda \frac{\partial \psi}{\partial R} = 0 \qquad \text{(I-25)}$$

Substituting the expressions for A and ψ in these equations, we get the two equations

$$2\pi R + \lambda\pi R^2 = 0$$

$$4\pi R + 2\pi H + \lambda 2\pi RH = 0 \qquad \text{(I-26)}$$

These two equations and Equation (I-23) are the three equations which need to be solved for H, R, and λ. In a simple problem such as this, the equations can be solved exactly. From the first of the equations of (I-26), we get

$$\lambda = -\frac{2}{R}$$

which, upon substituting in the second of the equations of (I-26), gives us

$$H = 2R$$

which, upon substituting in Equation (I-23), yields the earlier answer

$$R = R_{\text{opt}} = \left(\frac{V}{2\pi}\right)^{1/3}$$

The Method of Steepest Ascent/Descent

A numerical procedure for finding the extremum which is relatively easy to program is the method of steepest ascent/descent. The procedure can be illustrated with the help of Figure I4. The figure shows some of the constant-value contours of an objective function $U = U(x, y)$ of two variables x and y, and the expected location of its extremum. The procedure, like all numerical iterative procedures, requires that we choose a starting point. From this starting point we move in the direction of the maximum slope. After moving a certain distance we stop and change the direction to that of the maximum slope from that point, and so on, until the extremum is reached.

The direction of the steepest slope from a given point is found as shown in Figure I5. From the starting point (x_0, y_0), we find the slopes in the x and y directions. The distance moved in each of these directions, Δx and Δy, is proportional to the respective slopes; i.e.,

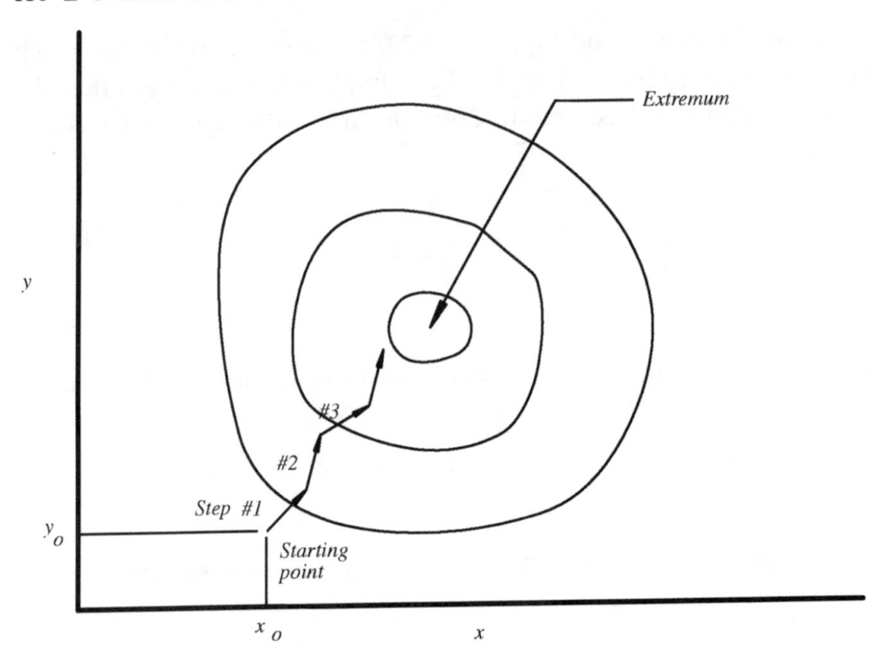

FIGURE I4. Constant-*U* contours.

$$\Delta x = k\left(\frac{\partial U}{\partial x}\right)_0 \qquad \text{and} \qquad \Delta y = k\left(\frac{\partial U}{\partial y}\right)_0 \tag{I-27}$$

where k is the same constant in each case. Thus, the next point (x_1, y_1) is given by

$$x_1 = x_0 + \Delta x \qquad \text{and} \qquad y_1 = y_0 + \Delta y \tag{I-28}$$

The process is then repeated, using (x_1, y_1) as the new starting point.

At each step we need to know how far to move, i.e., the value of k. This can be found by iterating at each step, until U does not change. Another method is the analytical process. The objective function at the new values of x and y is

$$U = U\left\{\left[x_0 + k\left(\frac{\partial U}{\partial x}\right)_0\right], \left[y_0 + k\left(\frac{\partial U}{\partial y}\right)_0\right]\right\} \tag{I-29}$$

Now, to find k, put

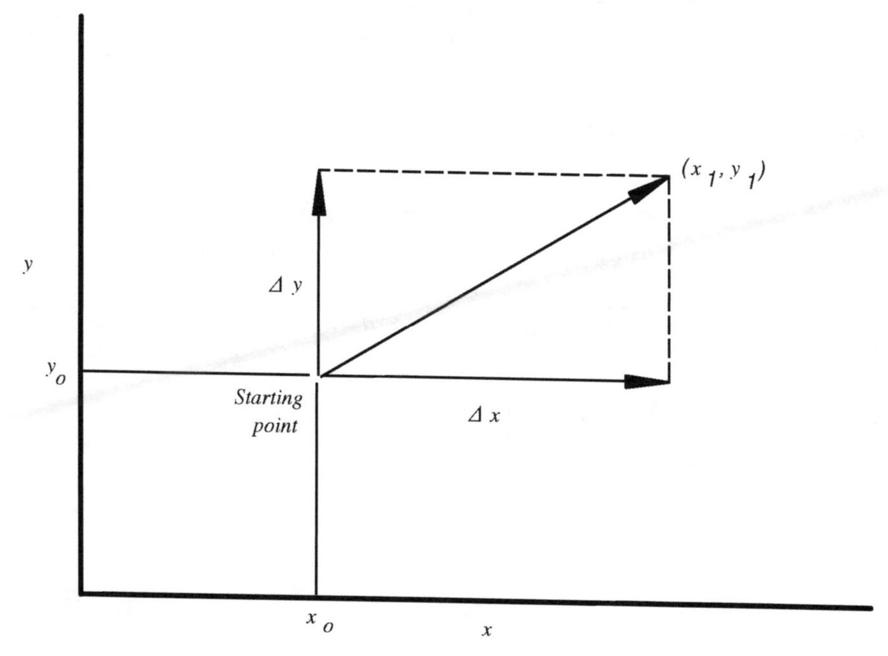

FIGURE I5. Direction of the steepest slope.

$$\frac{dU}{dk} = 0 \tag{I-30}$$

and solve for k. This can be a complex equation to solve.

The stepwise process is continued until two successive values of U do not show significant change.

LINEAR PROGRAMMING

An optimization problem in which the objective function, the functional constraints, and the regional constraints are linear functions of the variables is called a *linear programming problem*. In general, a linear programming problem is stated as follows:

Maximize the criterion function U:

$$U = c_1x_1 + c_2x_2 + \cdots + c_nx_n \tag{I-31}$$

subject to the functional constraints (which are linear inequalities)

$$\begin{aligned} a_{11}x_1 + a_{12}x_2 + \cdots + a_{1n}x_n &\leq b_1 \\ a_{21}x_1 + a_{22}x_2 + \cdots + a_{2n}x_n &\leq b_2 \\ \cdots\cdots\cdots\cdots\cdots\cdots\cdots\cdots \\ a_{m1}x_1 + a_{m2}x_2 + \cdots + a_{mn}x_n &\leq b_m \end{aligned} \tag{I-32}$$

and subject to the regional constraints

$$\begin{aligned} x_1 &\geq 0 \\ x_2 &\geq 0 \\ \cdots\cdots \\ x_n &\geq 0 \end{aligned} \tag{I-33}$$

Thus, all variables are restricted to positive or zero values by the regional constraints and limited to certain domains by the functional constraints. Linear programming problems occur in production/profit situations.

As an example, consider that a company makes x units of one item and y units of another item per day. Let the profit be \$4 for each unit of the first item and \$2 for each unit of the second item. Thus, the total daily profit is given by

$$U = 4x + 2y \tag{I-34}$$

subject to the regional constraints

$$x \geq 0 \quad \text{and} \quad y \geq 0 \tag{I-35}$$

The production facilities are limited in such a way that if only one of the items is produced at a time, then 8 units can be produced of the first item, or 4 units of the second item. This can be expressed by the functional constraint

$$x + 2y \leq 8 \tag{I-36}$$

It is required to find the values of x and y which yield the maximum profit U.

The data for this example are shown in Figure I6. The line defined by the equality in Equation (I-36) is drawn to show the limits of x and y. The other limits are the x and y axes, as defined by Equation (I-35). This triangular area is therefore the domain of feasible solutions, indicated by the shading in Figure I6.

The contours for constant U in linear programming problems are straight lines. The line for $U = 0$, defined by Equation (I-34), passes through the origin, as shown in the figure. As U increases, this line moves to the right and up. Therefore, we need to find for what maximum value of U the line would still stay in the allowable domain. The figure shows that the line for $U = 8$ passes through the top apex at $x = 0$ and $y = 4$, and that the line for $U = 32$ passes through the right apex at $x = 8$ and $y = 0$.

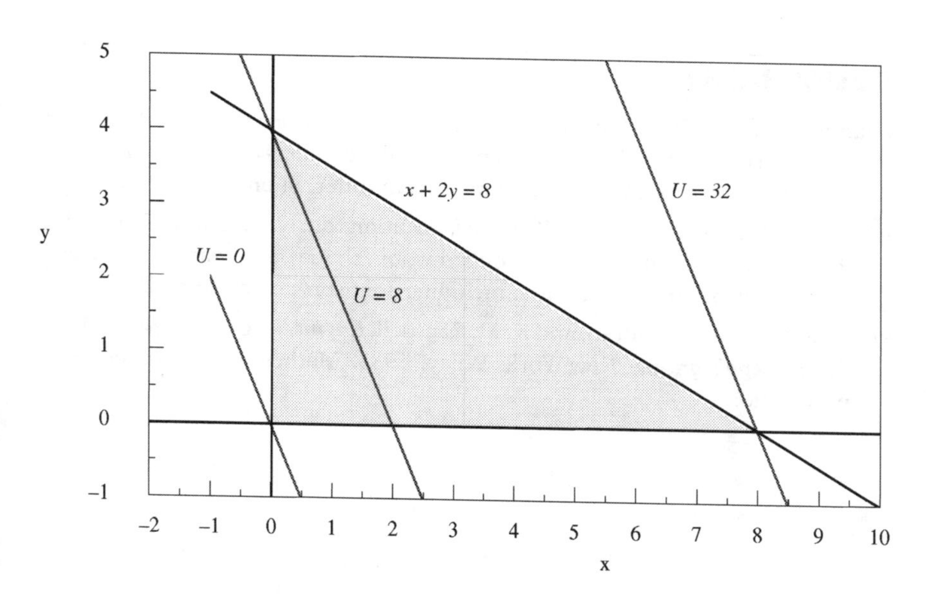

FIGURE I6. Linear programming example.

The required solution is therefore defined by the point $x = 8$ and $y = 0$, for which $U = 32$. The fact that the maximum occurs at a "corner" point is typical of linear programming problems. For problems involving more than two variables, it becomes difficult to visualize the situation. Commercial software packages are available for solving linear programming problems.

SUMMARY

The optimization methods described in this appendix can be applied only after the problem is reduced to the form of mathematical equations. This happens only at the detail design stage. Engineering analysis and design expertise is required to bring the problem to this stage. After the numerical results are obtained, it requires further engineering expertise to put those results to use. Before proceeding to find the optimum parameter values, one should gain insight into the problem and confidence in its mathematical model by doing a parameter search.

BIBLIOGRAPHY

Chapra, S. C., and R. P. Canale. *Numerical Methods for Engineering*. New York: McGraw-Hill, 1985. A discussion of numerical methods—e.g., equation solving, integration, regression—with examples from different engineering fields.

Johnson R. C. *Mechanical Design Synthesis*. Huntington, N.Y.: Krieger Publishing, 1978. Has chapters on optimum configuration, shape optimization, and classical methods of optimum design, including Johnson's own method.

Reklaitis, G. V., A. Ravindran, and K. M. Ragsdell. *Engineering Optimization Methods and Applications*. New York: Wiley, 1983. Mathematical techniques of optimization.

About the Author

Mahendra S. Hundal received his BSME degree from Osmania University in India in 1954. He then worked at the Tata Steel Company in Jamshedpur for six years, a major part of that time as a design engineer of heavy steel mill equipment. During this period he was granted leave to work as a design engineer at Lurgi-Chemie, a manufacturer of iron ore processing plants, in Frankfurt, Germany.

In late 1960 he enrolled in the graduate program at the University of Wisconsin–Madison, where he obtained his M.S. degree in 1962 and Ph.D. in 1964, both in mechanical engineering. From 1964 to 1967 he was Assistant Professor at San Diego State University. During the summers he gained experience at John Deere (Moline, Illinois) and at Solar and Rohr (San Diego area). He won the SAE Ralph R. Teetor Award in 1967.

In 1967 he moved to the University of Vermont in Burlington as Associate Professor and attained the rank of Professor in 1977. He has been active in a number of fields, including vibrations, noise, biomechanics, nuclear power plant simulation, and design. He was a visiting Professor at University of Oulu in Finland in 1976, and a Fulbright scholar at Munich University of Technology in 1987 to 1988. Under an NSF grant in Design Theory and Methodology, he developed methods for computer-aided conceptual design. He has authored over 120 papers, including book chapters on designing to cost. He was Chairman of the Department of Mechanical Engineering at the University of Vermont from 1980 to 1982 and again from 1991 to 1996.

As a member of ASME and its Design for Manufacturability committee (of the Design Engineering Division), he has organized sessions at the National Design Engineering Conference on topics of cost, environment, and time-driven design. He has presented papers at the International Conference on Engineering Design meetings in Boston (1987), Budapest (1988), Dubrovnik (1990), Zurich (1991), The Hague (1993), Prague (1995), and Tampere (1997).

Index

E